STUDY GUIDE FOR MORTIMER'S
CHEMISTRY

Sixth Edition

Donald W. Shive
Muhlenberg College

Louise E. Shive
Northhampton Area Senior High School

WADSWORTH PUBLISHING COMPANY
BELMONT, CALIFORNIA
A DIVISION OF WADSWORTH, INC.

Printed in the United States of America

1 2 3 4 5 6 7 8 9 10--90 89 88 87 86

ISBN 0-534-05675-X

CONTENTS

PREFACE

To the student:

This self-study guide is designed to accompany the sixth edition of *Chemistry*, by Charles E. Mortimer. The organization of the guide follows the text chapter by chapter, and thus the guide is easy to use after you have attended class and read the text. It is designed to supplement and reinforce the content of the text in several major ways which we will now discuss.

The section entitled Objectives in each chapter of the guide outlines the objectives of the corresponding chapter of the text and lists key terms used in the text. Each key term is followed by a specific text reference where the term is described.

A section entitled Units, Symbols, Mathematics follows. It includes appropriate remedial mathematical procedures which will help solve problems in the chapter, and it also lists symbols, along with their meanings, which are used in the chapter. We suggest that you pay particular attention to this section since we find that basic mathematics and the symbols used in chemistry often limit understanding the material.

The next section, Exercises, contains diverse unclassified problems which are appropriate to the chapter.

The following section, Answers to Exercises, provides the answers to each problem in the left-hand column and details of the solution and explanations on the right.

Finally, a Self-Test helps you gauge your competence. The answers to the self-tests can be found in the back of the guide.

We sincerely hope that this guide is helpful and that it makes your study of chemistry more enjoyable.

To the teacher:

While this study guide is primarily designed to assist students without your intervention, you can use it successfully to make initial assignments, to prescribe remedial work, and to emphasize certain concepts. For instance, you might want students to learn the chemical symbols for the elements early in the course. This guide, in its Units, Symbols, Mathematics sections, gradually presents the elements and their symbols. By Chapter 9, 38 common elements have been included. You might also consider assigning other exercises in the guide which are appropriate for your students to study. Especially, those students who are having trouble in the course can be directed to helpful additional material and detailed problem solutions in the guide.

Our hope, overall, is that this guide helps your students better understand chemistry.

INTRODUCTION

CHAPTER

1

OBJECTIVES   I. You should be able to demonstrate your knowledge of the following terms by defining them, describing them, or giving
specific examples of them:

alchemy [1.1]
analytical chemistry [1.1]
biochemistry [1.1]
centi- [1.3]
chemical change [1.2]
chemical symbol [1.2]
chemistry [introduction]
compound [1.2]
conversion factor [1.5]
density [1.5]
element [1.2]
empirical [1.1]
Greek period (600 B.C. to 300 B.C.) [1.1]
heterogeneous mixture [1.2]
homogeneous mixture [1.2]
inorganic chemistry [1.1]
International System of Units (SI) [1.3]
    base units [1.3]
    derived units [1.3]
    supplementary units [1.3]
kilo- [1.3]
law of conservation of mass [1.1]
law of definite proportions [1.2]
mass [1.2]
metric system [1.3]
matter [1.2]
micro- [1.3]
milli- [1.3]
mixture [1.2]

modern chemistry period (1790 to -) [1.1]
organic chemistry [1.1]
percentage [1.5]
phlogiston [1.1]
phlogiston period (1650 to 1790) [1.1]
phase [1.2]
physical change [1.2]
physical chemistry [1.1]
practical arts period (- to 600 B.C.) [1.1]
pure substance [1.2]
rate [1.5]
scientific notation [1.4; Appendix A]
significant figures [1.4]
solution [1.2]
substance [1.2]
transmutation [1.1]
weight [1.2]

II. You should be able to write numbers in scientific notation. [Appendix A]

III. You should be able to determine and work with the proper number of significant figures.

IV. You should study the instruction manual that came with your calculator so that you can perform calculations using conversion factors quickly and efficiently--noting only the final answer on paper and not writing down answers to intermediate steps.

V. You should be able to derive conversion factors and use them in performing calculations.

UNITS,
SYMBOLS,
MATHEMATICS

I. You should start building your understanding of SI notation and learning some of the prefixes that are used in conjunction with these standard units. For example, the measurement of length is the meter, m. Some of the standard prefixes used with the meter are listed in Table 1.1 along with the meaning of each symbol.

II. It is often inconvenient to work with the standard form of very large or very small numbers. For example, the multiplication involving the numbers 701,000 and 0.00000077 can easily be done incorrectly if we lose track of the zeros. Scientific notation is used to simplify calculations

(a) Move the decimal point so that there is a single digit (not zero) to its left; for example,

0.00000077

(b) Count the number of digits between the original and new decimal point positions to determine the magnitude of the exponent of 10; for example,

$$0.0000007\underset{1234567}{\odot}7$$

(c) Determine the sign of the exponent of 10 by the direction from the original to the new decimal point position (right is negative, left is positive); for example,

$$0.0000007\odot7$$

⟶

The direction indicates a negative sign. Thus,

0.00000077 equals 7.7×10^{-7}.

TABLE 1.1   Prefixes Used in SI Notation

Prefix		Multiplier	Length Measurements[a]	
			Name	Symbol and Equivalent
pico-	p	10^{-12}	picometer	1 pm = 1×10^{-12} m
nano-	n	10^{-9}	nanometer	1 nm = 1×10^{-9} m
micro-	μ	10^{-6}	micrometer	1 μm = 1×10^{-6} m
milli-	m	10^{-3}	millimeter	1 mm = 1×10^{-3} m
		1	meter	1 m = 1 m
kilo-	k	10^{3}	kilometer	1 km = 1×10^{3} m
mega-	M	10^{6}	megameter	1 Mm = 1×10^{6} m

[a]The centimeter (1 cm = 1×10^{-2} m) is acceptable but not preferred usage.

Properly used, scientific notation clearly indicates the number of significant figures. When 0.00000077 is expressed in scientific notation, only the two sevens are retained. The zeros preceding the sevens were expressed by the magnitude of of the exponent. Thus, the zeros are not significant. Only the two sevens are significant. General rules concerning zeros can be summarized as follows:

(a) Reading a number from left to right, the first significant digit encountered is the first non-zero digit. Zeros to the left of the first non-zero digit are not significant.

(b) Zeros to the right of the last non-zero digit are significant when a decimal point is included in the number. If a decimal point is not included, the zeros are not significant.

(c) Zeros between non-zero digits are significant.

III. The conversion factor method should be used in solving scientific problems. You may already use such a method, but you may not have formalized your thoughts in precisely this way. For example, if you were to determine the number of eggs in 2 dozen, you would immediately answer 24. The problem can be stated: "How many eggs equal two dozen?", or mathematically:

? eggs = 2 dozen.

Including the conversion factor, 12 eggs per dozen, the solution is

$$? \text{ eggs} = 2 \text{ dozen} \left(\frac{12 \text{ eggs}}{1 \text{ dozen}} \right) = 24 \text{ eggs}.$$

Note that the label "dozen" cancels from the equation and the label "eggs" remains. Identical labels are canceled from the numerator and the denominator in exactly the same way numbers are canceled.

Similarily to determine the total mass of tires discarded each year if the average tire weighs 30. pounds and 200. million are discarded per year, you would phrase the question: "How many pounds are there in 200,000,000 tires?" This can be expressed as

$$? \text{ lb} = 2.00 \times 10^8 \text{ tires}.$$

Each conversion factor must be an equivalency. Here, since one tire equals 30. pounds, the solution is

$$? \text{ lb} = 2.00 \times 10^8 \text{ tires} \left(\frac{30. \text{ lb}}{1 \text{ tire}} \right) = 6.0 \times 10^9 \text{ lb}$$

Using the original information and the fact that 2.20 lb = 1.00 kg, compute the number of kilograms of tires discarded each year:

$$? \text{ kg} = 2.00 \times 10^8 \text{ tires} \left(\frac{30. \text{ lb}}{1 \text{ tire}} \right) \left(\frac{1.00 \text{ kg}}{2.20 \text{ lb}} \right) = 2.7 \times 10^9 \text{ kg}$$

If 25 percent of the rubber is recoverable for use as blacktop material on roads, how many grams of tires could be recycled?

$$? \text{ g recycled}$$

$$= 2.00 \times 10^8 \text{ tires} \left(\frac{30. \text{ lb}}{1 \text{ tire}} \right) \left(\frac{1000 \text{ g}}{2.20 \text{ lb}} \right) \left(\frac{25 \text{ g recycled}}{100 \text{ g}} \right)$$

$$= 6.8 \times 10^{11} \text{ g recycled}$$

IV. Chemical symbols are used routinely in chemistry. It would be wise to begin learning these symbols for the more common elements now. Start with the following-- we will gradually build the list in subsequent chapters so learning them is not painful.

1. silver, Ag
2. hydrogen, H
3. sodium, Na
4. calcium, Ca
5. oxygen, O
6. nitrogen, N
7. copper, Cu

EXERCISES   I. Write the following numbers in scientific notation and indicate the number of significant figures in each. Check the answers in the next section.

Scientific Notation		Number of Significant Figures
_____	1. 751 meters	_____
_____	2. 781,000.00 kilometers	_____
_____	3. 781.000 kilometers	_____
_____	4. 0.050 grams	_____
_____	5. 0.000745 grams	_____
_____	6. 3 liters	_____
_____	7. 3.0 million liters	_____
_____	8. 7.0010 kilograms	_____
_____	9. 2.57 centimeters	_____
_____	10. 0.0057 micrograms	_____
_____	11. 1.00 liter	_____
_____	12. 574.4 kilograms	_____
_____	13. 100. degrees Celsius	_____
_____	14. 1.00 milliliter	_____

II. Write the abbreviations for the units associated with all numbers given in Section I of these exercises.

1. _____   8. _____

2. _____   9. _____

3. _____   10. _____

4. _____   11. _____

5. _____   12. _____

6. _____   13. _____

7. _____   14. _____

III. Perform the following operations and report the answer to the proper number of significant figures and with the proper symbols.

1.   10.2 m × 179.8 m =

2.   10.1 cm + 1.672 cm =

3.   5.70 cm × 6.40 cm × 7.320 cm =

4.   5.73 g ÷ 7.64 g =

5.   14.72 g + 735.0 kg =

6.   (0.00278 cm) × (0.00136 cm)

7.   75.6 g ÷ 5.3 cm^3 =

8. 7.3×10^{15} atoms + 6.8×10^{18} atoms =

9. 7.3×10^{15} atoms + 6.87453×10^{18} atoms =

10. $\left(\dfrac{4.56 \text{ g}}{7.32 \text{ g}} \right) \times 100\%$ =

IV. Write the symbols for each of the following elements.

1. hydrogen   _____   5. oxygen   _____

2. sodium   _____   6. copper   _____

3. nitrogen   _____   7. silver   _____

4. calcium   _____

V. From the following list of terms, choose the one that best completes each sentence.

a. alchemy   l. matter

b. analytical chemistry   m. organic chemistry

c. biochemistry   n. phlogiston

d. chemical change   o. phase

e. compound   p. physical change

f. conversion factor   q. practical arts period

g. empirical   r. rate

h. Greek period   s. significant figures

i. base units   t. solution

j. law of conservation   u. substance
    of mass
                               v. transmutation
k. mass

1. Developments based on only practical experience are _____.

2. The concept that proposes that one element can be changed into another is called _____.

3. _____ was assumed to be a constituent of any substance that could undergo combustion.

4. _____ is the principal branch of chemistry that is concerned with the qualitative and quantitative identification of substances.

5. _____ states that mass is not destroyed during a chemical reaction.

6. _____ is a measure of the quantity of matter.

7. A _____ is a substance composed of two or more elements in fixed proportions.

8. A physically distinct portion of matter that is uniform throughout in composition and properties is called a _____.

9. The _____, 1cm/0.01m, will allow you
    to convert 7.2 meters into centimeters.

10. Burning coal is a _____.

ANSWERS TO   I. The answers to section I of the exercises are given without
EXERCISES   comment:

Scientific Notation		Number of Significant Figures
7.51×10^2 meters	1. 751 meters	3
7.8100000×10^5 kilometers	2. 781,000.00 kilometers	8
7.81000×10^2 kilometers	3. 781.000 kilometers	6
5.0×10^{-2} grams	4. 0.050 grams	2
7.45×10^{-4} grams	5. 0.000745 grams	3
3 liters	6. 3 liters	1
3.0×10^6 liters	7. 3.0 million liters	2
7.0010 kilograms	8. 7.0010 kilograms	5
2.57 centimeters	9. 2.57 centimeters	3
5.7×10^{-3} micrograms	10. 0.0057 micrograms	2
1.00 liter	11. 1.00 liter	3
5.744×10^2 kilograms	12. 574.4 kilograms	4
1.00×10^2 degrees Celsius	13. 100. degrees Celsius	3
1.00 milliliter	14. 1.00 milliliter	3

II. The correct answers are given on the left, and alternate
    acceptable expressions are given to the right.

1. 751 m

2. 781,000.00 km   781.00000 Mm

3. 781.000 km

4. 0.050 g   50. mg

5. 0.000745 g   745 µg

6. 3 L

7. 3.0 ML

8. 7.0010 kg

9. 2.57 cm   25.7 mm

10. 0.0057 µg   5.7 ng

11. 1.00 L

12. 574.4 kg

13. 100.$^{\circ}$C

14. 1.00 mL

III. Calculations and Significant Figures

1. 1.83×10^3 m^2

An answer of 1833.96 is obtained on the calculator after proper multiplication. However, the answer should be reported to only three significant figures:

1.83×10^3 m^2

Other methods of reporting the answer are incorrect for the following reasons:

1833.96 m^2 too many significant figures

1.83×10^3 no units identifying the number

2. 11.8 cm or 118 mm

An answer of 11.772 is obtained on the calculator and should be reported as 11.8 cm or 118 mm.

3. 267 cm^3

A numerical answer of 267.0336 is obtained on the calculator.

4. 0.750

BE VERY CAREFUL! Calculators usually display more than the number of significant figures needed, but most models do not display significant zeros to the right of the decimal point unless they are followed by a non-zero digit. For instance, if you multiply 100.00 times 1.0000 on your calculator, the displays are:

(enter 100.00)

(press the multiplication key and enter 1.0000)

(press the equals key and read)

Note that the final answer should be reported to five significant figures; that is, 100.00. The extra zeros must be added. Similarily, the display of 5.73 ÷ 7.64 is

$$\boxed{0.75}$$

In reporting the answer a zero must be added to the right.

5. 735.0 kg

First change the units so that they are the same:

$14.72 \text{ g} + 7.350 \times 10^5 \text{ g}$

Add them on the calculator to obtain

$7.3501472 \times 10^5 \text{ g}$

Report the answer to the proper number of significant figures:

$7.350 \times 10^5 \text{ g}$ or 735.0 kg

6. preferably:

$3.78 \times 10^{-6} \text{cm}^2$

also acceptable:

0.00000378 cm^2

BE VERY CAREFUL! We noted in problem 5 of this section that calculators do not always display significant zeros. In this exercise we see that the problem of insufficient digits can plague us in other ways. Calculators usually display only seven digits after the decimal (a few even drop additional digits from memory); so, the answer to this question may appear on your calculator as:

$$\boxed{0.0000037}$$

This display shows only two significant digits. To lessen the problem of dropped digits, many calculators change to scientific notation automatically:

$$\boxed{3.78080\text{-}06}$$

If your calculator does not, you should force the change by converting one of the numbers used in the calculation to scientific notation. For instance:

(enter 0.00278 as 2.78×10^{3})

$$\boxed{3.78\text{-}3}$$

(press the multiplication key and enter 0.00136)

$$\boxed{0.00136}$$

(press the equals key and read)

$$\boxed{3.7808\text{-}6}$$

More than the needed number of significant figures are now available.

7. 14 g/cm^3

A density of 14 grams per cubic centimeter is obtained. Because this is division, the answer is limited to two significant figures by the factor with the smallest number of significant figures, 5.3 cm^3.

8. 6.8×10^{18} atoms

The 7.3×10^{15} atoms (alternatively written as 0.0073×10^{18} atoms) is not a significant addition to the much larger quantity 6.8×10^{18} atoms.

9. 6.8818×10^{18} atoms

The answer is rounded off to four decimal places because, in addition, the fewest number of decimal places limits the number of significant figures:

$$\begin{array}{r} 6.87453 \times 10^{18} \text{ atoms} \\ + \quad 0.0073 \times 10^{18} \text{ atoms} \\ \hline 6.88183 \times 10^{18} \text{ atoms} \end{array}$$

10. 62.3%   A unitless fraction multiplied by
                              the exact number 100 is a percentage
                              or parts per hundred.

IV. Names and Symbols of Elements

1. H   hydrogen

2. Na   sodium

3. N   nitrogen

4. Ca   calcium

5. O   oxygen

6. Cu   copper

7. Ag   silver

V. The correct answers to the sentence-completion questions
   are on the left. The sentences on the right show the
   answers in context.

1. empirical   1. Developments based on only practical
                          experience are empirical.

2. transmutation   2. The concept that proposes that one element
                          can be changed into another is called
                          transmutation.

3. phlogiston   3. Phlogiston was assumed to be a constituent of
                          any substance that could undergo combustion.

4. analytical   4. Analytical chemistry is the principal branch
    chemistry   of chemistry that is concerned with the
                          qualitative and quantitative identification
                          of substances.

5. law of   5. The law of conservation of mass states that
    conservation   mass is not destroyed during a chemical
    of mass   reaction.

6. mass   6. Mass is a measure of the quantity of matter.

7. compound   7. A compound is a substance composed of two or
                          more elements in fixed proportions.

8. phase

8. A physically distinct portion of matter that is uniform throughout in composition and properties is called a phase.

9. conversion factor

9. The conversion factor, 1cm/0.01m, will allow you to convert 7.2 meters into centimeters.

10. chemical change

10. Burning coal is a chemical change.

SELF-TEST

Complete the test in 30 minutes.

I. Answer each of the following:

_____ 1. The process of water changing into steam is called
   (a) freezing   (c) a physical change
   (b) a chemical change   (d) fusion

_____ 2. The number 0.070020 has how many significant figures?
   (a) 4   (b) 5   (c) 6   (d) 7

_____ 3. The number 0.070020 should be written in scientific notation as
   (a) 7.002×10^2 (c) 7.0020×10^{-2}
   (b) 7.002×10^{-2} (d) 7.0020×10^2

_____ 4. The number 700. should be written in scientific notation as
   (a) 7×10^2 (c) 7.00×10^2
   (b) 7.0×10^2 (d) cannot be determined

_____ 5. One microgram is equal to
   (a) 10^3 g   (c) 10^6 g
   (b) 10^{-3} g   (d) 10^{-6} g

_____ 6. One kilometer equals
   (a) 10^3 m   (c) 10^6 m
   (b) 10^{-3} m   (d) 10^{-6} m

_____ 7. The number 700.0 should be written in scientific notation as
   (a) 7×10^2 (c) 7.000×10^2
   (b) 7.00×10^2 (d) 7.000×10^3

_____ 8. The process of gasoline burning in an automobile cylinder is called
   (a) a physical change   (c) vaporization
   (b) a chemical change   (d) boiling

9. The sum of the number of 71.742, 6.0, and 21.3413 is
   (a) 99.0833   (c) 99.1
   (b) 99.0   (d) 1.0×10^2

10. The number 199.969, rounded off to three significant figures, should be written as
    (a) 200   (c) 2×10^2
    (b) 199   (d) 2.00×10^2

11. The number 0.74 has how many significant figures?
    (a) 1   (c) 3
    (b) 2   (d) 4

12. SI units are
    (a) units of the International System
    (b) Standard International units
    (c) Substituted International units
    (d) Scientific International units

13. Multiplication of the number 23.6 by the number 7.50×10^3 yields
    (a) 1.77×10^3 (c) 1.7×10^5
    (b) 1.77×10^5 (d) 177×10^3

14. A cow produces milk from ingested foodstuffs. This process can be called
    (a) a liquefaction   (c) a physical change
    (b) a chemical change   (d) a biological marvel

15. During what period in the development of chemistry was the theory of transmutation first proposed?
    (a) practical arts   (c) alchemy
    (b) Greek   (d) phlogiston

II. Work the following problems.

1. President Reagan, in his first major address to the nation, stated that a stack of $1,000.00 bills stacked four inches high equals a million dollars. How thick in centimeters is a single $1,000.00 bill? Use the conversion factor 1.00 inches equals 2.54 cm.

2. What is the density of a piece of material if a cube of it measuring 1.24 cm on a side weighs 2.36 g? Give your answer in g/cm^3.

3. In 1980 the United States used 35.7 quads of oil. What weight in kilograms of oil was consumed during that year? One quad equals one quadrillion Btu's; one quadrillion equals 1,000,000,000 millions; one barrel yields 5.8 million Btu's and weighs 306 pounds; one pound equals 453.59 g.

INTRODUCTION TO
ATOMIC THEORY

CHAPTER

2

OBJECTIVES   I. You should be able to demonstrate your knowledge of the
following terms by defining them, describing them, or giving
specific examples of them:

actinides [2.7]
alkali metal [2.7]
alpha particle, α [2.7]
atom [2.7]
atomic mass unit, u [2.9]
atomic number, Z [2.6]
atomic weight [2.9]
beta particle, β [2.5]
binding energy [2.9]
cathode ray [2.2]
coulomb [2.2]
electron [2.2]
family [2.7]
gamma radiation, γ [2.5]
group [2.7]
halogen [2.7]
ion [2.6]
isotope [2.6]
lanthanide [2.7]
law of conservation of mass [2.1]
law of definite proportions [2.1]
law of multiple proportions [2.1]
mass number, A [2.6]
mass spectrometer [2.8]
metal [2.7]
metaloid [2.7]
neutron [2.4]
noble gas [2.7]
nonmetal [2.7]

nucleon [2.6]
nucleus [2.6]
period [2.7]
periodic law [2.7]
positive rays [2.3]
proton [2.3]
radioactivity [2.5]
semimetal [2.7]
unit electrical charge, e [2.2]

II. You should understand the nuclear atom and be able to identify the number of protons, neutrons, and electrons in an isotope given the mass number, atomic number, and charge on the isotope. Likewise, given the number of protons, neutrons, and electrons, you should be able to identify the element, its charge, mass number, and atomic number.

III. You should be able to determine the number of protons, neutrons, and electrons in any isotope of any element.

IV. You should be able to calculate atomic weights from masses and relative abundances of isotopes.

V. You should understand Dalton's atomic theory and the laws of chemical composition.

UNITS,
SYMBOLS,
MATHEMATICS

I. Since chemical symbols are used routinely in chemistry, it would be wise to continue learning some of the symbols that are used to represent the more common elements. We suggest that you memorize the following list, which includes elements listed in Chapter 1.

1. potassium, K	10. neon, Ne
2. silver, Ag	11. copper, Cu
3. hydrogen, H	12. aluminum, Al
4. sodium, Na	13. sulfur, S
5. calcium, Ca	14. magnesium, Mg
6. oxygen, O	15. fluorine, F
7. chlorine, Cl	16. bromine, Br
8. nitrogen, N	17. iron, Fe
9. carbon, C	18. helium, He

II. The following symbols are used in this chapter, and you should be familiar with them.

- α is the symbol used to represent the alpha particle.

- A identifies the mass number of an element. A always equals the sum of the number of protons and neutrons in a nucleus of the element.

- β is the symbol used to represent beta radiation.

- C is the abbreviation for coulomb, the SI unit of charge.

- e is the symbol used to represent the unit electrical charge; that is, the charge of a single electron.

- γ is the symbol representing gamma radiation.

- g is the abbreviation for gram, the SI unit of mass.

- m stands for the mass of a particle.

- q is a symbol used to represent charge on a particle.

- u is the abbreviation for the unified atomic mass unit. By definition, one-twelfth the mass of ^{12}C equals exactly one u.

- Z identifies the atomic number of an element. Z always equals the number of protons in the nucleus of the element. The atomic number is always placed at the lower left corner of the symbol for the element: for example, in $^{12}_{6}C$ the atomic number of carbon is identified as 6.

EXERCISES

I. Choose the best answer to each of the following. You should consult a periodic table of the elements as needed.

1. Vanadium has an atomic number of

   a. 49
   b. 50
   c. 51

   d. 52
   e. none of these

2. Vanadium has a mass number of

   a. 49
   b. 50
   c. 51

   d. 52
   e. cannot tell from the information given

3. Silver has a mass number of

   a. 46
   b. 47
   c. 48

   d. 49
   e. cannot tell from the information given

4. Which of the following particles has the lowest mass?

   a. electron
   b. proton
   c. neutron

   d. He^{2+}
   e. an alpha particle

5. Which of the following has no electrical charge?

    a. electron   d. He^{2+}
    b. proton   e. an alpha particle
    c. neutron

6. Which of the following has a negative charge?

    a. electron   d. He^{2+}
    b. proton   e. an alpha particle
    c. neutron

7. Which of the following positive ions is deflected most in an electrical field?

    a. He^{+} c. Ne^{+}
    b. He^{2+} d. Ne^{2+}

8. The value of e, the unit electrical charge, was determined by

    a. Dalton   d. Newton
    b. Millikan   e. Leucippus
    c. Faraday

9. The atomic symbol for potassium is

    a. P   d. Ps
    b. Po   e. none of these
    c. K

10. Calcium, Ca, has how many electrons?

    a. 0   d. 60
    b. 20   e. cannot tell from the
    c. 40   information given

11. Which of the following is a metal?

    a. Ag   d. C
    b. Cl   e. none of these
    c. O

12. Which of the following is a nonmetal?

    a. Na   d. K
    b. Ca   e. none of these
    c. O

13. The elements Br, F, and Cl are members of

     a. a group   d. nonmetals
     b. a family   (e.) all of these
     c. the halogens

14. The elements Ne, O, Br, and N are members of

     a. a group   (d.) nonmetals
     b. a family   e. all of these
     c. the halogens

15. The elements Li, C, O, and N are members of

     (a.) a group   d. nonmetals
     b. a family   (e.) none of these
     c. the halogens

16. The elements K, Fe, Mg, and Ca are members of

     a. a group   d. nonmetals
     b. a family   (e.) none of these
     c. the halogens

17. Which of the following elements has the largest atomic
     mass?

     a. K   (d.) Br
     b. Fe   e. Ca
     c. Mg

18. The elements ^{79}Br and ^{81}Br

     a. are isotopes
     b. have different masses
     c. have the same number of protons
     d. have different numbers of neutrons
     (e.) all of the above

19. The following person did not believe that atoms
     existed

     a. Aristotle   d. Leucippus
     b. Dalton   (e.) all believed that atoms
     c. Democritus   existed

20. When atoms disintegrate by radioactive decay they can
     emit

     a. alpha radiation   (d.) all of the above
     b. gamma radiation   e. none of the above
     c. beta radiation

II. Complete the following table. Use a periodic table of elements to help identify them.

Name	Symbol	Z (Atomic #)	A (Atomic mass.)	Protons	Neutrons	Electrons
_____	H	1	1	1	0	1
_____	H$^+$	1	3	1	2	0
_____	Li	3	6	3	3	3
_____	C	6	12	6	8	_____
_____	_____	_____	_____	8	8	10
_____	S	_____	_____	_____	20	_____
_____	Pd	_____	110	_____	_____	_____
_____	Cs	_____	132	_____	_____	54
_____	Te^{2-}	52	128	_____	68	_____
_____	_____	_____	132	_____	80	52
_____	Xe	_____	_____	_____	80	_____
_____	_____	52	125	_____	_____	54
_____	I$^-$	_____	125	_____	_____	54
_____	_____	76	184	_____	_____	76
_____	Cf	_____	_____	_____	153	98

III. Using the periodic table, identify the element that is described by the statement:

_____   1. $Z = 13$

_____   2. The atomic weight is 74.9 u.

_____   3. The atomic number is 50.

_____   4. The mass number is 109 and the number of neutrons is 62.

_____   5. There are 16 electrons in the neutral atom.

_____   6. Two isotopes exist, one with a mass of 10.013 u and one with a mass of 11.009 u.

_____

7. There are 18 electrons in the ion that has a single negative charge, X^-.

_____

8. There are 22 electrons in the ion that has two positive charges.

IV. You have learned to calculate atomic weights from the masses of isotopes. Work the following problems to test your expertise:

Let x = abundance ^{11}B

$1 - x = {}^{10}B$

$x(11.00931) + (1-x)(10.0129) =$

$11.00931 x + 10.0129 - 10.0129 x = 10.811$

$0.99641 x = .7981$

$x = .80097$

$^{10}B \ 19.9\% \quad ^{11}B \ 80.1\%$

$^{10}B \ 10.01294$

$^{11}B \ 11.00931$

10.811

1. Boron consists of two naturally occurring isotopes. One isotope, ^{10}B, has a mass of 10.01294 u and the other, ^{11}B, has a mass of 11.00931 u. If the atomic weight of boron is 10.811, what percent of each of the two isotopes is naturally occurring boron?

2. Chlorine consists of two naturally occurring isotopes, ^{35}Cl and ^{37}Cl. The ^{35}Cl isotope is more abundant than the ^{37}Cl isotope (75.77% vs. 24.23%). Calculate the approximate atomic weight of naturally occurring chlorine. Assume that the mass of ^{35}Cl is 34.969 u and that of ^{37}Cl is 36.966 u.

$0.7577(34.969) = 26.496$

$0.2423(36.966) = 8.9569$

atomic weight = 35.45

ANSWERS TO EXERCISES

I. Multiple choice questions:

1. e

Vanadium has an atomic number of 23.

2. e

Vanadium has an atomic weight of 50.9415. You cannot determine the mass number without knowing the specific isotope.

3. e

Same reason as given in problem #2.

4. a

Refer to Table 2.1 in your text.

5. c

Refer to Table 2.1 in your text.

6. a

Refer to Table 2.1 in your text.

7. b

The deflection is proportional to the value of q/m.
The largest value of q/m is for the He^{2+} ion.

8. b

Millikan did it in his famous oil drop experiment.

9. c

Potassium, K.

10. b

The neutral calcium atom has the same number of electrons as protons, 20.

11. a

12. c

13. e

14. d

15. e   They are members of a period.   18. e

16. e   They are all metals.   19. a

17. d   Bromine, Br   20. d

II. Problems involving the use of the periodic table of the elements

Name	Symbol	Z	A	Protons	Neutrons	Electrons
hydrogen	H	1	1	1	0	1
hydrogen ion	H^+	1	3	1	2	0
lithium	Li	3	6	3	3	3
carbon	C	6	14	6	8	6
oxygen ion	O^{2-}	8	16	8	8	10
sulfur	S	16	36	16	20	16
palladium	Pd	46	110	46	64	46
cesium ion	Cs^+	55	132	55	77	54
tellurium ion	Te^{2-}	52	120	52	68	54
tellurium	Te	52	132	52	80	52
xenon	Xe	54	134	54	80	54
tellurium ion	Te^{2-}	52	125	52	73	54
iodine ion	I^-	53	125	53	72	54
osmium	Os	76	184	76	108	76
californium	Cf	98	251	98	153	98

III. Identification of elements

1. Al, aluminum

2. As, arsenic

3. Sn, tin   The symbol for atomic number is Z.

4. Ag, silver   The atomic number is obtained by subtracting the number of neutrons from the mass number, $109 - 62 = 47$. The atomic number is 47.

5. S, sulfur   The number of electrons in a neutral atom equals the number of protons in that atom; therefore, $Z = 16$.

6. B, boron   Boron is the only element that has an atomic weight lying between 10.013 u and 11.009 u.

7. Cl, chlorine   The ion has one more electron than the neutral atom; therefore, the neutral atom has 17 electrons; $Z = 17$.

8. Cr, chromium   The ion has a 2+ charge, i.e., two electrons, each of which has a negative charge, have been removed. The neutral atom should have 24 electrons; $Z = 24$.

IV. Mathematical problems

1. 19.903% ^{10}B
   80.097% ^{11}B

Set x = fraction of ^{10}B, the isotope with a mass of 10.01294 u. Then $1 - x$ = fraction of ^{11}B, the isotope with a mass of 11.00931 u. The atomic weight of naturally occurring boron is the weighted average of these isotopes; therefore,

$$10.01294(x) + 11.00931(1 - x) = 10.811$$

Multiplying terms,

$$10.01294x + 11.00931 - 11.00931x = 10.811$$

Combining terms,

$$x = 0.19903$$
$$1 - x = 0.80097$$

Since a percentage is 100 times a corresponding fraction, the values expressed as percentages are 19.903% and 80.097%. Thus, naturally occurring boron is a mixture composed of 19.903% ^{10}B and 80.097% ^{11}B.

2. 35.45 u   Using the information given in the problem, we obtain an answer of 35.45 u:

0.7577(34.963) + 0.2423(36.966) = 35.45 u

This answer is very close to the actual atomic weight, 35.453 u.

SELF-TEST   Complete the test in 15 minutes.

Fill in each space provided.

✓1. The fundamental subatomic particles are the
     __protons__, the __neutrons__, and
     the __electrons__.

✓2. The fundamental subatomic particle that has no charge
     is the __neutrons__.

✗ 3. The radiation that is emitted from naturally
     radioactive elements and has no charge is
     __alpha particle__.   *gamma particle.*

✓4. Atoms with the same atomic number but different mass
     numbers are called __isotopes__.

✓5. Lithium is a mixture of two naturally occurring
     isotopes: 7.40% of the mixture is ^{6}Li, an isotope that
     has a mass of 6.0169 u; 92.60% of the mixture is ^{7}Li,
     an isotope that has a mass of 7.0182 u. The atomic
     weight of lithium is __6.944__.

✓6. The isotope used as the reference standard for
     the unified atomic mass scale is __carbon 12__.

✓7. The element that contains 82 protons is
     __Lead (Pb)__.   You may check a periodic table
     of the elements.

✓8. The halogen with the lowest atomic mass is
     __Flourine__.

✓9. An element has a nucleus consisting of 14 neutrons
     and 13 protons; its mass number is __27__.

✓10. If the element described in problem 9 is made into an
     ion with a charge of 3+, the ion will contain
     __10__ electrons.

periods X 11.   The horizontal rows of elements in the periodic table
are called ___family___. (group)

✓ 12.   An atom that does not contain an equal number of
protons and electrons is an ___ion___.

gamma particle. 13.   A highly energetic form of radiation that is similar
to X rays and emitted from radioactive elements is
___Beta rays___.

✓ 14.   Most of the mass of an atom is concentrated in the
___nucleus___.

✓ 15.   A neutral copper atom contains 29 electrons. How many
protons are in the nucleus? ___29 protons___

5. $0.074 (6.0169) = 0.44525$

$0.9260 (7.0182) = 6.4988$

6.944

$\frac{12}{15} = 80\%$

STOICHIOMETRY, PART I: CHEMICAL FORMULAS

CHAPTER
3

OBJECTIVES

I. You should be able to demonstrate your knowledge of the following terms by defining them, describing them, or giving specific examples of them:

anion [3.1]
Avogadro's number [3.4]
cation [3.1]
chemical formula [3.1]
diatomic molecule [3.1]
empirical formula [3.2]
formula weight [3.3]
ion [3.1]
mole [3.4]
molecular formula [3.1]
molecular weight [3.3]
monatomic ion [3.1]
percentage composition [3.5]
polyatomic ion [3.1]
simplest formula [3.2]
stoichiometry [introduction]
structural formula [3.1]

II. You should be able to write formulas for ionic compounds given the ions and their charges.

III. You should be able to name the elements that occur in nature as diatomic molecules.

IV. You should be able to perform calculations to obtain numbers of moles and molecules, percent composition, empirical formulas, and molecular formulas.

UNITS,
SYMBOLS,
MATHEMATICS

I. Certain conversion factors will be used frequently in this and subsequent chapters. These factors are summarized here with the corresponding units:

Mass percent, %, has units of (g/g)100. Molecular weight and formula weight usually have units of g/mol, but may have units of atomic mass units, u, or no units at all.

Avogadro's number has the value of 6.02205×10^{23} and usually has units of either atoms/mol or molecules/mol.

II. Continue to learn the names and symbols of common elements:

1. potassium, K
2. silver, Ag
3. hydrogen, H
4. sodium, Na
5. calcium, Ca
6. oxygen, O
7. chlorine, Cl
8. nitrogen, N
9. carbon, C
10. lithium, Li
11. boron, B
12. neon, Ne
13. copper, Cu
14. aluminum, Al
15. sulfur, S
16. magnesium, Mg
17. fluorine, F
18. bromine, Br
19. iron, Fe
20. helium, He
21. berylium, Be
22. silicon, Si

EXERCISES

I. Answer the following questions.

1. Write formulas for the compounds formed between Na^+ ion and

   a. Cl^- ion
   b. O^{2-} ion
   c. N^{3-} ion

2. Write formulas for the compounds formed between Ca^{2+} ion and

   a. NO_3^- ion
   b. CO_3^{2-} ion
   c. PO_4^{3-} ion

3. Determine the formula weight of each compound in question 2.

4. Write the empirical formula for each of the following molecular formulas.

   a. C_6H_6
   b. B_2H_6
   c. Sb_2O_5
   d. $C_{34}H_{36}O_6N_2$

5. Determine the molecular formulas of the compounds for which the empirical formula and molecular weight are given.

    a. HPO_3, 160 g/mol

    b. B_5H_7, 122 g/mol

    c. $C_4H_5NO_2$, 297 g/mol

II. Work the following problems.

1. Calculate the number of moles of atoms in a 2.50 g sample of nickel.

2. How many atoms are contained in the 2.50 g sample of nickel?

3. What is the mass of 2.41×10^{24} molecules of oxygen gas?

4. There are 3.01×10^{22} molecules in a 3.90 g sample of DMSO. What is the molecular weight of DMSO?

5. Most iron is obtained from Fe_2O_3. What is the percentage of iron (by mass) in Fe_2O_3?

6. Gibbsite, $Al_2O_3 \cdot 3H_2O$, is a naturally occurring material from which aluminum is produced.* What percentage of aluminum is in gibbsite?

7. Chemical analyses are often performed to determine the percent composition of a pure material, and from this information the empirical formula can be determined. An organic material is analyzed and found to contain 75.92% C, 17.71% N, and 6.37% H. Determine the empirical formula of the compound.

8. Tritopine, an alkaloid isolated from opium, has been found to be 74.0% C, 7.90% H, 14.0% O, and 4.10% N by weight. What is the empirical formula? The molecular weight is known to be 682 g/mol. What is the molecular formula?

---

*Many minerals and other materials are combinations of two or more compounds in specific ratios. Gibbsite is a combination of aluminum oxide, Al_2O_3, and water in a 1:3 ratio.

ANSWERS TO   I. Questions
EXERCISES

1. a. NaCl   One negative charge on the chloride ion is balanced by one
                positive charge on one sodium ion.

   b. Na_2O Since the oxide ion has two negative charges, it must be
                balanced by two positive charges supplied by two sodium
                ions.

   c. Na_3N Three negative charges are balanced by three positive
                charges.

2. a. $Ca(NO_3)_2$ Two negative charges are needed to balance the two positive
                charges on the calcium ion.

   b. $CaCO_3$ Each ion has a charge of two.

   c. $Ca_3(PO_4)_2$ There must be equal numbers of positive and negative charges
                for a balanced formula. Three ions with positive two
                charges and two ions with negative three charges balance
                each other with a total charge of six positive charges and
                six negative charges.

3. a. 164.1   The atomic mass of one Ca atom is   40.1
                The mass of two N atoms is 2(14), or: 28.0
                The mass of six oxygen atoms is 6(16)   96.0
                The total formula weight is: 164.0

   b. 100.1   Ca   40.1
                       C   12.0
                 + 3 O   48.0
                             100.1

   c. 310.3   3 Ca   120.3
                     2 P   62.0
                 + 8 O   128.0
                             310.3

4. a. CH   The ratio of 6:6 reduces to the simplest whole number
                ratio of 1:1.

   b. BH_3 Find the simplest whole number ratio by dividing each
                subscript by the smallest subscript. In B_2H_6 the smallest
                subscript is 2.   (2/2) = 1 and (6/2) = 3.

   c. Sb_2O_5 The formula cannot be reduced further since 5 divided by
                2 is not a whole number.

   d. $C_{17}H_{18}O_3N$ Each subscript is divided by 2, the smallest subscript.

5. a. $H_2P_2O_6$ The formula weight of HPO_3 is 80 g/mol. The molecular weight is given as 160 g/mol $((160/80) = 2)$, which means it takes two formula weights to make one molecular weight.

   b. $B_{10}H_{14}$ The formula weight of B_5H_7 is 61.0; $122/61 = 2$.

   c. $C_{12}H_{15}N_3O_6$ $297/99 = 3$

II. Calculations

1. 4.26×10^{-2} mol   State the problem mathematically:

   $? \text{ mol} = 2.50 \text{ g Ni}$

   Then use the conversion factor, the atomic weight of nickel, to solve the problem:

   $? \text{ mol} = 2.50 \text{ g Ni} \left(\dfrac{1 \text{ mol}}{58.7 \text{ g Ni}} \right) = 0.0426 \text{ mol}$

   $= 4.26 \times 10^{-2} \text{ mol}$

2. 2.56×10^{22} atoms   State the problem mathematically:

   $? \text{ atoms} = 2.50 \text{ g Ni}$

   Then solve using appropriate conversion factors:

   $? \text{ atoms} = 2.50 \text{ g Ni} \left(\dfrac{1 \text{ mol}}{58.7 \text{ g Ni}} \right) \left(\dfrac{6.02 \times 10^{23} \text{ atoms}}{1 \text{ mol}} \right)$

   $= 2.56 \times 10^{22} \text{ atoms}$

3. 128 g   Remember that oxygen is one of the seven diatomic elements. Thus one mole of oxygen gas weighs 32.0 g. State the problem mathematically:

   $? \text{ g } O_2 = 2.41 \times 10^{24} \text{ molecules } O_2$

   Then use the appropriate conversion factors to solve the problem.

   $? \text{ g } O_2 = 2.41 \times 10^{24} \text{ molecules } O_2 \left(\dfrac{1 \text{ mole } O_2}{6.02 \times 10^{23} \text{ molecules}} \right)$

   $\left(\dfrac{32.0 \text{ g } O_2}{\text{mole } O_2} \right) = 1.28 \times 10^2 \text{ g } O_2 = 128 \text{ g } O_2$

4. 78.0 g/mol

Since a molecular weight is simply the number of grams constituting one mole of material, the problem can be stated, "How many grams does one mole of DMSO weigh?"

? g = 1.00 mol DMSO

Solve by using the conversion factor given in the problem, 3.01×10^{22} molecules in 3.90 g, and Avogadro's number:

$$? \; g = 1.00 \; \cancel{mol} \; DMSO \left(\frac{3.90 \; g}{3.01 \times 10^{22} \; \cancel{molecules}} \right) \left(\frac{6.02 \times 10^{23} \; \cancel{molecules}}{1 \; \cancel{mol}} \right)$$

$$= 78.0 \; g \; DMSO$$

5. 69.943% Fe

The formula indicates the ratio of moles. From this the ratio of masses must be calculated to determine percent composition. In 1 mole of Fe_2O_3 there would be how many grams of Fe and how many grams of O?

$$? \; g \; Fe = 1 \; mol \; Fe_2O_3 \left(\frac{55.847 \; g \; Fe}{1 \; mol \; Fe} \right) \left(\frac{2 \; mol \; Fe}{1 \; mole \; Fe_2O_3} \right)$$

$$= 111.694 \; g \; Fe$$

$$? \; g \; O = 1 \; mol \; Fe_2O_3 \left(\frac{15.9994 \; g \; O}{1 \; mol \; O} \right) \left(\frac{3 \; mol \; O}{1 \; mol \; Fe_2O_3} \right)$$

$$= 47.9982 \; g \; O$$

The mass of 1 mole of Fe_2O_3 is the sum of the masses of the component parts, 159.692 g of Fe_2O_3 per 1 mol of Fe_2O_3. Percent composition is the mass of a component divided by the mass of the whole times 100. Therefore:

$$\% \; Fe \; in \; Fe_2O_3 = \left(\frac{111.694 \; g \; Fe}{159.692 \; g \; Fe_2O_3} \right) 100\%$$

$$= 69.9434\% \; Fe$$

6. 34.59019% Al

First calculate the mass of 1 mol of gibbsite:

$$? \; g \; Al = 1 \; mol \; Al_2O_3 \cdot 3H_2O \left(\frac{2 \; mol \; Al}{mol \; Al_2O_3 \cdot 3H_2O} \right) \left(\frac{26.98154 \; g \; Al}{1 \; mol \; Al} \right)$$

$$= 53.96308 \; g \; Al / 1 \; mol \; Al_2O_3 \cdot 2H_2O$$

$$? \; g \; O = 1 \; mol \; Al_2O_3 \cdot 3H_2O \left(\frac{6 \; mol \; O}{mol \; Al_2O_3 \cdot 3H_2O} \right) \left(\frac{15.9994 \; g \; O}{1 \; mol \; O} \right)$$

$$= 95.9964 \; g \; O / 1 \; mol \; Al_2O_3 \cdot 3H_2O$$

$$? \text{ g H} = 1 \text{ mol Al}_2O_3 \cdot 3H_2O \left(\frac{6 \text{ mol H}}{\text{mol Al}_2O_3 \cdot 3H_2O} \right) \left(\frac{1.0079 \text{ g H}}{1 \text{ mol H}} \right)$$

$$= 6.0474 \text{ g H/1 mol Al}_2O_3 \cdot 3H_2O$$

The mass of 1 mol of gibbsite is the sum of the three
values just calculated: 53.96308 g + 6.0474 g + 95.9964 g
= 156.0069 g. Then calculate the percentage of aluminum:

$$\% \text{ Al} = \left(\frac{\text{mass Al}}{\text{mass Al}_2O_3 \cdot 3H_2O} \right) 100\%$$

$$= \left(\frac{53.96308 \text{ g}}{156.0069 \text{ g}} \right) 100\%$$

$$= 34.59019\% \text{ Al}$$

7. C_5H_5N

Determine the ratio of moles of atoms in the molecule.
Assume for convenience that there are 100 grams of
material. From the percentages given in the problem,
there would be 75.92 g of C, 17.71 g of N, and 6.37 g
of H in the sample.

$$? \text{ mol C} = 75.92 \text{ g C} \left(\frac{1 \text{ mol C}}{12.011 \text{ g C}} \right)$$

$$= 6.321 \text{ mol C}$$

$$? \text{ mol N} = 17.71 \text{ g N} \left(\frac{1 \text{ mol N}}{14.007 \text{ g N}} \right)$$

$$= 1.264 \text{ mol N}$$

$$? \text{ mol H} = 6.37 \text{ g H} \left(\frac{1 \text{ mol H}}{1.008 \text{ g H}} \right)$$

$$= 6.319 \text{ mol H}$$

The compound has the formula $C_{6.321}N_{1.264}H_{6.3119}$, but
this is not an acceptable representation, since atoms
combine in whole-number ratios. To reduce any formula
to the nearest whole-number ratio, divide by the
smallest number. Thus, the empirical formula is

$$\frac{C_{6.321} \quad N_{1.264} \quad H_{6.319}}{1.264 \quad 1.264 \quad 1.264} = C_5NH_5$$

For this compound the empirical formula is the same as
the molecular formula. The molecule is pyridine, which
has the structure

8. empirical
   formula:
   $C_{21}H_{27}O_3N$;
   molecular
   formula:
   $C_{42}H_{54}O_6N_2$

Assume for convenience that there is a 100 g sample, and determine the mole ratios:

$$? \text{ mol C} = 74.0 \text{ g C}\left(\frac{1 \text{ mol C}}{12.01 \text{ g C}}\right)$$

$$= 6.16 \text{ mol C}$$

$$? \text{ mol H} = 7.90 \text{ g H}\left(\frac{1 \text{ mol H}}{1.008 \text{ g.H}}\right)$$

$$= 7.84 \text{ mol H}$$

$$? \text{ mol O} = 14.0 \text{ g O}\left(\frac{1 \text{ mol O}}{16.00 \text{ g O}}\right)$$

$$= 0.875 \text{ mol O}$$

$$? \text{ mol N} = 4.10 \text{ g N}\left(\frac{1 \text{ mol N}}{14.01 \text{ g N}}\right)$$

$$= 0.293 \text{ mol N}$$

The nearest whole-number ratio for this compound is $C_{21}H_{27}O_3N$, which has a formula weight of 341 g/mol. The molecular formula is found by determining what multiple the weight of the empirical formula is of the actual molecular weight:

$$\frac{\text{actual molecular weight}}{\text{weight of empirical formula}} = ?$$

$$\frac{682 \text{ g/mol}}{341 \text{ g/mol}} = 2$$

Thus, the molecular formula is twice the empirical formula, and the molecular formula is

$$C_{42}H_{54}O_6N_2$$

SELF-TEST

Complete the test in 20 minutes.

1. Name the diatomic elements.

2. Which of the following is an empirical formula?

        (a) C_2H_2 (c) C_2H_4

        (b) C_2H_3 (d) C_3H_6

3. Calculate the number of atoms in a 3.20 g sample of copper.

4. In which of the following compounds is the mass percentage of calcium greatest?
    (a) $Ca_3(PO_4)_2$ (c) CaI_2
    (b) CaO   (d) $CaCO_3$

5. A certain organic chemical is shown by analyses to contain by weight 43.90% carbon, 3.05% hydrogen, 9.76% oxygen, and 43.29% chlorine. What is the empirical formula of the compound?

STOICHIOMETRY, PART II: CHEMICAL EQUATIONS

OBJECTIVES

I. You should be able to demonstrate your knowledge of the following terms by defining them, describing them, or giving specific examples of them:

actual yield [4.4]
balanced equation [4.1]
chemical equation [4.1]
concentration [4.5]
limiting reactants [4.3]
molarity [4.5]
percent yield [4.4]
product [4.1]
reactant [4.1]
solute [4.5]
solvent [4.5]
theoretical yield [4.4]

II. You should be able to balance chemical equations and use them in stoichiometric calculations.

III. You should be able to predict the products for the combustion of compounds containing carbon, hydrogen, sulfur, and nitrogen.

IV. You should be able to perform calculations to obtain concentrations.

V. You should be able to compute percent yields from actual yields and theoretical yields. You should be able to calculate theoretical yields.

UNITS,
SYMBOLS,
MATHEMATICS

II. Continue to learn the names and symbols of common elements:

1. potassium, K
2. silver, Ag
3. hydrogen, H
4. sodium, Na
5. calcium, Ca
6. oxygen, O
7. chlorine, Cl
8. nitrogen, N
9. carbon, C
10. lithium, Li
11. boron, B
12. phosphorus, P

13. neon, Ne
14. copper, Cu
15. aluminum, Al
16. sulfur, S
17. magnesium, Mg
18. fluorine, F
19. bromine, Br
20. iron, Fe
21. helium, He
22. berylium, Be
23. silicon, Si
24. iodine, I

EXERCISES

I. Balance the following chemical equations.

1. $Al_4C_3 + H_2O \longrightarrow Al(OH)_3 + CH_4$

2. $ThO_2 + HF \longrightarrow ThF_4 + H_2O$

3. $FeCr_2O_4 + C \longrightarrow Fe + Cr + CO$

4. $XeF_6 + H_2O \longrightarrow XeO_3 + HF$

5. $I_2O_5 + CO \longrightarrow I_2 + CO_2$

6. $Na + SF_6 \longrightarrow Na_2S + NaF$

7. $Si_4H_{10} + O_2 \longrightarrow SiO_2 + H_2O$

8. $C_6H_6O + O_2 \longrightarrow$ predict the products

9. $S + O_2 \longrightarrow$ predict the products

10. $NH_3 + O_2 \longrightarrow$ predict the products

II. Work the following problems.

1. Commercially iron is obtained from the reduction of hematite, Fe_2O_3:

   $Fe_2O_3 + CO \rightarrow Fe + CO_2$

   How many grams of iron can be obtained from 5.24 g of Fe_2O_3?

2. If only 2.04 g of Fe is obtained from the sample described in problem 1 of this section, what is the percent yield of the reduction process?

3. How many grams of iron can be obtained by the reaction of 2.78 g of CO with excess Fe_2O_3? (See problem 1 of this section.)

4. The carbon monoxide needed for producing iron by the reaction given in problem 1 of this section is obtained by the reaction of oxygen with carbon, which is in the form of coke:

$$C + O_2 \rightarrow CO$$

How much iron can be obtained from 1.47 g of C?

5. Refer to problem 1 of this section and answer the following:

   (a) How much iron can be obtained from 4.02 g of Fe_2O_3 and 1.78 g of CO?

   (b) What are the masses of all compounds after the reaction is complete?

6. Ethyl alcohol, C_2H_5OH, is a product of the fermentation of sugars. In the presence of the enzyme invertase, sucrose undergoes conversion into invert sugar.*

$$C_{12}H_{22}O_{11} + H_2O \xrightarrow{\text{invertase}} 2C_6H_{12}O_6$$

The invert sugar is converted into ethyl alcohol and carbon dioxide by the enzyme zymase. Usually only half the sucrose is converted into alcohol, a 50.0 percent yield.

$$C_6H_{12}O_6 \xrightarrow{\text{zymase}} 2C_2H_5OH + 2CO$$

How much ethyl alcohol, C_2H_5OH, can be obtained from 171 g of sucrose, $C_{12}H_{22}O_{11}$?

7. A total of 4.00 g of $AgNO_3$ is dissolved in a small quantity of water and diluted to 1.00 L. What is the molarity of the final $AgNO_3$ solution?

8. How many grams of $K_2Cr_2O_7$ are needed to prepare 1.00 L of 0.100 M $K_2Cr_2O_7$?

9. Calculate the molarity of 60.0 mL of solution of $KMnO_4$ which contains 1.90 g of $KMnO_4$.

---

*A catalyst such as an enzyme is not written in the balanced equation of a reaction; rather, it is written over the reaction arrow of the equation. Zymase increases the rate of conversion of sucrose to alcohol such that the conversion proceeds at a reasonable rate.

10. How many grams of $K_2Cr_2O_7$ are needed to prepare 60. mL of a 0.15 M $K_2Cr_2O_7$ solution?

11. How many grams of KCl can react with 10.0 mL of 0.050 M $AgNO_3$ to form AgCl?

12. What volume of 0.200 M NaOH is required to react with 25.0 mL of 0.100 M HCl? The equation for the reaction is HCl + NaOH $\longrightarrow$ NaCl + H_2O

13. What is the molarity of the acetic acid, CH_3COOH, in vinegar if 32.4 mL of 0.250 M NaOH reacts with a 10.0 mL sample of vinegar?

    CH_3COOH + NaOH $\longrightarrow$ CH_3COONa + H_2O

III. Work the following problems. These are more challenging than those in part II of this chapter.

1. A piece of iron weighs 15.74 g. Assume each atom of iron is a perfect sphere with a radius of 117 pm. How great a distance would this iron span if the atoms were arranged one against another in a single line?

2. A 0.00745 g sample containing carbon, hydrogen, and chlorine is burned in an oxygen rich atmosphere. The products of the combustion react with added chemicals converting the chlorine to chloride ion, Cl^-. It takes exactly 5.01 ml of 0.0140 M silver nitrate to completely precipitate the chloride:

    $Ag^+ + NO_3^- + Cl^- \longrightarrow AgCl + NO_3^-$

    What was the % chloride in the original sample?

3. A 7.84 g sample of a mixture of $KClO_3$ and KCl is heated until the $KClO_3$ decomposes to KCl and O_2 gas. The unbalanced equation is

    $KClO_3 \longrightarrow KCl + O_2$

    The decomposed sample weighs 6.92 g. How many grams of KCl were in the original 7.84 g sample? How many grams of KCl are in the sample after the heating process?

4. What is the molarity of chloride, Cl^-, in 250. ml of solution which contains 7.00 g KCl and 7.00 g NaCl?

ANSWERS TO   I. Balancing equations
EXERCISES

1. $Al_4C_3 + 12H_2O \longrightarrow 4Al(OH)_3 + 3CH_4$

> A balanced equation contains the same number of each kind of atom on both sides of the arrow. It is usually easiest to leave hydrogen and oxygen atoms for last and begin with the other atoms.

$Al_4C_3 + H_2O \longrightarrow Al(OH)_3 + CH_4$

> There are four Al atoms on the left and only one on the right. This should be balanced by putting the coefficient 4 in front of $Al(OH)_3$.

$Al_4C_3 + H_2O \longrightarrow 4Al(OH)_3 + CH_4$

> There are three C atoms on the left and one on the right. Balance the C atoms by placing the coefficient 3 in front of CH_4.

$Al_4C_3 + H_2O \longrightarrow 4Al(OH)_3 + 3CH_4$

> There are two H atoms on the left and 24 H atoms on the right [12 in $3Al(OH)_3$ and 12 in $3CH_4$]. Balance the H atoms by putting the coefficient 12 in front of H_2O.

$Al_4C_3 + 12H_2O \longrightarrow 4Al(OH)_3 + 3CH_4$

> There are 12 O atoms on both sides of the arrow. The equation is now balanced.

2. $ThO_2 + HF \longrightarrow ThF_4 + H_2O$

3. $FeCr_2O_4 + C \longrightarrow Fe + Cr + CO$

4. $XeF_6 + H_2O \longrightarrow XeO_3 + HF$

5. $I_2O_5 + CO \longrightarrow I_2 + CO_2$

6. $Na + SF_6 \longrightarrow Na_2S + NaF$

7. $Si_4H_{10} + O_2 \longrightarrow SiO_2 + H_2O$

8. $C_6H_6O + O_2 \longrightarrow CO_2 + H_2O$

> Near the end of section 4.1 in your text is a list of the products of complete combustion of compounds in oxygen. The compound in this question contains carbon and hydrogen, thus CO_2 and H_2O are produced with combustion.

9. $S + O_2$ SO_2

10. $4NH_3 + 3O_2$ $2N_2 + 6H_2O$

II. Chemical Calculations

1.   3.67 g Fe   First make sure that the chemical equation is balanced. The equation as written in the problem is not balanced. The balanced equation

$$Fe_2O_3 + 3CO \rightarrow 2Fe + 3CO_2$$

states that 1 mol Fe_2O_3 and 3 mol CO react to give 2 mol Fe and 3 mol CO_2. The molecular weights of all compounds can be determined from atomic weights. The molecular weights are: Fe_2O_3, 159.692 g/mol; CO, 28.010 g/mol; Fe, 55.847 g/mol; and CO_2, 44.010 g/mol. From this information it is obvious that the chemical equation also says that 159.692 g Fe_2O_3 and (3 × 28.010) g CO react to yield (2 × 55.847) g Fe and (3 × 44.010) g CO_2:

$$Fe_2O_3 + 3CO \rightarrow 2Fe + 3CO_2$$
1 mol Fe_2O_3 + 3 mol CO → 2 mol Fe + 3 mol CO_2
159.692 g Fe_2O_3 + (3 × 28.010) g CO
    → (2 × 55.847) g Fe + (3 × 44.010) g CO_2
159.692 g Fe_2O_3 + 84.030 g CO
    → 111.694 g Fe + 132.030 g CO_2

Remember these types of relationships; such relationships are necessary for solving problems involving chemical reactions. Conversion factors can be used to solve the problem:

$$? \text{ g Fe} = 5.24 \text{ g } Fe_2O_3 \left(\frac{(2 \times 55.85) \text{ g Fe}}{159.7 \text{ g } Fe_2O_3} \right)$$

$$= 3.67 \text{ g Fe}$$

As long as we keep track of the units, we can use other conversion factors:

$$? \text{ g Fe} = 5.24 \text{ g } Fe_2O_3 \left(\frac{1 \text{ mol } Fe_2O_3}{159.7 \text{ g } Fe_2O_3} \right) \left(\frac{2 \text{ mol Fe}}{1 \text{ mol } Fe_2O_3} \right) \left(\frac{55.85 \text{ g Fe}}{1 \text{ mol Fe}} \right)$$

$$= 3.67 \text{ g Fe}$$

2.   55.6%   From the definition of percent yield and the calculated theoretical yield calculated in problem 1 of tnis section:

$$\text{percent yield} = \left(\frac{\text{actual yield}}{\text{theoretical yield}}\right)100\%$$

$$= \left(\frac{2.04 \text{ g}}{3.67 \text{ g}}\right)100\%$$

$$= 55.6\%$$

3. 3.70 g Fe

Since all the CO reacts and some Fe_2O_3 is left over, the original amount of CO limits the amount of Fe that can be produced:

$$? \text{ g Fe} = 2.78 \text{ g CO} \frac{(2 \times 55.85) \text{ g Fe}}{(3 \times 28.01) \text{ g CO}}$$

$$= 3.70 \text{ g Fe}$$

4. 4.56 g Fe

First make sure that all equations are balanced:

$2C + O_2 \rightarrow 2CO$

2 mol C $\longrightarrow$ 2 mol CO

(2×12.011) g C $\longrightarrow$ (2×28.010) g CO

Using the information in the preceding equation and the one for Fe_2O_3 reduction (see problem 8 of this section), one can write many series of conversion factors. Some of the following are more efficient than others, but your concern should be to find a method that is logical to you:

$$? \text{ g Fe} = 1.47 \text{ g C}\left(\frac{(2 \times 28.01) \text{ g CO}}{(2 \times 12.01) \text{ g C}}\right)\left(\frac{(2 \times 55.85) \text{ g Fe}}{(3 \times 28.01) \text{ g CO}}\right)$$

$$= 4.56 \text{ g Fe} \qquad or$$

$$? \text{ g Fe} = 1.47 \text{ g C}\left(\frac{1 \text{ mol C}}{12.01 \text{ g C}}\right)\left(\frac{2 \text{ mol Fe}}{3 \text{ mol C}}\right)\left(\frac{55.85 \text{ g Fe}}{1 \text{ mol Fe}}\right)$$

$$= 4.56 \text{ g Fe} \qquad or$$

$$? \text{ g Fe} = 1.47 \text{ g C}\left(\frac{1 \text{ mol C}}{12.01 \text{ g C}}\right)\left(\frac{1 \text{ mol CO}}{1 \text{ mol C}}\right)\left(\frac{2 \text{ mol Fe}}{3 \text{ mol CO}}\right)\left(\frac{55.85 \text{ g}}{1 \text{ mol Fe}}\right)$$

$$= 4.56 \text{ g Fe}$$

5. (a) 2.37 g Fe

In this problem the quantities of two reagents are specified. Most probably they are not mixed in the exact reaction ratio. Therefore, one of them will not react completely. In order to work the problem, the reagent that reacts completely must be identified, since it limits the amount of product that can be obtained:

Method 1:

With an electronic calculator the problem can be solved very quickly by assuming that first one and then the other reagent is limiting. If Fe_2O_3 were limiting,

$$? \text{ g Fe} = 4.02 \text{ g } Fe_2O_3\left(\frac{(2 \times 55.85) \text{ g Fe}}{(1 \times 159.69) \text{ g } Fe_2O_3}\right)$$

$$= 2.81 \text{ g Fe}$$

If CO were limiting,

$$? \text{ g Fe} = 1.78 \text{ g CO}\left(\frac{(2 \times 55.85) \text{ g Fe}}{(3 \times 28.01) \text{ g CO}}\right)$$

$$= 2.37 \text{ g Fe}$$

It is obvious that only 2.37 g of Fe could be produced. Since all the CO is consumed to produce this amount of iron, the remaining Fe_2O_3 cannot react. The reactant that would yield the least amount of product is the limiting reagent.

Method 2:

First calculate the number of moles of each reactant as detailed in your text:

$$? \text{ mol } Fe_2O_3 = 4.02 \text{ g } Fe_2O_3\left(\frac{1 \text{ mol } Fe_2O_3}{159.69 \text{ g } Fe_2O_3}\right)$$

$$= 0.02517 \text{ mol } Fe_2O_3$$

$$? \text{ mol} = 1.78 \text{ g CO}\left(\frac{1 \text{ mol CO}}{28.01 \text{ g CO}}\right)$$

$$= 0.06355 \text{ mol CO}$$

The chemical equation says that

$$1 \text{ mol } Fe_2O_3 = 3 \text{ mol CO}$$

Compare the number of moles:

$$\left(\frac{0.02517 \text{ mol } Fe_2O_3}{1 \text{ mol } Fe_2O_3}\right) = 0.02517$$

$$\left(\frac{0.06355 \text{ mol CO}}{3 \text{ mol CO}}\right) = 0.02118$$

The CO is limiting because a lower proportionate amount is present. Solve the problem using the limiting reactant, CO:

$$? \text{ g Fe} = 0.06355 \text{ mol CO} \left(\frac{(2 \times 55.85) \text{ g Fe}}{3 \text{ mol CO}} \right) = 2.37 \text{ g Fe}$$

5. (b) 2.37 g Fe,
      0.00 g CO,
      0.64 g Fe_2O_3,
      2.80 g CO_2

After the reaction is complete, 2.37 g of Fe are formed and 0.00 g of CO remains. The mass of CO_2 and the mass of residual Fe_2O_3 need to be calculated:

$$? \text{ g CO}_2 = 0.06355 \text{ mol CO} \left(\frac{(3 \times 44.01) \text{ g CO}}{3 \text{ mol CO}} \right) = 2.80 \text{ g CO}_2$$

The mass of Fe_2O_3 remaining can be calculated:

$$? \text{ g Fe}_2O_3 \text{ remaining} = \text{initial g Fe}_2O_3 - \text{reacted g Fe}_2O_3$$

$$= 4.02 \text{ g Fe}_2O_3 - 0.06355 \text{ mol CO} \left(\frac{(1 \times 159.69) \text{ g Fe}_2O_3}{3 \text{ mol CO}} \right)$$

$$= 4.02 \text{ g Fe}_2O_3 - 3.38 \text{ g Fe}_2O_3$$

$$= 0.64 \text{ g Fe}_2O_3$$

6. 46.1 g C_2H_5OH

$$? \text{ g C}_2\text{H}_5\text{OH} = 171 \text{ g C}_{12}\text{H}_{22}\text{O}_{11} \left(\frac{1 \text{ mol C}_{12}\text{H}_{22}\text{O}_{11}}{342 \text{ g C}_{12}\text{H}_{22}\text{O}_{11}} \right)$$

$$\times \left(\frac{1 \text{ mol C}_{12}\text{H}_{22}\text{O}_{11} \text{ reacting}}{2 \text{ mol C}_{12}\text{H}_{22}\text{O}_{11} \text{ present}} \right) \left(\frac{2 \text{ mol C}_6\text{H}_{12}\text{O}_6}{1 \text{ mol C}_{12}\text{H}_{22}\text{O}_{11}} \right)$$

$$\times \left(\frac{2 \text{ mol C}_2\text{H}_5\text{OH}}{1 \text{ mol C}_6\text{H}_{12}\text{O}_6} \right) \left(\frac{46.1 \text{ g C}_2\text{H}_5\text{OH}}{1 \text{ mol C}_2\text{H}_5\text{OH}} \right)$$

$$= 46.1 \text{ g C}_2\text{H}_5\text{OH}$$

7. 0.0235 M

State problem mathematically and solve using conversion factors:

$$? \text{ mol AgNO}_3 = 1.00 \text{ L AgNO}_3\text{sol'n} \left(\frac{4.00 \text{ g AgNO}_3}{1.00 \text{ L AgNO}_3 \text{ sol'n}} \right)$$

$$\times \left(\frac{1 \text{ mol AgNO}_3}{169.9 \text{ g AgNO}_3} \right)$$

$$= 0.0235 \text{ mol}$$

There are 0.0235 mol $AgNO_3$ in 1.00 L. Thus the concentration is 0.235 mol $AgNO_3$/L or 0.0235 M.

The equation states that one mole NaOH reacts with one mole HCl. Thus

$$? \text{ mole NaOH} = 2.50 \times 10^{-3} \text{ mole HCl} \left(\frac{1 \text{ mole NaOH}}{1 \text{ mole HCl}}\right)$$

$$= 2.50 \times 10^{-3} \text{ mole HCl}$$

Find the volume of NaOH using the number of moles and the concentration.

$$? \text{ mL NaOH} = 2.50 \times 10^{-3} \text{ mole NaOH} \left(\frac{1000 \text{ mL NaOH solution}}{0.200 \text{ mole NaOH}}\right)$$

$$= 12.5 \text{ mL NaOH solution}$$

13.   0.812 *M*   Determine the number of moles of acetic acid that react with the NaOH.

$$? \text{ moles CH}_3\text{COOH} = 32.4 \text{ mL NaOH}$$

$$\left(\frac{0.150 \text{ mole NaOH}}{1000 \text{ mL NaOH}}\right) \left(\frac{1 \text{ mole CH}_3\text{COOH}}{1 \text{ mole NaOH}}\right)$$

$$= 8.10 \times 10^{-3} \text{ mole CH}_3\text{COOH}$$

The concentration units are moles per liter. The problem states that 10.0 mL was used in the reaction.

$$? \text{ } M \text{ CH}_3\text{COOH}$$

$$= \left(\frac{8.10 \times 10^{-3} \text{ mole CH}_3\text{COOH}}{10.0 \text{ mL CH}_3\text{COOH solution}}\right)$$

$$\left(\frac{1000 \text{ mL}}{1 \text{ mL}}\right)$$

$$= 0.812 \text{ } M \text{ CH}_3\text{COOH}$$

III.   More-Challenging Problems

1. 3.97×10^{13} m   First calculate the number of atoms in the sample:

$$? \text{ atoms} = 15.74 \text{ g Fe} \left(\frac{6.022 \times 10^{23} \text{ atoms}}{55.85 \text{ g Fe}}\right)$$

$$= 1.697 \times 10^{23} \text{ atoms}$$

Then calculate the distance. Note that each atom has a diameter of 234 pm and thus contributes that amount to the total distance.

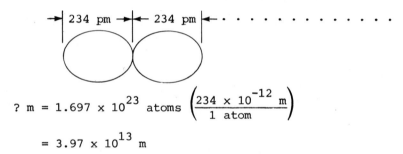

$$? \text{ m} = 1.697 \times 10^{23} \text{ atoms} \left(\frac{234 \times 10^{-12} \text{ m}}{1 \text{ atom}} \right)$$

$$= 3.97 \times 10^{13} \text{ m}$$

This is greater than the distance between the earth and the sun.

2. 33.4%

First determine the number of grams of chloride precipitated by the silver nitrate. This is the same amount as the amount of chlorine in the original sample.

$$? \text{ g Cl} = 0.00501 \text{ L AgNO}_3 \text{ sol'n} \left(\frac{0.0140 \text{ mol AgNO}_3}{1.00 \text{ L AgNO}_3 \text{ sol'n}} \right)$$

$$\times \left(\frac{1 \text{ mol Cl}}{1 \text{ mol AgNO}_3} \right) \left(\frac{35.45 \text{ g Cl}}{1 \text{ mol Cl}} \right)$$

$$= 2.49 \times 10^{-3} \text{ g Cl}$$

Now calculate the percent chloride:

$$? \% \text{ Cl} = \left(\frac{2.49 \times 10^{-3} \text{ g Cl}}{7.45 \times 10^{-3} \text{ g sample}} \right) 100$$

$$= 33.4\%$$

3. 5.49 g KCl
   originally

First balance the equation:

$$2 \text{KClO}_3 \longrightarrow 2 \text{KCl} + 3 \text{O}_2$$

Determine the number of moles of $KClO_3$ which were initially present from the mass of oxygen liberated during the heating process. The mass of oxygen liberated is determined by the weight loss of the sample:

$$7.84 \text{ g} - 6.92 \text{ g} = 0.92 \text{ g O}_2 \text{ liberated}$$

Then complete the calculation:

$$? \text{ g KClO}_3 = 0.92 \text{ g O}_2 \left(\frac{1 \text{ mol O}_2}{32.0 \text{ g O}_2}\right)\left(\frac{2 \text{ mol KClO}_3}{3 \text{ mol O}_2}\right)$$

$$\left(\frac{122.5 \text{ g KClO}_3}{1 \text{ mol KClO}_3}\right)$$

$$= 2.35 \text{ g KClO}_3$$

The original mass of KCl is the mass of the original sample minus the mass of $KClO_3$:

$$7.84 \text{ g sample} - 2.35 \text{ g KClO}_3 = 5.49 \text{ g KCl}$$

6.92 g KCl finally

The entire 6.92 g sample remaining after heating is KCl.

4. 0.856 M Cl⁻

Determine the number of moles of Cl^- supplied by each reagent, KCl and NaCl:

$$? \text{ mol Cl}^- = 7.00 \text{ g KCl} \left(\frac{1 \text{ mol Cl}^-}{74.55 \text{ g KCl}}\right)$$

$$= 0.0939 \text{ mol Cl}^-$$

$$? \text{ mol Cl}^- = 7.00 \text{ g NaCl} \left(\frac{1 \text{ mol Cl}^-}{58.45 \text{ g NaCl}}\right)$$

$$= 0.120 \text{ mol Cl}^-$$

The total moles of Cl^- is determined by summing those from each reagent:

$$\text{total mol Cl}^- = 0.0939 \text{ mol Cl}^- + 0.120 \text{ mol Cl}^- = 0.214 \text{ mol Cl}^-$$

Now determine the number of moles per liter of solution, i.e., the molarity:

$$? \text{ mol Cl}^- = 1.00 \text{ L sol'n} \left(\frac{0.214 \text{ mol Cl}^-}{0.250 \text{ L sol'n}}\right) = 0.856 \text{ mol Cl}^-$$

SELF-TEST

Complete the test in 20 minutes.

1. The compound S_4N_3Cl can be formed by the reaction

$$3S_4N_4 + 2S_2Cl_2 \rightarrow 4S_4N_3Cl$$

How many grams of S_4N_3Cl can be formed with .500 g of S_4N_4 and 3.00 g of S_2Cl_2?

2. How many grams of acetylene, H_2C_2, can be produced by
    mixing 1.00 g calcium carbide, CaC_2, with 1.00 g water?
    What is the sum of the weight of all reactants and
    products after the reaction is completed? The unbalanced
    chemical equation is

    $$CaC_2 + H_2O \longrightarrow Ca(OH)_2 + H_2C_2$$

3. Magnesium hydroxide is used as an antacid. It reacts
    with stomach acid as shown in the following unbalanced
    equation.

    $$HCl + Mg(OH)_2 \longrightarrow MgCl_2 + H_2O$$

    If a 0.57 g antacid tablet reacts completely with 11 mL
    of 1.0 M HCl, what is the percent $Mg(OH)_2$ in the antacid
    tablet?

THERMOCHEMISTRY

CHAPTER

5

OBJECTIVES   I. You should be able to demonstrate your knowledge of the
                    following terms by defining them, describing them, or
                    giving specific examples of them:

        acceleration [5.1]
        bond energy [5.7]
              average bond energy [5.7]
              bond dissociation energy [5.7]
        calorie, cal [5.2]
        calorimeter [5.3]
        Celsius temperature scale [5.2]
        centigrade, °C [5.2]
        endothermic [5.4]
        energy [5.1]
        enthalpy [5.4]
              enthalpy of formation [5.4]
              standard enthalpy of formation [5.4]
        exothermic [5.4]
        Fahrenheit temperature scale [5.2]
        heat [5.2]
        heat capacity [5.3]
        joule, J [5.1]
        law of Hess [5.5]
        polyatomic molecule [5.7]
        specific heat [5.2]
        standard state [5.6]
        temperature [5.2]
        thermochemistry [introduction]
        work [5.1]

II. You should be able to convert temperatures from the Fahrenheit to the Celsius scale.

III. You should be able to calculate enthalpies of reactions and heat capacities from data obtained with calorimeters.

IV. You should be able to use the law of Hess in thermo-chemical calculations

V. You should be able to use enthalpies of formation to calculate enthalpies of reactions.

VI. You should be able to estimate enthalpies of reactions from average bond energies.

UNITS,
SYMBOLS,
MATHEMATICS

I. The SI units and associated meanings of certain factors should be familiar to you:

(a) Specific heat has units of J/(g °C).
(b) Heat capacity has units of either J/°C or kJ/°C.
(c) The enthalpy of a reaction, ΔH, has units of J or kJ. This is the amount of heat either lost or gained by the reaction of the number of moles of each species in the balanced chemical equation. For example, for the thermochemical reaction

$$4NH_3(g) + 3O_2(g) \longrightarrow 2N_2(g) + 6H_2O(l) \quad \Delta H = -1531kJ$$

appropriate conversion factors would include

$$\frac{-1531 \text{ kJ}}{4 \text{ mol } NH_3} \quad \frac{-1531 \text{ kJ}}{3 \text{ mol } O_2} \quad \frac{-1531 \text{ kJ}}{2 \text{ mol } N_2} \quad \frac{-1531 \text{ kJ}}{6 \text{ mol } H_2O}$$

(d) ΔH is negative for exothermic reactions, i.e., those evolving heat.
(e) ΔH is positive for endothermic reactions, i.e., those consuming heat.
(f) ΔH is positive for breaking chemical bonds.

II. The following symbols are used in this chapter and you should be familiar with them.

- a is the symbol used in the equation $F = ma$ to represent acceleration.

- (aq) is the symbol used after a chemical formula to indicate that the chemical is dissolved in water.

- C is the symbol used to represent heat capacity in equations involving calorimetry. For example, C = (mass)(specific heat).

- cal is the symbol for the calorie, which is sometimes used to define heat energy. It is not an SI unit, but can be defined in terms of the SI unit joule:
1 cal = 4.184 joule (exactly)

- Δ is the capitalized Greek letter delta and is used before other symbols to represent the change in that value.

- H is the symbol used to represent enthalpy.

- ΔH is the symbol used to indicate change in enthalpy. For a chemical reaction it equals the enthalpy of the products minus the enthalpy of the reactants.

- ΔH_f^o is the symbol used to represent the standard enthalpy of formation of a compound.

- ΔH^o is the symbol used to indicate standard enthalpy changes during reactions of chemicals in their standard physical state.

- ^{o}F is the symbol for degrees Fahrenheit.

- (g) is a symbol used after a chemical formula to indicate that the chemical is gaseous.

- J is the symbol for the joule, the SI unit of work.

- $J/(g\,^oC)$ are the units for specific heat.

- kJ is the symbol for Kilojoule.

- (l) is a symbol used after a chemical formula to indicate that the chemical is a liquid.

- s is the abbreviation used for the unit of time, the second.

- Σ is the capitalized Greek letter sigma and it is used before other symbols to indicate the sum of all values.

- (s) is a symbol used after a chemical formula to indicate that the chemical is a solid.

- W is the symbol used to stand for work in equations such as W = (force)(distance).

III. Continue to learn the names and symbols of common elements.

1.	potassium, K	14.	neon, Ne
2.	silver, Ag	15.	copper, Cu
3.	hydrogen, H	16.	aluminum, Al
4.	sodium, Na	17.	sulfur, S
5.	calcium, Ca	18.	magnesium, Mg
6.	oxygen, O	19.	fluorine, F
7.	chlorine, Cl	20.	bromine, Br
8.	nitrogen, N	21.	iron, Fe
9.	carbon, C	22.	helium, He
10.	lithium, Li	23.	berylium, Be
11.	boron, B	24.	silicon, Si
12.	phosphorus, P	25.	iodine, I
13.	zinc, Zn	26.	nickel, Ni

EXERCISES   I. Use the data in the following table when needed to solve the thermochemical problems in this section. In the space provided write the balanced equation that corresponds to the listed enthalpy of formation.

ENTHALPIES OF FORMATION AT 25°C

Molecule	ΔH_f (kJ)	Equation
$CO(g)$	-110.5	$C + \frac{1}{2}O_2 \rightarrow CO$
$CO_2(g)$	-393.7	$C + O_2 \rightarrow CO_2$
$Fe_2O_3(s)$	-822.2	~~Fe_2O_3~~ $2Fe + \frac{3}{2}O_2 \rightarrow$
$HCl(g)$	-92.5	
$H_2O(g)$	-241.8	_____
$H_2O(l)$	-285.8	
$H_2SO_4(l)$	-811.3	_____
$I_2(g)$	+62.3	
$NH_3(l)$	-66.9	_____
$NO(g)$	+90.4	
$NO_2(g)$	+52.7	_____
$O_3(g)$	+142.3	

PCl_3 (g)	-306.3	_____
PCl_5 (g)	-399.2	
$POCl_3$ (g)	-592.0	_____
SO_2 (g)	-297.0	
SO_3 (g)	-395.0	_____

1. What is the enthalpy change at 25°C for the following reaction?

   $$SO_3 (g) + H_2O(l) \rightarrow H_2SO_4 (l)$$

2. What is the enthalpy change at 25°C for the following reaction?

   $$PCl_5 (g) + H_2O(g) \rightarrow POCl_3 (g) + 2HCl (g)$$

3. The enthalpy of combustion of pentane, C_5H_{12} (g), to CO_2 (g) and $H_2O(l)$ is -3536.3 kJ/mol. What is the enthalpy of formation of pentane at 25°C?

4. The specific heat of platinum is 0.136 J/(g°C). What is the heat capacity of 25.0 g of platinum?

5. How many joules of heat are required to raise the temperature of the sample of platinum in problem 4 from 20.00 to 21.50°C?

6. A 12.45 g sample of P_4O_{10} (s) reacted with a stoichio-metrically equivalent quantity of water in a vessel placed in a calorimeter containing 950.0 g of water. The temperature of the calorimeter and its contents increases from 22.815°C to 26.885°C. If the calorimeter has a heat-absorbing capacity equal to that of 165.0 g H_2O, what is the enthalpy change for the following reaction?

   $$P_4O_{10} (s) + 6HO (l) \rightarrow 4H_3PO_4 (aq)$$

   Assume the specific heat of water to be 4.184 J/(g°C).

7. Use the average bond energies from Table 3.2 of your text to calculate the value of ΔH for the reaction

   $$CH_4 (g) + 2O_2 (g) \longrightarrow O=C=O (g) + 2H_2O (g)$$

   Is this reaction endothermic or exothermic?

8. For the reaction

$$CH_4(g) + Br_2(g) \longrightarrow H-\overset{\overset{\displaystyle H}{|}}{\underset{\underset{\displaystyle H}{|}}{C}}-Br(g) + HBr(g) \qquad \Delta H = -25 \text{ kJ}$$

Calculate the C-Br bond energy using the data in Table 3.2 of your text and the value of ΔH for the reaction given in this problem.

9. Given

(a) $COCl_2(g) + H_2S(g) \longrightarrow 2HCl(g) + COS(g)$ $\Delta H = -78.71 \text{ kJ}$

(b) $COS(g) + H_2S(g) \longrightarrow H_2O(g) + CS_2(l)$ $\Delta H = 3.42 \text{ kJ}$

determine the heat of the reaction

$$COCl_2(g) + 2H_2S(g) \longrightarrow 2HCl(g) + H_2O(g) + CS_2(l)$$

10. Given

(a) $\tfrac{1}{4}P_4(s) + 3/2 \, Cl_2(g) \longrightarrow PCl_3(g)$ $\Delta H = -306.4 \text{ kJ}$

(b) $P_4(s) + 5O_2(g) \longrightarrow P_4O_{10}(g)$ $\Delta H = -2967.3 \text{ kJ}$

(c) $PCl_3(g) + Cl_2(g) \longrightarrow PCl_5(g)$ $\Delta H = -84.2 \text{ kJ}$

(d) $PCl_3(g) + \tfrac{1}{2}O_2(g) \longrightarrow Cl_3PO(g)$ $\Delta H = -285.7 \text{ kJ}$

calculate the value of ΔH for the reaction

$$P_4O_{10}(g) + 6PCl_5(g) \longrightarrow 10Cl_3PO(g)$$

11. Given that the N≡N bond energy in N_2 gas is 941 kJ/mol and that the O=O bond energy in O_2 gas is 494 kJ/mol, calculate the NO bond energy in NO and ONO. Explain any discrepancies. Use data from the table in this section of the Study Guide.

12. What quantity of heat is liberated from the combustion of 1.00 l of pentane, C_5H_{12}? Pentane has a density of 0.621 g/ml. Compare this to the quantity of heat liberated from the combustion of 1.00 l of carbon disulfide, CS_2. Carbon disulfide has a density of 1.263 g/ml. The thermochemical equations are

$$C_5H_{12}(l) + O_2(g) \longrightarrow 5CO_2(g) + 6H_2O(g) \qquad \Delta H = -3563.3 \text{ kJ}$$

$$CS_2(l) + 3O_2(g) \longrightarrow CO_2(g) + 2SO_2(g) \qquad \Delta H = -1075.2 \text{ kJ}$$

I. Thermochemical calculations
Equations

$C(graphite) + 1/2\ O_2(g) \rightarrow CO(g)$

$2Fe(s) + 3/2\ O\ (g) \rightarrow Fe_2O_3(s)$

$H_2(g) + 1/2\ O_2(g) \rightarrow H_2O(g)$

$H_2(g) + S(s) + 2O_2(g) \rightarrow H_2SO_4(l)$

$1/2\ N_2(g) + 3/2\ H_2(g) \rightarrow NH_3(l)$

$1/2\ N_2(g) + O_2(g) \rightarrow NO_2(g)$

$P(s) + 3/2\ Cl_2(g) \rightarrow PCl_3(g)$

$P(s) + 1/2\ O_2(g) + 3/2\ Cl_2(g) \rightarrow POCl_3(g)$

$S(s) + 3/2\ O_2(g) \rightarrow SO_3(g)$

1. $\Delta H = -130.5$ kJ   Sum any available equations including the enthalpies
such that the final equation and the enthalpy for that
equation are obtained. In the final equation $SO_3(g)$
is to the left of the reaction arrow, so an equation
with known enthalpy change is needed that contains
$SO_3(g)$ on the left. From the table choose the
equation and standard enthalpy of formation of $SO_3(g)$:

$S(s) + 3/2\ O_2(g) \rightarrow SO_3(g)$ $\Delta H = -395.0$ kJ

Invert the equation so that $SO_3(g)$ is on the left.
Inverting the equation changes the sign of the value
of ΔH:

$SO_3(g) \rightarrow S(s) + 3/2\ O_2(g)$ $\Delta H = +395.0$ kJ   (1)

In the final equation $H_2O(l)$ is also needed on the
left. From the table we choose the equation and
standard enthalpy of formation of $H_2O(l)$, invert the
equation, and change the sign of ΔH:

$H_2O(l) \rightarrow H_2(g) + 1/2\ O_2(g)$ $\Delta H = +285.8$ kJ   (2)

Sulfuric acid is needed on the right in the final
equation, so we choose the equation and standard
enthalpy of formation of $H_2SO_4(l)$:

$H_2(g) + S(s) + 2O_2(g) \rightarrow H_2SO_4(l)$ $\Delta H = -811.3$ kJ   (3)

Sum equations (1), (2), and (3), including the values:

$SO_3(g) \rightarrow S(s) + 3/2\ O_2(g)$ $\Delta H = +395.0$ kJ

$H_2O(1) \rightarrow H_2(g) + 1/2\ O_2(g)$ $\Delta H = +285.8$ kJ

$H_2(g) + S(s) + 2O_2(g) \rightarrow H_2SO_4(1)$ $\Delta H = -811.3$ kJ

$SO_3(g) + H_2O(1) \rightarrow H_2SO_4(1)$ $\Delta H = -130.5$ kJ

To find the enthalpy change of a reaction we need only find appropriate thermochemical data for reactions that, when mathematically combined, give the desired reaction.

For any reaction, ΔH equals the sum of the enthalpies of formation of the reactants. Each ΔH_f° must be multiplied by the number of moles that appear before the corresponding molecule in the balanced chemical equation.

In this problem,

$\Delta H = \Delta H_f^\circ$ (products) $- \Delta H_f^\circ$ (reactants)

$\quad = (-811.3\text{ kJ}) - (-395.0\text{ kJ} - 285.8\text{ kJ})$

$\quad = -130.5$ kJ

2. $\Delta H = -136.0$ kJ

$P(s) + 3/2\ Cl_2(g) + 1/2\ O_2(g) \rightarrow POCl_3(g)$ $\Delta H = -592.0$ kJ

$H_2(g) + Cl_2(g) \rightarrow 2HCl(g)$ $\Delta H = 2(-92.5)$ kJ

$PCl_5(g) \rightarrow P(s) + 5/2\ Cl_2(g)$ $\Delta H = +399.2$ kJ

$H_2O(g) \rightarrow H_2(g) + 1/2\ O_2(g)$ $\Delta H = +241.8$ kJ

$PCl_5(g) + H_2O(g) \rightarrow POCl_3(g) + 2HCl(g)$ $\Delta H = -136.0$ kJ

3. $\Delta H = -147.1$ kJ   First balance the equation:

$C_5H_{12}(g) + 8O_2(g) \rightarrow 5CO_2(g) + 6H_2O(1)$ $\Delta H = -3536.3$ kJ

From the law of Hess we know

$\Delta H = \Delta H_f^\circ$ (products) $- \Delta H_f^\circ$ (reactants)

$\Delta H_{combustion} = 5\Delta H_f^\circ(CO_2) + 6\Delta H_f^\circ(H_2O) - \Delta H_f^\circ(C_5H_{12}) - 8\Delta H_f^\circ(O_2)$

Rearrange the preceding equation to find the enthalpy of formation of pentane:

$\Delta H_f^\circ(C_5H_{12}) = 5\Delta H_f^\circ(CO_2) + 6\Delta H_f^\circ(H_2O) - 8\Delta H_f^\circ(O_2) - \Delta H_{combustion}$

Then substitute values for all known enthalpies:

$$\Delta H_f^\circ = 5(-393.7 \text{ kJ}) + 6(-285.8 \text{ kJ}) - 8(0) + 3536.2 \text{ kJ}$$

$$= -147.1 \text{ kJ}$$

4. 3.40 J/°C

The heat capacity of a substance, C, can be calculated using the relationship

$$C = (\text{mass})(\text{specific heat})$$

Substituting:

$$C = (25.0 \text{ g})(0.136 \text{ J/g°C}) = 3.40 \text{ J/°C}$$

5. 5.10 J

Since the heat capacity, C, of a given mass of a substance is the amount of heat required to raise the temperature of the mass by 1°C, the heat required to raise it 1.50°C would be 1.50 times the heat capacity:

$$q = C(t_2-t_1)$$

$$= (3.40 \text{ J/°C})(21.50°C - 20.00°C) = (3.40 \text{ J/°C})(1.50°C)$$

$$= 5.10 \text{ J}$$

6. ΔH
$= -433.0$ kJ/mol

Calculate the change in temperature:

$$\Delta T = 26.885°C - 22.815°C = +4.070°C$$

Thus, heat is evolved by the reaction:

$$? \text{ kJ} = (950.0 \text{ g } H_2O + 165.0 \text{ g } H_2O)(4.070°C)(4.184 \text{ J/g°C})$$

$$= 1.899 \times 10^4 \text{ J}$$

$$= 18.99 \text{ kJ}$$

The preceding amount of heat is liberated by the reaction of 12.45g of P_4O_{10}. The enthalpy change per mole of P_4O_{10} can be determined as follows:

$$? \text{ kJ/mol} = \left(\frac{18.99 \text{ kJ}}{12.45 \text{ g } P_4O_{10}}\right)\left(\frac{283.9 \text{ g } P_4O_{10}}{1 \text{ mol } P_4O_{10}}\right)$$

$$= -433.0 \text{ kJ}$$

7. -622 kJ
exothermic

Energy is absorbed (ΔH is positive) when bonds are broken, and energy is evolved (ΔH is negative) when bonds are formed.

Four C-H bonds are broken in CH_4: $\Delta H = 4(+414)$ kJ

$= +1656$ kJ

Two O=O bonds are broken in O_2: $\Delta H = 2(+494)$ kJ

$= +988$ kJ

Two C=O bonds are formed in CO_2: $\Delta H = 2(-707)$ kJ

$= -1414$ kJ

Four O-H bonds are formed in H_2O (two per H_2O):

$\Delta H = 4(-463)$ kJ

$= -1852$ kJ

The solution is obtained by summing:

$$CH_4 + O_2 \longrightarrow CO_2 + 2H_2O \qquad \Delta H = -622 \text{ kJ}$$

Heat is evolved in this reaction. The reaction is exothermic.

8. +268 kJ

ΔH for the reaction can be considered to be the sum of the energy required to break the bonds of the reactants and the energy released by the formation of the bonds of the products:

$\Delta H = -25$ kJ $= 4(+$ C-H bond energy) $+$ Br-Br bond energy

$+ 3(-$ C-H bond energy) $+ (-$ C-Br bond energy)

$+ (-$ H-Br bond energy)

-25 kJ $= 4(+414)$kJ $+ 193$ kJ $+ 3(-414)$kJ

$+ (-$ C-Br bond energy) $+ (-364)$kJ

-25 kJ $= (1656+193-1242-364)$kJ $-$ C-Br bond energy

C-Br bond energy $= 268$ kJ

9. -75.29 kJ

The law of Hess states that ΔH for a chemical reaction is constant, regardless of the number of steps. Thus thermochemical data is treated algebraically. The reaction of interest,

$$COCl_2(g) + 2H_2S(g) \longrightarrow 2HCl(g) + H_2O(g) + CS_2(l)$$

can be obtained by the addition of the two equations given. The heat of the reaction is obtained by adding the heats of reaction for the equations given:

$$COCl_2(g) + H_2S(g) \longrightarrow 2HCl(g) + COS(g) \quad \Delta H = -78.71 \text{ kJ}$$

$$COS(g) + H_2S(g) \longrightarrow H_2O(g) + CS_2(l) \qquad \Delta H = +3.42 \text{ kJ}$$

$$COCl_2(g) + 2H_2S(g) \longrightarrow 2HCl(g) + H_2O(g)$$
$$+ CS_2(l) \quad \Delta H = -75.29 \text{ kJ}$$

10. -610.1 kJ

Write the equations so that the sum yields the desired net equation.

P_4O_{10} is needed on the left; therefore, equation (b), the only source of P_4O_{10}, is reversed. The sign of ΔH is also reversed.

Six moles of PCl_5 are needed on the left also. Equation (c) is reversed and multiplied by 6. ΔH is multiplied by 6 and the sign is reversed also.

10 moles of Cl_3PO are needed on the right. Since equation (d) is the only source of Cl_3PO, it is multiplied by 10, as well as ΔH.

At this point look for molecules which do not appear in the desired equation.

$$P_4O_{10} \longrightarrow P_4 + 5O_2 \qquad\qquad \Delta H = +2967.3 \text{ kJ}$$
$$6PCl_5 \longrightarrow 6PCl_3 + 6Cl_2 \qquad\quad \Delta H = +505.2 \text{ kJ}$$
$$10PCl_3 + 5O_2 \longrightarrow 10Cl_3PO \qquad \Delta H = -2857.0 \text{ kJ}$$
$$P_4O_{10} + 6PCl_5 + 4PCl_3 \longrightarrow P4 + 6Cl_2 + 10Cl_3PO$$

The desired equation does not contain $4PCl_3$, P_4, or $6Cl_2$. The addition of equation (a) multiplied by four will cancel unwanted species:

$$P_4O_{10} \longrightarrow P_4 + 5O_2 \qquad\qquad \Delta H = +2967.3 \text{ kJ}$$
$$6PCl_5 \longrightarrow 6PCl_3 + 6Cl_2 \qquad\quad \Delta H = +505.2 \text{ kJ}$$
$$10PCl_3 + 5O_2 \longrightarrow 10Cl_3PO \qquad \Delta H = -2857.0 \text{ kJ}$$
$$P_4 + 6Cl_2 \longrightarrow 4PCl_3 \qquad\qquad \Delta H = -1225.6 \text{ kJ}$$
$$P_4O_{10} + 6PCl_5 \longrightarrow 10Cl_3PO \qquad \Delta H = -610.1 \text{ kJ}$$

11. in NO
    +627.1 kJ
    in ONO
    +455.9 kJ

For NO, the thermochemical equation for the enthalpy of formation may be written from the table:

$$\tfrac{1}{2}N_2(g) + \tfrac{1}{2}O_2(g) \longrightarrow NO(g) \qquad \Delta H = +90.4 \text{ kJ}$$

This process involves breaking $\tfrac{1}{2}$ mole of N_2 bonds and $\tfrac{1}{2}$ mole of O_2 bonds and forming one mole of NO bonds:

$$\tfrac{1}{2}N_2(g) \longrightarrow N(g) \qquad \Delta H = \tfrac{1}{2}(+941) \text{kJ}$$

$$\tfrac{1}{2}O_2(g) \longrightarrow O(g) \qquad \Delta H = \tfrac{1}{2}(+494) \text{kJ}$$

$$N(g) + O(g) \longrightarrow NO(g) \qquad \Delta H = -x$$

Here x equals the NO bond energy.
The sum of these three equations is the enthalpy of formation:

$$\tfrac{1}{2}N_2(g) + \tfrac{1}{2}O_2(g) \longrightarrow NO(g) \qquad \Delta H = +90.4 \text{kJ}$$

Therefore, the sum of the enthalpies in the first three equations must add up to the enthalpy of formation:

$$\tfrac{1}{2}(941\text{kJ}) + \tfrac{1}{2}(494\text{kJ}) - (x) = +90.4\text{kJ}$$

or

$$x = 627\text{kJ}$$

Similarly for ONO,

$$\tfrac{1}{2}N_2 \longrightarrow N \qquad \Delta H = \tfrac{1}{2}(941\text{kJ})$$

$$O_2 \longrightarrow 2O \qquad \Delta H = 494\text{kJ}$$

$$\underline{N + 2O \longrightarrow ONO \qquad \Delta H = -2x}$$

$$\tfrac{1}{2}N_2 + O_2 \longrightarrow ONO \qquad \Delta H = +52.7\text{kJ}$$

Solving for x:

$$52.7\text{kJ} = \tfrac{1}{2}(941\text{kJ}) + 494\text{kJ} - 2x$$

or
$$x = +456\text{kJ}$$

There is a stronger NO bond in NO than in ONO. The structures bear this out since NO has a double bond between the nitrogen and oxygen atoms and ONO has an average of only one and one-half bonds between the nitrogen and oxygen atoms:

N=O   O=N-O

12. for C_5H_{12}
   $3.07 \times 10^4 kJ$
   for CS_2
   $1.79 \times 10^4 kJ$

First ask the question in mathematical form:

$$? \text{ kJ} = 1.00 \text{ l } C_5H_{12}$$

Then use conversion factors to solve the problem:

$$? \text{ kJ} = 1.00 \text{ l } C_5H_{12} \left(\frac{621 \text{ g}}{1.00 \text{ l}}\right)\left(\frac{1 \text{ mol}}{72.1 \text{ g}}\right)\left(\frac{-3563.3 \text{ kJ}}{1 \text{ mol } C_5H_{12}}\right)$$

$$= 3.07 \times 10^4 kJ$$

For carbon disulfide, proceed in the same way:

$$? \text{ kJ} = 1.00 \text{ l } CS_2 \left(\frac{1.263 \text{ g}}{1.00 \text{ l}}\right)\left(\frac{1 \text{ mol}}{76.1 \text{ g}}\right)\left(\frac{-1075.2 \text{ kJ}}{1 \text{ mol}}\right)$$

$$= 1.79 \times 10^4 kJ$$

More heat is evolved in the combustion of the pentane than in the combustion of the carbon disulfide. The percent increase can easily be calculated:

$$\% = \left(\frac{3.07 \times 10^4 kJ - 1.79 \times 10^4 kJ}{1.79 \times 10^4 kJ}\right) \times 100\%$$

$= 71.5\%$ more heat from pentane than from carbon disulfide.

SELF-TEST   I. Complete this test in 50 minutes

1. The specific heat of aluminum at 20°C is $0.895 J/(g°C)$. How many joules will be needed to raise the temperature of a 3.45 g sample of aluminum from 20.00°C to 21.23°C?

2. How many grams of aluminum were burned by the reaction

$$2Al(s) + 3/2 \, O_2(g) \longrightarrow Al_2O_3(s) \qquad \Delta H = -1669.8 \text{ kJ}$$

when a calorimeter with a heat capacity of 3.78kJ/°C and its contents of 0.876 kg of water experience a temperature increase of 0.496°C?

3. Given the following thermochemical equation and the enthalpy of formation values from the table in this chapter of the *Study Guide*, determine the enthalpy of formation of $OSCl_2(l)$.

$$OSCl_2(l) + H_2O(l) \longrightarrow SO_2(g) + 2HCl(g) \qquad \Delta H = +10.3 \text{ kJ}$$

4. Use average bond energies from Table 5.2 of your text
    to calculate the enthalpy of formation of HF (g).

5. Given the following thermochemical equations:

$BCl_3 (g) + 3H_2O (l) \longrightarrow H_3BO_3 (s) + 3HCl (g)$ $\Delta H = -112.5$ kJ

$B_2H_6 (g) + 6Cl_2 (g) \longrightarrow 2BCl_3 (g) + 6HCl (g)$ $\Delta H = -1376.0$ kJ

$B_2H_6 (g) + 6H_2O (l) \longrightarrow 2H_3BO_3 (s) + 6H_2 (g)$ $\Delta H = -493.4$ kJ

determine the enthalpy of formation of HCl (g).

6. Calculate the enthalpy change of the reaction

$BaO (s) + CO_2 (g) \rightarrow BaCO_3 (s)$

from the following standard enthalpies of formation at
298 K: BaO (s), -558.6 kJ/mol; CO_2 (g), 393.50 kJ/mol;
and $BaCO_3$ (s), 1216.7 kJ/mol.

THE ELECTRONIC STRUCTURE OF ATOMS

CHAPTER

6

OBJECTIVES

I. You should be able to demonstrate your knowledge of the following terms by defining them, describing them, or giving specific examples of them:

aufbau method [6.7]
Balmer series [6.2]
diamagnetic substance [6.6]
electromagnetic radiation [6.1]
electronic configuration [6.6]
energy level [6.2, 6.5]
excited state [6.2]
exclusion principle [6.5]
frequency [6.1]
ground state [6.2]
Hund's rule [6.6]
magnetic orbital quantum number, m_l [6.5]
magnetic spin quantum number, m_s [6.5]
orbit [6.2]
orbital [6.4]
paramagnetic substance [6.6]
periodic law [6.3]
photon [6.1]
Planck's constant [6.1]
principal quantum number, n [6.5]
quantum [6.1]
representative element [6.9]
spectrum [6.2]
subshell [6.5]
subsidiary quantum number, l [6.5]
transition element [6.7, 6.9]
uncertainty principle [6.4]
valence electrons [6.6]
wavelength [6.1]

II. You should understand the relationship between energy and frequency, $E = h\upsilon$ or $E = hc/\lambda$, and be able to work problems relating to these equations.

III. You should understand the Bohr theory.

IV. You should be able to write a complete set of quantum numbers for each electron of any element.

V. You should be able to write the electronic notation of each element in the periodic table and of any monatomic ion. You should also be able to predict the number of unpaired electrons and the magnetic properties of each of these species.

UNITS,
SYMBOLS,
MATHEMATICS

I. The following symbols are used in this chapter and you should be familiar with them.

- c stands for the speed of light, 2.9979×10^8 m/s

- h stands for Planck's constant of proportionality.

- Hz is the abbreviation used for the SI unit of frequency, the Hertz, and has units of reciprocal seconds, 1/s.

- l is used to identify the subsidiary quantum number.

- λ stands for the wavelength of light.

- m_l is used to identify the magnetic orbital quantum number.

- m_s is used to identify the magnetic spin quantum number.

- n is used to identify the principal quantum number.

- υ stands for the frequency of light.

- s is the abbreviation for second, the SI unit of time.

- s, p, d, f, g are notations used to represent the electronic subshells in atoms.

II. You should continue to memorize chemical symbols and should know the following by now.

1. potassium, K
2. silver, Ag
3. hydrogen, H
4. sodium, Na
5. calcium, Ca
6. oxygen, O
7. chlorine, Cl
8. nitrogen, N
9. carbon, C
10. lithium, Li
11. boron, B
12. phosphorus, P
13. zinc, Zn
14. arsenic, As
15. neon, Ne
16. copper, Cu
17. aluminum, Al
18. sulfur, S
19. magnesium, Mg
20. fluorine, F
21. bromine, Br
22. iron, Fe
23. helium, He
24. berylium, Be
25. silicon, Si
26. iodine, I
27. nickel, Ni
28. chromium, Cr

EXERCISES

I. Answer each of the following with *TRUE* or *FALSE*. If a statement is false, correct it. Use this section as a guide for further study.

1. The energy of a quantum of radiation is directly proportional to its wavelength. The proportionality constant, h, is Planck's constant.

2. Electrons are distributed in atoms so that a maximum number is paired.

3. Paramagnetic materials interact with magnetic fields.

4. The frequency of a quantum of radiation is inversely proportional to its wavelength.

5. Electrons associated with an atom can exist only in discrete energy levels.

6. It is possible for an electron to have the set of quantum numbers $n = 1$, $l = 1$, $m_l = +1$, $m_s = +\frac{1}{2}$.

7. It is possible for an electron to have the set of quantum numbers $n = 2$, $l = 0$, $m_l = 0$, $m_s = +\frac{1}{2}$.

8. It is possible for an electron to have the set of quantum numbers $n = 2$, $l = 1$, $m_l = +1$, $m_s = +\frac{1}{2}$.

9. It is possible for an electron to have the set of quantum numbers $n = 4$, $l = 3$, $m_l = -2$, $m_s = +\frac{1}{2}$.

10. A d subshell can contain a maximum of six electrons.

11. A noble gas is unreactive chemically because the electronic configuration of its outer shell consists of completely filled s and p subshells.

12. A diamagnetic material has unpaired electrons.

II. Using the periodic table, identify the element that is described by the statement:

N ✓

1. The electronic notation for the neutral atom is $1s^2 2s^2 2p^3$.

(N)i

2. The electronic notation for the neutral atom is $1s^2 2s^2 2p^6 3s^2 3p^6 3d^8 4s^2$.

Kr

3. The electronic notation for the neutral atom is $1s^2 2s^2 2p^6 3s^2 3p^6 3d^{10} 4s^2 4p^6$.

Na$^+$

4. The electronic notation for the ion that has a single positive charge, X$^+$, is $1s^2 2s^2 2p^6$.

~~Ge~~ Cu$^+$

5. The electronic notation for the ion that has a single positive charge, X$^+$, is $1s^2 2s^2 2p^6 3s^2 3p^6 3d^{10}$.

6. The quantum numbers of the last electron added to the neutral atom according to the aufbau method are $n = 2$, $l = 1$, $m_l = 0$, $m_s = +\frac{1}{2}$. C

7. The quantum numbers of the last electron added to the neutral atom according to the aufbau method are $n = 3$, $l = 1$, $m_l = 0$, $m_s = -\frac{1}{2}$. Cl

8. The quantum numbers of the last electron added to the neutral atom according to the aufbau method are $n = 4$, $l = 2$, $m_l = 0$, $m_s = -\frac{1}{2}$. Pd

III. Write the electronic notations for the following:

1. Ca, calcium

2. Zn, zinc $1s^2 2s^2 2p^6 3s^2 3p^6 4s^2 3d^{10}$

3. Cs, cesium $1s^2 2s^2 2p^2 3s^2 3p^6 4s^2 3d^{10} 4p^6 5s^2 4d^{10} 5p^6 6s^1$

4. Kr, krypton $1s^2 2s^2 2p^6 3s^2 3p^6 3d^8 4s^2$

5. S, sulfur $1s^2 2s^2 2p^6 3s^2 3p^4$

IV. Write the four quantum numbers of the last electron added to the element, the differentiating electron:

1. O, oxygen $n=2$ $l=1$ $m_l = +1, 0, -1$ m_s

2. K, potassium

3. V, vanadium

4. As, arsenic

5. Ni, nickel

6. Au, gold

V. Fill in the blank with the best answer to each question.

1. Substances that contain unpaired electrons are drawn into a magnetic field. They are called _____.

2. Diamagnetic substances are _____ by a magnetic field.

3. Electrons in the outermost shells of atoms are called the _____ electrons.

4. Elements which, by the aufbau method of adding electrons, have a d electron as the differentiation electron are called _____ elements.

5. The elements found at the end of each period in group 0 are the _____.

6. The scientist who is well known for the principle that the exact position and exact momentum of a body as small as an electron cannot be determined simultaneously is _____.

7. The quantum number that identifies the shell in which the electron is found is _____.

8. The scientist who proposed the principle that no two electrons in the same atom may have identical sets of all four quantum numbers is _____.

9. Elements that have either an s or a p differentiating electron are _____.

10. The electromagnetic radiation that is similar to X rays but can be more energetic is _____ radiation.

11. How many seconds would it take light to travel 4,000 miles? (1 mile = 1.609 km) _____

12. What is the frequency of a microwave if the wavelength in vacuum is 5.00 cm? _____

13. When electrons in atoms are in the condition of the lowest possible energy, they are in the _____ state.

14. The theory that proposes that light can be emitted in small particles called photons is the _____ theory.

15.   The electronic configuration of copper is an exception to the aufbau order because of the _____ orbitals.

ANSWERS TO
EXERCISES

I. Principles of atomic structure

1. False

Energy, E, is directly proportional to frequency, ν.

$$E = h\nu$$

The constant of proportionality is called Planck's constant, h. Energy, however, is inversely proportional to wavelength, λ, since

$$E = hc/\lambda$$

in which c is the speed of light.

2. False

The number of electrons that is unpaired is a maximum, as stated in Hund's rule of maximum multiplicity.

3. True

4. True

The relationship between frequency and wavelength is

$$\nu = c/\lambda$$

in which ν is the frequency in hertz or /s; λ is the wavelength in cm; and c is the speed of light in a vacuum, in units of cm/s.

5. True

6. False

If $n = 1$, l can only be zero.

7. True

For $n = 2$, the following sets of quantum numbers are possible:

$l = 0$, $m_l = 0$, $m_s = +\frac{1}{2}$ or $-\frac{1}{2}$
$l = 1$, $m_l = +1$, $m_s = +\frac{1}{2}$ or $-\frac{1}{2}$
$l = 1$, $m_l = 0$, $m_s = +\frac{1}{2}$ or $-\frac{1}{2}$
$l = 1$, $m_l = -1$, $m_s = +\frac{1}{2}$ or $-\frac{1}{2}$

In general, for any value of n the possible values of l and m_l are

$l = (n - 1)$, $(n - 2)$, . . ., 0
$m_l = +1$, $+(l - 1)$, . . ., 0, . . ., $(l - 1)$, $-l$

8. True

9. True

10. False

A maximum of 10 electrons can be added to a d subshell. The d orbitals are being filled in each series of transition elements. The possible quantum numbers of electrons in d orbitals are

$l = 2$; $m_l = +2$, +1, 0, -1, or -2; $m_s = +\frac{1}{2}$ or $-\frac{1}{2}$

11. True

12. False   In diamagnetic materials all electrons are paired.

II. Identification of the Elements

1. N, nitrogen   Each superscript indicates the number of electrons in the subshell. The $1s$ subshell has 2 electrons, the $2s$ subshell has 2 electrons, and the $2p$ subshell has 3 electrons. A total of 7 electrons indicates an atom with $Z = 7$, nitrogen. Check the periodic table in the back of the study guide.

2. Ni, nickel   The total number of electrons in the atom is 28; $Z = 28$. Observe the position of nickel in the periodic table. The order of filling the atomic orbitals is shown up to $Z = 28$, and the corresponding electronic notation is given to the left of the table.

$1s^2$

$2s^2 2p^6$

$3s^2 3p^6$

$4s^2 3d^8$

1s																	1s
2s	2s											2p	2p	2p	2p	2p	2p
3s	3s								Ni			3p	3p	3p	3p	3p	3p
4s	4s	3d	3d	3d	3d	3d	3d	3d	3d								

3. Kr, krypton   The sum of the superscripts from the electronic notation is 36; $Z = 36$.

4. Na, sodium $Z = 11$

5. Cu, copper $Z = 29$

6. C, carbon   Since n, the principal quantum number, is 2, the element must be in the second row of the periodic table. Since $l = 1$, the element is one in which the p subshell is being filled. The convention used in your text suggests that the p orbitals are filled in the order given in Table 6.1 of the study guide.

TABLE 6.1   Order of Filling the *p* Orbitals

			Element	
Order	m_l	m_s	$n = 2, l = 1$	$n = 3, l = 1$
1	+1	$+\frac{1}{2}$	B	Al
2	0	$+\frac{1}{2}$	C	Si
3	−1	$+\frac{1}{2}$	N	P
4	+1	$-\frac{1}{2}$	O	S
5	0	$-\frac{1}{2}$	F	Cl
6	−1	$-\frac{1}{2}$	Ne	At

7. Cl, chlorine   Elements in which the last electron is added to the
   $n = 3$ shell are indicated by the shaded area of the
   periodic table.

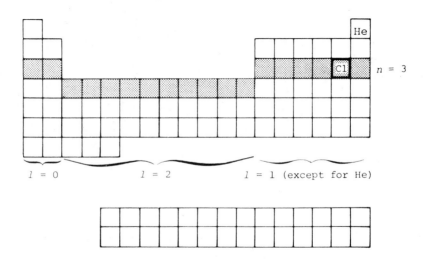

8. Pd, palladium   Elements in which the last electron is added to the $n = 4$
   shell are indicated by the shaded area of the periodic
   table.   The convention used in your text suggests that
   the *d* orbitals ($l = 2$) are filled in the order given in
   Table 6.2 of the study guide.

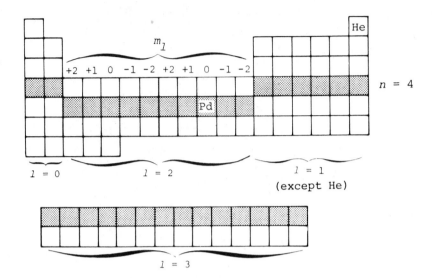

TABLE 6.2   Order of Filling the d Orbitals

			Element	
Order	m_l	m_s	$n = 4,\ l = 2$	$n = 3,\ l = 2$
1	+2	$+\frac{1}{2}$	Y	Sc
2	+1	$+\frac{1}{2}$	Zr	Ti
3	0	$+\frac{1}{2}$	Nb	V
4	-1	$+\frac{1}{2}$	Mo	Cr
5	-2	$+\frac{1}{2}$	Tc	Mn
6	+2	$-\frac{1}{2}$	Ru	Fe
7	+1	$-\frac{1}{2}$	Rh	Co
8	0	$-\frac{1}{2}$	Pd	Ni
9	-1	$-\frac{1}{2}$	Ag	Cu
10	-2	$-\frac{1}{2}$	Cd	Zn

III. Electronic configurations

1. Ca: $1s^2 2s^2 2p^6 3s^2 3p^6 4s^2$ Reading the periodic table from left to right, we note that the electronic notation for Ca is $1s^2 2s^2 2p^6 3s^2 3p^6 4s^2$. The arrangement of the periodic table corresponds to the aufbau filling order.

$1s^2$

$2s^2 2p^6$

$3s^2 3p^6$

$4s^2$

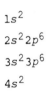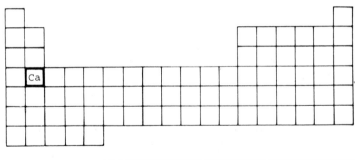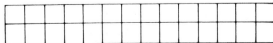

2. Zn: $1s^2 2s^2$ $2p^6 3s^2 3p^6 3d^{10}$ $4s^2$ Always write a notation in order of increasing principal quantum number; begin with the *s* subshell, then write the *p* subshell, the *d* subshell, and finally the *f* subshell of that shell. The notation for Zn is $1s^2 2s^2 2p^6 3s^2 3p^6 3d^{10} 4s^2$.

$1s^2$

$2s^2 2p^6$

$3s^2 3p^6$

$4s^2 3d^{10}$

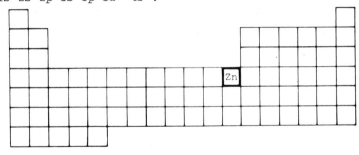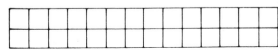

3. Cs: $1s^2 2s^2$ $2p^6 3s^2 3p^6 3d^{10}$ $4s^2 4p^6 4d^{10}$ $5s^2 5p^6 6s^1$ The electronic notation for the configuration of Cs is $1s^2 2s^2 2p^6 3s^2 3p^6 3d^{10} 4s^2 4p^6 4d^{10} 5s^2 5p^6 6s^1$.

$1s^2$

$2s^2 2p^6$

$3s^2 3p^6$

$4s^2 3d^{10} 4p^6$

$5s^2 4d^{10} 5p^6$

$6s^1$

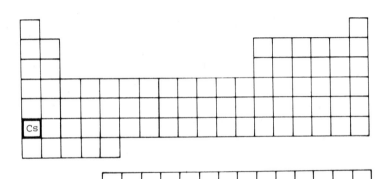

4. Kr: $1s^2 2s^2$
$2p^6 3s^2 3p^6 3d^{10}$
$4s^2 4p^6$

5. S: $1s^2 2s^2$ The electronic configuration of S is $1s^2 2s^2 2p^6 3s^2 3p^4$.
$2p^6 3s^2 3p^4$

IV. Quantum numbers

1. O: $n = 2$,   The last electron added is the fourth one in the $2p$ sub-
$l = 1$,   shell. Therefore, $n = 2$ and $l = 1$. The quantum numbers
$m_l = +1$, m_l and m_s by the convention used in your text are +1 and
$m_l = -\frac{1}{2}$ $-\frac{1}{2}$, respectively (see Table 6.1 of the study guide). The
set of quantum numbers that corresponds to the last
electron added is

$$n = 2, \ l = 1, \ m_l = +1, \ m_s = -\frac{1}{2}$$

The periodic table in the study guide may be helpful in
the determination of the quantum numbers of the last
electron added in the building of elements when the aufbau
method is used.

2. K: $n = 4$,
$l = 0$,
$m_l = 0$
$m_s = +\frac{1}{2}$

3. V: $n = 3$,
$l = 2$,
$m_l = 0$,
$m_s = +\frac{1}{2}$

4. As: $n = 4$,
$l = 1$,
$m_l = -1$,
$m_s = +\frac{1}{2}$

5. Ni: $n = 3$,
$l = 2$,
$m_l = 0$,
$m_s = -\frac{1}{2}$

6. Au: $n = 5$,
$l = 2$,
$m_l = -1$,
$m_s = -\frac{1}{2}$

IV. Fill in the blank with the best answer to each question.

1. Substances that contain unpaired electrons are drawn into a magnetic field. They are called paramagnetic.

2. Diamagnetic substances are repelled by a magnetic field.

3. Electrons in the outermost shells of atoms are called the valence electrons.

4. Elements which, by the aufbau method of adding electrons, have a d electron as the differentiation electron are called transition elements.

5. The elements found at the end of each period in group 0 are the nobel gases.

6. The scientist who is well known for the principle that the exact position and exact momentum of a body as small as an electron cannot be determined simultaneously is Heisenberg.

7. The quantum number that identifies the shell in which the electron is found. n

8. The scientist who proposed the principle that no two electrons in the same atom may have identical sets of all four quantum numbers is Pauli.

9. Elements that have either an s or a p differentiating electron are representative.

10. The electromagnetic radiation that is similar to X rays but can be more energetic is gamma radiation.

11. How many seconds would it take light to travel 4,000 miles? (1 mile = 1.609 km)   0.02 sec

12. What is the frequency of a microwave if the wavelength in vacuum is 5.00 cm? 6.00×10^9 Hz

13. When electrons in atoms are in the condition of the lowest possible energy, they are in the ground state.

14. The theory that proposes that light can be emitted in small particles called photons is the quantum theory.

15. The electronic configuration of copper is an exception to the aufbau order because of the half-filled orbitals.

SELF-TEST   Complete the test in 20 minutes.

I. Choose the best answer to each of the following questions.

  1. Planck's constant, h, times the frequency of
electromagnetic radiation equals

 $h\nu = ?$

    a. the wavelength of the radiation
    b. the energy of the radiation
    c. the velocity of the radiation
    d. the amplitude of the radiation

 2. If an element has an electronic structure in which
all the electrons are paired, it is

    a. paramagnetic
    b. diamagnetic
    c. ferromagnetic
    d. none of these

3. The electronic notation for barium, Ba, is

    a. $1s^2\ 2s^2\ 2p^6\ 3s^2\ 3p^6\ 3d^{10}\ 4s^2\ 4p^6\ 4d^{10}\ 5s^2\ 5p^6$
       $6d^1\ 6s^2$

    b. $1s^2\ 2s^2\ 2p^6\ 3s\ 3p^6\ 3d^{10}\ 4s^2\ 4p^6\ 4d^{10}\ 5s^2\ 5p^6\ 6s^2$

    c. $1s^2\ 2s^2\ 2p^6\ 3s^2\ 3p^6\ 3d^{10}\ 4s^2\ 4p^6\ 4d^{10}\ 5s^2\ 5p^6$
       $6s^2\ 6p^1$

    d. $1s^2\ 2s^2\ 2p^6\ 3s^2\ 3p^6\ 3d^{10}\ 4s^2\ 4p^6\ 4d^{10}\ 5s^2\ 5p^6$
       $6s^2\ 6d^1$

 4. The differentiating electron of a sulfur atom has which
of the following sets of quantum numbers?
    a. $n = 3,\ l = 1,\ m_l = +1,\ m_s = +\frac{1}{2}$
    b. $n = 3,\ l = 1,\ m_l = -1,\ m_s = +\frac{1}{2}$
    c. $n = 3,\ l = 1,\ m_l = +1,\ m_s = -\frac{1}{2}$
    d. $n = 3,\ l = 1,\ m_l = -1,\ m_s = -\frac{1}{2}$

5. Which of the following sets of quantum numbers
represents an impossible arrangement?
    a. $n = 2,\ l = 0,\ m_l = 0,\ m_s = -\frac{1}{2}$
    b. $n = 7,\ l = 4,\ m_l = -1,\ m_s = +\frac{1}{2}$
    c. $n = 3,\ l = 1,\ m_l = +1,\ m_s = -\frac{1}{2}$
    d. $n = 3,\ l = -1,\ m_l = -1,\ m_s = -\frac{1}{2}$

6. Which of the following is the most paramagnetic?
   a. K^+ b. Zn^{2+} c. Cu^{2+} d. Fe^{3+}

7. The wave mechanical approach to atomic structure permits the calculation of
   a. a volume about the nucleus in which an electron of specified energy will most probably be found
   b. the most probable radius of an orbit that an electron of specified energy will follow
   c. the most probable position of an electron of specified energy at a given time, t
   d. the most probable spin value that will be associated with an electron of specified energy.

8. As electrons move from a ground state to an excited state,
   a. an emission spectrum results
   b. energy is absorbed
   c. heat is liberated
   d. light is emitted

9. The intensity of a spectral line observed in an atomic emission spectrum can be directly related to
   a. the difference in energy of the energy levels involved in the electron transition that gives rise to the line
   b. the number of electrons undergoing the transition that gives rise to the line
   c. the number of energy levels involved in the transition that gives rise to the line
   d. the speed with which an electron undergoes a transition from one energy level to another

10. The wavelength at which a spectral line is observed in an atomic emission spectrum is inversely related to
    a. the difference in energy of the energy levels involved in the electron transition that gives rise to the line
    b. the number of electrons undergoing the transition that gives rise to the line
    c. the number of energy levels involved in the transition that gives rise to the line
    d. the speed with which an electron undergoes a transition from one energy level to another.

11. The quantum number that designates the spin of an electron is
    a. n b. l c. m_l d. m_s

_____ A ✓ 12. Scandium
a. is a transition element
b. has an atomic number of 45
c. has 21 electrons in the Sc^{2+} ion
d. has a completely filled $n = 3$ subshell

_____ D ✓ 13. The hertz

a. is the SI unit of frequency
b. has units of per second
c. is abbreviated Hz
d. all of the above

C   ✓ 14. The larger the energy of electromagnetic radiation,

a. the higher the frequency
b. the shorter the wavelength
c. all of the above
d. none of the above

b   ✓ 15. With increasing distance from the nucleus

a. the radius of the electron shell increases
b. the energy of an electron in the shell increases
c. less energy must be supplied to move the electron
    to the next higher principal level
d. all of the above

B or a ✓ 16. When an electron falls from an excited state to a lower
energy state it

a. goes to the K shell
b. can emit a quantum of electromagnetic radiation
c. is further from the nucleus
d. all of the above

C   ✗ 17. Consider the de Broglie equation $\lambda = h/mv$

a. larger masses have a shorter wavelength than
    smaller masses traveling at the same velocity
b. equal masses have the same wavelength no matter
    what the velocity
c. masses at rest have a very small wavelength
d. all are true

b   ✓ 18. The magnetic spin quantum number is denoted by

a. m_l
b. m_s
c. l
d. n

19.   Boundary surface diagrams are used to represent

 a.   orbitals
 b.   electron orbits
 c.   the limits within which the electron is confined
 d.   all of the above

20.   The subsidiary quantum number

 a.   is denoted by the symbol l
 b.   always has a value which is less than the value of
       the principle quantum number for any given electron
 c.   has a corresponding notation of s, p, d, f, g, ...
 d.   all of the above

PROPERTIES OF ATOMS AND THE IONIC BOND

CHAPTER

7

OBJECTIVES I. You should be able to demonstrate your knowledge of the following terms by defining them, describing them, or giving specific examples of them:

anion [7.1]
atomic radius [7.1]
Born-Haber cycle [7.5]
cation [7.4]
d^{10} ion [7.6]
$d^{10}s^2$ ion [7.6]
electron affinity [7.3]
enthalpy of sublimation [7.5]
ion [7.1, 7.4, 7.6]
ionic bonding [introduction, 7.4]
ionic radius [7.7]
ionic reaction [7.7]
ionization energy [7.2]
isoelectronic [7.4]
lattice energy [7.5]
noble-gas ion [7.4, 7.5]
s^2 ion [7.6]
shielding [7.1]

II. Using the periodic table of the elements, you should be able to predict relative sizes of atoms and ions and relative magnitudes of ionization energies and electron affinities.

III. You should be able to construct and use Born-Haber cycles for thermochemical calculations.

IV. You should be able to write formulas for anions, cations, and simple ionic compounds.

V. You should be able to write notations for the ground-state electronic configurations of ions.

VI. You should know the formula and charge for each of the ions listed in Table 7.5 in your text if your instructor requests that you do so. You should be able to write molecular formulas for compounds formed from the ions and name those compounds.

UNITS, SYMBOLS, MATHEMATICS

I. The following symbols have been used in the chapter. You should be familiar with all of them.

- ΔH is the symbol used to identify an enthalpy change.

- ΔH_{diss} is the symbol used to identify the enthalpy change associated with a bond dissociation.

- $\Delta H_{elec\ af}$ is the symbol used to identify the enthalpy change associated with the process in which an electron is added to an atom.

- ΔH_f is the symbol used to identify the enthalpy change associated with the formation of a compound from the necessary pure elements in their most stable state.

- $\Delta H_{ion\ en}$ is the symbol used to identify the enthalpy change associated with the removal of electrons from the atom in its ground state.

- $\Delta H_{lat\ en}$ is the symbol used to identify the enthalpy change associated with the condensation of gaseous ions into ionic crystals.

- ΔH_{subl} is the symbol used to identify the enthalpy change associated with the direct conversion of a solid into a gas.

- eV is the abbreviation for electron volt, a unit of energy.

- kJ is the abbreviation for kilojoules, an SI unit of energy equal to 10^3 joules.

- n is used to identify the principal quantum number.

- pm is the abbreviation for picometer, an SI unit of length equal to 10^{-12} meter.

- s, p, d, f, g are notations used to represent electronic subshells of atoms.

- Z identifies the atomic number of an element.

II. You should memorize the symbols for the elements listed in Table 7.1.

TABLE 7.1   Common Elements and Their Symbols

Symbol	Element	Symbol	Element
Al	aluminum	Hg	mercury
Ar	argon	I	iodine
As	arsenic	K	potassium
B	boron	Li	lithium
Be	beryllium	Mg	magnesium
Br	bromine	Mn	manganese
C	carbon	N	nitrogen
Ca	calcium	Na	sodium
Cl	chlorine	Ne	neon
Cr	chromium	Ni	nickel
Cu	copper	O	oxygen
F	fluorine	P	phosphorus
Fe	iron	S	sulfur
H	hydrogen	Sn	tin
He	helium	Zn	zinc

EXERCISES   I. Answer the following:

1. Which has the larger radius?

Mg
Cs
Ni

a. Mg or Si

b. Li or Cs

c. Ni or Zn

d. Ti or Cr

e. Mg or Mg^{2+}

f. Cl or Cl$^-$

2. Which has the larger first ionization energy?

Bi
Al
O

a. Ba or Bi

b. Al or Tl

c. C or O

d. Br or Kr

e. P or O

3. Which has the more negative value for the first electron affinity?

F

a. B or F

b. Cl or S

II. Answer each of the following with true or false. If a statement is false, correct it.

T

1. An atom of any group V A element has five valence electrons.

T

2. The ionization energy of a noble gas is very high compared to that of any of the other elements in the same period.

_____T_____

3. The first electron affinity of a halogen is low compared to that of other elements in the same period. That is, less energy is evolved when an electron is added to a halogen atom than when an electron is added to any other element in the same period.

_____F_____

4. Ionization energy is the amount of energy required to add an electron to an isolated gaseous atom in its ground state.

_____T_____

5. Na^+ and Ne have the same electronic structure.

_____T_____

6. The second ionization energy of sodium should be larger than the second ionization energy of magnesium.

_____F_____

7. Purely ionic bonds are formed by two atoms sharing electrons.

III. Perform the following calculations.

1. Calculate the lattice energy of potassium chloride from the following data. For potassium the ionization energy is 414 kJ/mol, and the enthalpy of sublimation is 88 kJ/mol. For chlorine the dissociation energy is 243 kJ/mol and the electron affinity is -348 kJ/mol. The enthalpy of formation of KCl is -435 kJ/mol.

2. Use the following data to calculate the first electron affinity of chlorine. The enthalpy of formation of $MgCl_2$ is -642 kJ/mol, and the lattice energy of $MgCl_2$ is -2530 kJ/mol. The enthalpy of sublimation of Mg is +152 kJ/mol. The first ionization energy of Mg is +738 kJ/mol, and the second ionization energy of Mg is +1450 kJ/mol. The bond energy of the Cl_2 bond is +244 kJ/mol Cl_2.

IV. Write the notation for the electronic configuration of the following ions.

1. Zn^{2+}

2. S^{2-}

3. Cs^+

V. Write formulas for the following simple binary compounds.
A more complete description of nomenclature is contained
in Chapter 8.

1. cesium chloride
2. zinc oxide
3. magnesium sulfide
4. sodium cyanide
5. barium sulfate

VI. Name the following simple binary compounds.

1. LiBr
2. MgS
3. NiO
4. $CaCO_3$
5. KNO_3

ANSWERS TO   I. Atomic properties as related to chemical bonding.
EXERCISES

1. a. Mg   There is generally a decrease in atomic radius across a
                period from left to right.

   b. Cs   There is generally an increase in atomic radius in a
                group from top to bottom.

   c. Zn   In the transition series inner electrons of the atoms
                screen the outer electrons from the nuclear charge.
                Toward the end of the series the atomic radius actually
                increases with increasing atomic number.

   d. Ti   The screening effect of inner electrons is not so pro-
                nounced in the beginning of a transition series, and
                the atomic radius decreases with increasing atomic number.

   e. Mg   The magnesium ion, Mg^{2+}, has the same nuclear charge as
                the magnesium atom but two fewer electrons. The radius
                of the magnesium ion is therefore less than that of the
                magnesium atom.

   f. Cl⁻   The chloride ion, Cl^-, has the same nuclear charge as
                the chlorine atom but one additional electron. The
                radius of the chloride ion is therefore larger than that
                of the chlorine atom.

2. a. Bi   There is generally an increase in ionization energy
                across a period from left to right.

   b. Al   There is generally a decrease in ionization energy from
                top to bottom.

   c. O   See the explanation given for part (a) of this question.

d. Kr   Krypton is a noble gas and has 'a full outer shell of electrons. A large amount of energy would be required to remove an electron and form a less stable ion.

e. O   See the explanations given for parts (a) and (b) of this question.

3. a. F
   b. Cl   Electron affinity generally increases across a period from left to right; however, there are significant exceptions. See Table 7.2 of the study guide.

TABLE 7.2   Electron Affinities (kJ/mol.)*

H −73								He (+21)
Li −60	Be (+240)	B −27	C −122	N 0	O −141	F −328		Ne (+29)
Na −53	Mg (+230)	Al −43	Si −134	P −72	S −200	Cl −349		Ar (+35)
K −48	Ca (+156)	Ga −29	Ge −116	As −77	Se −195	Br −325		Kr (+39)
Rb −47	Sr (+168)	In −29	Sn −121	Sb −101	Te −190	I −295		Xe (+41)
Cs −45	Ba (+52)	Tl −29	Pb −35	Bi −91	Po −183	At −270		Rn (+41)

* Values in parentheses are estimated.

II. True-false questions.

1. true
2. true
3. true
4. false   Ionization energy is the energy required to remove the most loosely held electron.

5. true
6. true
7. false   Ionic bonding occurs through the transfer of electrons.

III. Chemical calculations.

1. $\Delta H = -711$ kJ   The lattice energy is the energy associated with the process.

$$K^+(g) + Cl^-(g) \rightarrow KCl(s) \qquad \Delta H_{lat\ en} = ?$$

A Born-Haber cycle can be used to determine the value of $\Delta H_{lat\ en}$ of KCl. From the law of Hess we know that

$$\Delta H_{lat\ en} = \Delta H_f(KCl) - \Delta H_f(K^+) - \Delta H_f(Cl^-)$$

The value of $\Delta H_f(KCl)$ is given, but we must determine $\Delta H_f(K^+)$ and $\Delta H_f(Cl^-)$ from a series of reactions, since ΔH_f pertains to the formation of a species from elements in standard states. Add the ionization energy of potassium, $\Delta H_{ion\ en} = 414$ kJ, which pertains to the equation

$$K(g) \rightarrow K^+(g) + e^-$$

and the enthalpy of sublimation, $\Delta H_{sub} = 88$ kJ, which pertains to the equation

$$K(s) \rightarrow K(g)$$

to obtain a ΔH_f of $K^+(g)$, $\Delta H_f(K^+) = 502$ kJ, which pertains to the equation

$$K(s) \rightarrow K^+(g) + e^-$$

Similarly, we add the electron affinity of chlorine, $\Delta H_{elec\ af} = -348$ kJ, which pertains to the equation

$$Cl(g) + e^- \rightarrow Cl^-(g)$$

and half the dissociation energy of Cl_2, $1/2\ \Delta H_{diss} = 122$ kJ, which pertains to the equation

$$1/2\ Cl_2(g) \rightarrow Cl(g)$$

to obtain $\Delta H_f(Cl^-)$, $\Delta H_f(Cl^-) = -226$ kJ, which pertains to the equation

$$1/2\ Cl_2(g) + e^- \rightarrow Cl^-(g)$$

To obtain $\Delta H_{lat\ en}$ we add the following equations and enthalpies:

$$
\begin{array}{ll}
K(s) + 1/2\ Cl_2(g) \rightarrow KCl(s) & \Delta H_f(KCl) = -435\ \text{kJ} \\
K^+(g) + e^- \rightarrow K(s) & -\Delta H_f(K^+) = -502\ \text{kJ} \\
\underline{Cl^-(g) \rightarrow 1/2\ Cl_2(g) + e^-} & \underline{-\Delta H_f(Cl^-) = 226\ \text{kJ}} \\
K^+(g) + Cl^-(g) \rightarrow KCl(s) & \Delta H_{lat\ en}(KCl) = -711\ \text{kJ}
\end{array}
$$

2. −348 kJ/mol   A Born-Haber cycle can be used to determine the first electron affinity for chlorine:

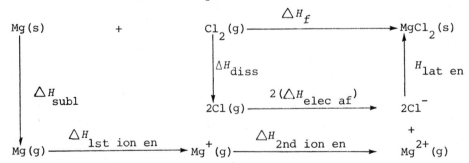

The heat of formation equals the sum of the other trans-
formations indicated in the diagram:

$$\Delta H_f = \Delta H_{subl} + \Delta H_{1st\ ion\ en} + \Delta H_{2nd\ ion\ en}$$
$$+ \Delta H_{diss} + 2(\Delta H_{elec\ af}) + \Delta H_{lat\ en}$$

Rearranging:

$$\Delta H_{elec\ af} = \tfrac{1}{2}(\Delta H_f - \Delta H_{subl} - \Delta H_{1st\ ion\ en}$$
$$- \Delta H_{2nd\ ion\ en} - \Delta H_{diss} - \Delta H_{lat\ en})$$

Substituting:

$$\Delta H_{elec\ af} = \tfrac{1}{2}\ [-642\ kJ/mol - (+152\ kJ/mol)$$
$$- (+738\ kJ/mol) - (+1450\ kJ/mol)$$
$$- (+244\ kJ/mol) - (-2530\ kJ/mol)]$$
$$= \tfrac{1}{2}(-696\ kJ/mol) = -348\ kJ/mol$$

IV. Electronic configuration.

1. Neutral Zn has the electronic configuration
$$1s^2\ 2s^2\ 2p^6\ 3s^2\ 3p^6\ 3d^{10}\ 4s^2$$

The two outermost electrons are removed so that Zn^{2+}
is formed. The electronic configuration for Zn^{2+} is
$$1s^2\ 2s^2\ 2p^6\ 3s^2\ 3p^6\ 3d^{10}$$

2. Neutral S has the electronic configuration
$$1s^2\ 2s^2\ 2p^6\ 3s^2\ 3p^4$$

Two electrons are added to the outer shell to give a
charge of 2-. The electronic configuration for S^{2-} is
$$1s^2\ 2s^2\ 2p^6\ 3s^2\ 3p^6$$

3. The outer electron of Cs is removed to give a +1 charge.
The electronic configuration of Cs^+ is
$$1s^2\ 2s^2\ 2p^6\ 3s^2\ 3p^6\ 3d^{10}\ 4s^2\ 4p^6\ 4d^{10}\ 5s^2\ 5p^6$$

V. Nomenclature

1. CsCl
2. ZnO
3. MgS
4. NaCN
5. $BaSO_4$

VI. Nomenclature

1. lithium bromide
2. magnesium sulfide
3. nickel(II) oxide
4. calcium carbonate
5. potassium nitrate

SELF-TEST   Work the following in 20 minutes.

1. Which of the following elements has the greatest electron
   affinity?
   a. Na   c. I
   b. O   d. Cl

2. Which of the following elements has the lowest first
   ionization energy?
   a. Na   c. I
   b. O   d. Cl

3. Which of the following ions has the largest ionic radius?
   a. Na^+ c. Ca^{2+}
   b. K^+ d. Ga^{3+}

4. Determine the dissociation energy of F_2 from the following
   data. The enthalpy of formation of KF is -563 kJ/mol and
   the lattice energy of KF is -823.5 kJ/mol. The enthalpy
   of sublimation of K is +90 kJ/mol and the first ionization
   energy of K is +415 kJ/mol. The first electron affinity
   of F is -322 kJ/mol.

5. Write the notation for the electronic configuration of
   Cu^+, Cu^{2+}, N^{3-}, and Cl^-.

THE COVALENT BOND

CHAPTER 8

OBJECTIVES
I. You should be able to demonstrate your knowledge of the following terms by defining them, describing them, or giving specific examples of them:

adjacent charge rule [8.5]
binary compound [8.7]
covalent bond [8.1]
 double bond [8.1]
 multiple bond [8.1]
 single bond [8.1]
 triple bond [8.1]
diatomic [8.1]
dipole moment [8.2]
electronegativity [8.3]
formal charge [8.4]
ion distortion [8.2]
Lewis structure [8.1, 8.5]
nomenclature [8.7]
partial ionic character [8.2, 8.3]
polar covalent bond [8.2]
resonance [8.6]

II. Given the names or chemical formulas of common binary compounds, you should be able to write the corresponding chemical formulas or names.

III. You should be able to predict relative magnitudes of electronegativities using the method of ion distortion, polarization, and electronegativity differences.

IV. You should be able to draw Lewis structures of molecules and ions and determine the formal charge of each atom in such structures.

V. You should be able to predict which bond in a given set of compounds has the most covalent character.

VI. You should be able to calculate the partial ionic character of a bond from dipole moment data and bond distance data. You should also be able to determine the polarity of a bond from the electronegativities of the elements that are bonded.

UNITS, SYMBOLS, MATHEMATICS

I. The following symbols have been used in this chapter. You should be familiar with all of them.

- C is the abbreviation for coulomb, the SI unit of charge.

- D stands for dipole moment.

- δ is the Greek letter delta and is used to indicate either a partial negative charge, δ^-, or a partial positive charge, δ^+.

- kJ is the abbreviation for kilojoules.

- kJ/mol is the abbreviation for kilojoules per mole of material.

- m is the abbreviation for meter, the SI unit of length.

II. The chemical symbols listed in Table 8.1 should be familiar to you.

TABLE 8.1   Common Elements and Their Symbols

Symbol	Element	Symbol	Element
Al	aluminum	Hg	mercury
Ar	argon	I	iodine
As	arsenic	K	potassium
B	boron	Li	lithium
Be	beryllium	Mg	magnesium
Br	bromine	Mn	manganese
C	carbon	N	nitrogen
Ca	calcium	Na	sodium
Cl	chlorine	Ne	neon
Cr	chromium	Ni	nickel
Cu	copper	O	oxygen
Fe	iron	Pb	lead
Ge	germanium	S	sulfur
H	hydrogen	Sn	tin
He	helium	Zn	zinc

EXERCISES

I. Write the formula of each of the compounds in the space provided:

_____
_____
_____
_____
_____
_____
_____
_____

1. sulfur trioxide
2. trioxygen difluoride
3. disulfur decafluoride
4. dinitrogen trioxide
5. dinitrogen tetrafluoride
6. iodine heptafluoride
7. xenon hexafluoride
8. carbon dioxide

II. Write the name of each of the following compounds.

_____
_____
_____
_____
_____
_____
_____
_____
_____
_____
_____
_____

1. NF_3
2. CCl_4
3. N_2O_5
4. I_2O_5
5. IF_7
6. S_2F_{10}
7. SiF_4
8. N_2O_3
9. OF_2
10. O_2F_2
11. O_3F_2
12. Br_2O_7

III. Answer the following:

1. Using the data in Table 8.2 and Figure 8.1 of the study guide, predict the degree of polarity of a bond between each of the following pairs of atoms. Arrange them in order of increasing polarity. What is the % ionic character of each bond?

_____
_____
_____
_____
_____
_____

a. C and I
b. Al and Cl
c. K and Cl
d. S and Cl
e. Cs and S
f. K and O

TABLE 8.2   Relative Electronegativities

H 2.2							He –
Li 1.0	Be 1.6	B 2.0	C 2.6	N 3.0	O 3.4	F 4.0	Ne –
Na 0.9	Mg 1.3	Al 1.6	Se 1.9	P 2.2	S 2.6	Cl 3.2	Ar –
K 0.8	Ca 1.0	Ga 1.8	Ge 2.0	As 2.2	Se 2.6	Br 3.0	Kr –
Rb 0.8	Sr 0.9	In 1.8	Sn 2.0	Sb 2.1	Te 2.1	I 2.7	Xe –
Cs 0.8	Ba 0.9	Tl 2.0	Pb 2.3	Bi 2.0	Po 2.0	At 2.2	Rn –

2. Using the concept of anion deformation, predict which one of the two compounds in each of the following pairs has the bond with the greater amount of covalent character.

    a. $SBCl_3$ or $BiCl_3$
    b. $CrCl_2$ or $CrCl_3$
    c. $HgCl_2$ or HgI_2
    d. BeO or BeS
    e. ZnS or CdS
    f. Tl_2O or Tl_2O_3
    g. MgO or CaO
    h. SO_2 or SeO_2

3. Using any appropriate concept, predict which one of the compounds or ions in each of the following pairs is more covalently bonded.

    a. H_2O or CO_2
    b. HCl or HBr
    c. NO or CaO
    d. BaO or MgO
    e. PbO or PbO_2
    f. PO_4^{3-} or SO_4^{2-}

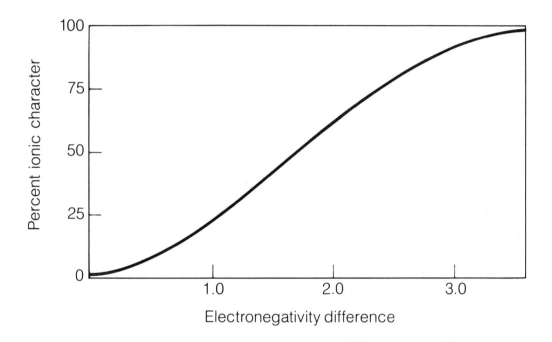

Figure 8.1   Percent ionic character of a bond plotted against
the difference in the electronegativities of the two bonded
atoms.

IV. Draw the best Lewis structure of each of the following and include the formal charges. The formulas of the more complex, multi-element molecules are written to indicate the general atomic arrangement of the molecule. Thus, N_2F_2 is written as FNNF to show that the two nitrogen atoms are joined by at least one chemical bond and that each nitrogen atom forms at least one bond with a single fluorine atom.

A formula which is written CH_4 indicates that the carbon atom forms at least one bond with each of the four hydrogen atoms. A formula which is written H_3COCH_3 indicates that each carbon atom forms at least one bond with each of three hydrogen atoms and at least one bond with the oxygen atom.

1. O_2

2. O_3

3. CH_4

4. N_3^-

5. OF_2

6. $H_2PO_2^-$

7. FNNF

8. C_2^{2-}

9. BF_4^-

10. NH_4^+

11. NO_3^-

12. SO_4^{2-}

13. N_2

14. ClO^-

15. NNO

16. O_2SF_2

17. H_2NNH_2

18. $SnCl_4$

19. OH^-

20. O_2NONO_2

21. $ONNO^{2-}$

22. H_3COCH_3

23. H_2CO

24. XeO_3

V. Draw resonance structures of a molecule of each of the following compounds. Include formal charges.

1. $OCCl_2$: This compound, called phosgene or carbonyl chloride, is a poisonous gas. Carbon is the central atom in the compound.

2. SO_3: Sulfur trioxide is an atmospheric contaminant that is responsible for imparting the characteristic blue color to smoke. It is the acid anhydride of sulfuric acid, i.e., SO_3 reacts with water to form H_2SO_4:

$$SO_3(g) + H_2O(aq) \rightarrow H_2SO_4(aq)$$

3. OCCCO: Little chemistry is reported for this rare gas, which is called tricarbon dioxide.

4. ClCN: Chlorine cyanide or cyanogen chloride is the name of this poisonous, volatile liquid.

5. NO_3^-: This extremely common anion is the nitrate ion. Nitrogen is the central atom.

6. O_3: This compound, ozone, is a major reactant in photochemical smog production and is formed by the action of ultraviolet light on oxygen in the upper atmosphere.

7. NNO: The chemical name of this compound is dinitrogen monoxide or nitrous oxide. The compound is the anesthetic commonly referred to as "laughing gas." It is also used as a propellant gas in whipped cream cans.

ANSWERS TO EXERCISES

I. Formulas of binary compounds

1. SO_3
2. O_3F_2
3. S_2F_{10}
4. N_2O_3
5. N_2F_4
6. IF_7
7. XeF_6
8. CO_2

It is customary to write the less electronegative element first and the more electronegative element last. The ending of the more electronegative element (i.e., the last one in the name or formula) is changed to -ide.

II. Names of covalent binary compounds

1. nitrogen trifluoride

2. carbon tetrachloride

3. dinitrogen pentaoxide

The prefixes used to identify the number of atoms of the same element in a covalent binary compound are given below:

Prefix	Number
mono-	1
di-	2

4. iodine pentaoxide

5. iodine heptafluoride

6. disulfur decafluoride

7. silicon tetrafluoride

8. dinitrogen trioxide

9. oxygen difluoride

10. dioxygen difluoride

11. trioxygen difluoride

12. dibromine heptaoxide

Prefix	Number
tri-	3
tetra-	4
penta-	5
hexa-	6
hepta-	7
octa-	8
nona-	9
deca-	10

III. Atomic properties as related to chemical bonding

1.   The descriptive terms that follow are the authors' personal descriptions of the trend from covalent to ionic bonds. There are no specific terms to indicate the degree of polarity that a bond possesses, but the trend from nonpolar to polar, a, d, b, e, c, and f should be reflected.

a. nonpolar

a. Covalent, nonpolar: The electronegativity difference = 2.7 - 2.6 = 0.1, and the bond is predicted to be covalent and almost nonpolar. The bond is approximately 1% ionic.

d. somewhat polar

d. Covalent, somewhat polar: The electronegativity difference = 3.2 - 2.6 = 0.6, and the bond is predicted to be covalent and somewhat polar. The bond is approximately 8% ionic.

b. moderately polar

b. Covalent, moderately polar: The electronegativity difference = 3.2 - 1.6 = 1.6, and the bond is predicted to be covalent and moderately polar. The bond is approximately 45% ionic.

e. highly polar

e. Covalent, highly polar: The electronegativity difference = 2.6 - 0.8 = 1.8, and the bond is predicted to be covalent and highly polar. The bond is approximately 54% ionic.

c.
ionic

c. Ionic: The electronegativity difference = 3.2 - 0.8
    = 2.4, and the bond is predicted to be ionic. The bond
    is approximately 75% ionic.

f.
very
ionic

f. Very ionic: The electronegativity difference = 3.4 - 0.8
    = 2.6, and the bond is predicted to be ionic. The bond
    is approximately 81% ionic.

2.

The circles in this section are drawn to scale in order to
show relative ionic sizes. Remember that only anion distortion is considered and that the more distorted the anion
the more covalent the bond.

a. $SbCl_3$

Since Sb^{3+} is smaller than Bi^{3+}, Sb^{3+} has a larger concentration
of charge. As a result, Sb^{3+} attracts the electron cloud
of the chlorine ion more than the Bi^{3+}.

The larger the cation charge concentration the more
distorted the anion and the more covalent the bond.

b. $CrCl_3$

Cr^{3+} has a larger concentration of charge than Cr^{2+}; therefore Cr^{3+} attracts the anion more than Cr^{2+}.

Larger cation charge   Smaller cation charge
  concentration   concentration
More distorted anion   Less distorted anion
More covalent   Less covalent

c. HgI_2

The electrons in the larger I^- ion can be more easily
attracted by Hg^{2+}.

Larger anion   Smaller anion
More distortion of anion   Less distortion of anion
More covalent   Less covalent

d. BeS   The sulfide anion is larger than the oxide anion and is
                  therefore more easily distorted than the oxide anion.

Be^{2+} S^{2-} Be^{2+} O^{2-}

Larger anion   Smaller anion
More easily distorted   Less easily distorted
  anion   anion
More covalent bond   Less covalent bond

e. ZnS   See the explanation given for part (a) of this problem.

f. Tl$_2$O$_3$ See the explanation given for part (b) of this problem.

g. MgO   See the explanation given for part (a) of this problem.

h. SO$_2$ See the explanation given for part (a) of this problem.

3. a. CO$_2$ The relative electronegativity difference for an OH bond
                  is 1.2 and that for the CO bond is 0.8. Therefore, CO$_2$
                  is more covalent than H$_2$O. The concept of anion deforma-
                  tion is not easily used in this case because the relative
                  sizes of the cations are unknown and, therefore, the
                  relative concentration of charge on the cation cannot be
                  determined.

   b. HBr   The relative electronegativity difference for the HCl bond
                  is 1.0 and that for the HBr bond is 0.8. Therefore, HBr
                  is more covalent. Since Br$^-$ is the larger anion, the
                  electrons can be more easily attracted by H$^+$.

   c. NO   The answer is easily predicted from relative electronega-
                  tivity differences. The concept of anion distortion
                  cannot be used easily because the relative sizes of the
                  Ca^{2+} and N^{2+} cations are not known.

   d. MgO   Both concepts easily give the same answer.

   e. PbO$_2$ The concept of anion distortion can be used to predict
                  that PbO$_2$ is more covalent than PbO. Because electrone-
                  gativity differences are the same in each case, that
                  concept does not work.

   f. SO$_4$$^{2-}$ The SO bond in SO$_4$$^{2-}$ should be more covalent. The
                  prediction can be made by both the relative electronegativity
                  and the anion distortion concepts.

IV. Lewis structures

Follow the steps outlined in the text:

1. Find the total number of valence electrons in the atoms forming the molecule.

2. Find the total number of electrons that are needed to fill the valence shell of all atoms in the molecule.

3. Compute the number of electrons that must be shared to obtain filled valence shells in all atoms (value from step 2 minus value from step 1).

4. Determine the number of bonds. Each pair of electrons that need to be shared represents one bond (one-half the value determined in step 3).

5. Draw the structure with the proper number of bonds. First place a single bond between bonded atoms; if more bonds are needed, add multiple bonds. There may be several possible positions for multiple bonding. Be sure to draw them all.

6. Add extra electrons in order to fill all valence shells.

7. Compute the formal charges for each atom in all possible structures.

8. (A special bit of advice.) Confirm the correct structure by guaranteeing that:
   a. each atom contains a filled valence shell.
   b. all valence electrons have been used.
   c. the lowest possible formal charges exist.
   d. formal charges of the same sign do not exist on adjacent atoms.

1. $\ddot{O}=\ddot{O}$

1. 12 electrons are available.
2. 16 electrons are needed. Eight for each oxygen atom.
3. 4 electrons must be shared.
4. 2 bonds
5. O$=$O
6. $:\!\ddot{O}=\ddot{O}\!:$
7. $:\!\ddot{O}=\ddot{O}\!:$
8. Check

2. $:\!\ddot{O}=\overset{\textstyle\oplus}{\ddot{O}}-\overset{\textstyle\ominus}{\ddot{O}}\!:$

   or

   $:\!\overset{\textstyle\ominus}{\ddot{O}}-\overset{\textstyle\oplus}{\ddot{O}}=\ddot{O}\!:$

1. 18 electrons are available.
2. 24 electrons are needed.
3. 6 electrons must be shared.
4. 3 bonds
5. O$=$O$-$O or O$-$O$=$O
6. $:\!\ddot{O}=\ddot{O}-\ddot{O}\!:$
7. $:\!\ddot{O}=\overset{\textstyle\oplus}{\ddot{O}}-\overset{\textstyle\ominus}{\ddot{O}}\!:$
8. Check

3.
```
      H
      |
  H—C—H
      |
      H
```

4. $\ddot{N}=N=\ddot{N}$

   $\ominus$ $\oplus$ $\ominus$

1. 16 electrons are available, 5 from each N atom and 1 from the formation of the anion.
2. 24 electrons are needed.
3. 8 electrons must be shared.
4. 4 bonds.
5. N=N=N or N≡N—N or N—N≡N
6. :N=N=N: or :N≡N—N̈: or :N̈—N≡N:
7. :N̈=N=N̈: or :N≡N—N̈: or :N̈—N≡N:

   $\ominus$ $\oplus$ $\ominus$ $\oplus$ $\circled{2-}$ $\circled{2-}$ $\oplus$

8. The structures

   :N̈—N≡N: and :N≡N—N̈:

   $\circled{2-}$ $\oplus$ $\oplus$ $\circled{2-}$

   are improbable because of the large formal charge on a single nitrogen atom.

5. :F̈—Ö—F̈:

6.
```
     ..   ⊖
    :O:
     |
 H—P—H
     |⊕
    :O:
     ..
        ⊖
```

7. :F̈—N=N̈—F̈: The structures
   :F̈—N̈—N=F̈: and F̈=N—N̈—F̈:

   $\ominus$ $\oplus$ $\oplus$ $\ominus$

   are not probable, because the formal charges are not necessary.

9.
```
       ..
      :F:
       |
  ..   |   ..
 :F—B—F:
  ..   |⊖  ..
      :F:
       ..
```

10.
```
       H
       |⊕
   H—N—H
       |
       H
```

11.   The structure

is doubtful because there are only 6 electrons in the
valence shell of nitrogen and the formal charge of
nitrogen is larger than necessary.

   or

   [structure with :Ö:⊖ bonded to N⊕ with two oxygens]

   or

   [structure with :Ö:⊖ bonded to ⊕N with two oxygens]

12.   [sulfate structure: :Ö:⊖ bonded to S, with :Ö—S—Ö:⊖, ⊖ and (2+), :Ö:⊖ below]

13.   :N≡N:

14.   :Ö—Cl:
      ⊖

15.   :N≡N—Ö: The total number of valence electrons is 16.   The structure
      ⊕ ⊖

               N̈=N=Ö
               ⊖ ⊕

is less probable because there are formal charges of
opposite sign on adjacent atoms of the same element.
The structure

               :N̈—N≡O:
               2⊖ ⊕ ⊕

is improbable because there are formal charges of the same
sign on adjacent atoms.

16.   [structure: :Ö:⊖ bonded to S, with :F—S—F:, (2+), :Ö:⊖ below]

17.

$$\begin{array}{ccc} H & & H \\ | & & | \\ :N & \!\!-\!\! & N: \\ | & & | \\ H & & H \end{array}$$

18.

$$:\ddot{C}l:$$
$$|$$
$$:\ddot{C}l-Sn-\ddot{C}l:$$
$$|$$
$$:\ddot{C}l:$$

19. $:\overset{\ominus}{\underset{..}{O}}\!-\!H$

20. $:\ddot{O}: \qquad :\ddot{O}:$ The total number of valence electrons is 40.
The structures

are improbable because in each there are positive formal
charges on three adjacent atoms.

21. $\overset{\ominus}{:}\ddot{O}\!-\!\ddot{N}\!=\!\ddot{N}\!-\!\ddot{O}\overset{\ominus}{:}$ The total number of valence electrons is 24.   The
structures

$$:\ddot{O}\!-\!\ddot{N}\!-\!\ddot{N}\!=\!\ddot{O} \quad and \quad \ddot{O}\!=\!\ddot{N}\!-\!\ddot{N}\!-\!\ddot{O}:$$
$$\ominus \quad \ominus \qquad\qquad\qquad \ominus \quad \ominus$$

are improbable because adjacent formal charges exist.

22.

$$\begin{array}{ccc} H & & H \\ | & & | \\ H-C-\ddot{O}-C-H \\ | & \ddot{} & | \\ H & & H \end{array}$$

23.

$$\begin{array}{c} H \\ | \\ C\!=\!\ddot{O}: \\ | \\ H \end{array}$$

24.

$$:\ddot{O}:$$
$$|$$
$$:Xe-\ddot{O}:$$
$$|$$
$$:\ddot{O}:$$

V. Resonance structures
   See your text.

1.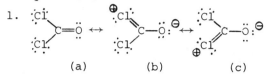

   (a)   (b)   (c)

   Three resonance structures (a, b, and c) can be drawn.
   The structure of a $OCCl_2$ molecule can be described as
   a resonance hybrid, or a weighted average, of the

   three structures. Structure (a) is more important
   than either (b) or (c), i.e., the actual structure
   of a $OCCl_2$ molecule contains more of the expected
   character structure (a) than of that of either (b)
   or (c). If you had been asked to draw a single Lewis
   structure to describe the molecule, structure (a)
   would be correct, because it most closely approximates
   the structure of the molecule.

2.

   In sulfur trioxide all S—O bonds are of equal length.
   The actual structure is an average of the three
   resonance forms, each contributing equally.

3. $\ddot{O}{=}C{=}C{=}C{=}\ddot{O}$ ↔ $:\overset{\oplus}{\ddot{O}}{\equiv}C{-}C{\equiv}C{-}\overset{\ominus}{\ddot{O}}:$ ↔ $:\overset{\ominus}{\ddot{O}}{-}C{\equiv}C{-}C{\equiv}O:^{\oplus}$

        (a)   (b)   (c)

   Structure (a) is the main contributor to the actual
   structure of tricarbon dioxide. A fourth structure
   containing triple bonds between each carbon and oxygen
   is very improbable.

4. $:\ddot{C}l{-}C{\equiv}N:$ ↔ $\ddot{C}l\overset{\oplus}{=}C{=}\ddot{N}^{\ominus}$

        (a)   (b)

   Structure (a) is a more important contributor to the
   actual structure than structure (b).
   Another structure

                    $:Cl{\equiv}C{-}\ddot{N}:$

                     ②⊕   ②⊖

   is very improbable.

5.

$$:\overset{..}{\underset{..}{O}}:^{\ominus} \quad \leftrightarrow \quad :O: \quad \leftrightarrow \quad :\overset{..}{\underset{..}{O}}:^{\ominus}$$

(a)   (b)   (c)

All three structures contribute equally to the actual structure of the nitrate ion.

6.

(a)   (b)

Structures (a) and (b) are expected to contribute equally to the actual structure.

7. $\overset{\ominus}{\overset{..}{N}}=\overset{\oplus}{N}=\overset{..}{\underset{..}{O}} \quad \leftrightarrow \quad :N\equiv\overset{\oplus}{N}-\overset{..}{\underset{..}{O}}:^{\ominus}$

(a)   (b)

Both are important contributors to the actual structure of nitrous oxide.   A third structure

$$^{(2)}\overset{..}{\underset{..}{N}}-N\equiv\overset{\oplus}{O}:$$

does not contribute appreciably.

SELF-TEST

Complete the test in 25 minutes.

Using a periodic chart, answer each of the following:

1. Which of the following is the best Lewis structure for NCCN? What is wrong with those structures that you did not choose?

a. $:N\equiv C—C\equiv N:$

b. $:\overset{..}{N}—C\equiv C—\overset{..}{N}:$

c. $\overset{..}{N}=C=C=N:$

d. $\overset{.}{N}=C=C=\overset{..}{N}$

e. $\overset{.}{N}—\overset{..}{C}—\overset{..}{C}—\overset{..}{N}$

f. $:N\equiv C=\overset{..}{C}—\overset{..}{N}:$

2. The thiocyanate ion, SCN⁻, is linear and the atoms are arranged in the order given. Draw Lewis structures for the three resonance forms of the ion *complete with formal charges*.

3. Which of the following elements has the greatest
   electronegativity?
   a. Na   c. Sn
   b. Ca   d. F

4. Which of the following elements has the smallest
   electronegativity?
   a. P   c. Sb
   b. S   d. Se

MOLECULAR GEOMETRY, MOLECULAR ORBITALS

CHAPTER 9

OBJECTIVES

I. You should be able to demonstrate your knowledge of the following terms by defining them, describing them, and giving specific examples of them:

antibonding orbital [9.4]
    sigma antibonding orbital [9.4]
    pi antibonding orbital [9.4]
axial positions [9.2]
bond order [9.4]
bonding orbital [9.4]
    sigma bonding orbital [9.4]
    pi bonding orbital [9.4]
bonding pair [9.2]
hybridization [9.3]
hybrid orbital [9.3]

    sp^3 hybrid orbital [9.3]

    sp^2 hybrid orbital [9.3]

    sp hybrid orbital [9.3]

    dsp^3 hybrid orbital [9.3]

    d^2sp^3 hybrid orbital [9.3]
equatorial positions [9.2]
lone pair [9.2]
nonbonding pair [9.2]
octahedron [9.2]
$p\pi - d\pi$ bond [9.6]
pi bond [9.4]
sigma bond [9.4]
square planar [9.2]
square pyramid [9.2]
tetrahedral [9.2]

trigonal bipyramid [9.2]
VSEPR theory [9.2]

II. You should be able to use VSEPR theory to predict molecular shapes.

III. You should be able to use molecular orbital hybridization to rationalize the geometry of covalent molecules.

IV. You should be able to use molecular orbital theory to draw energy-level diagrams of the diatomic molecules and ions formed from elements of the second period.

V. You should be able to rationalize observed molecular structures by drawing Lewis structures and discussing the molecular orbitals that are involved in the bonding.

UNITS,
SYMBOLS,
MATHEMATICS

I. The following symbols were used in this chapter. You should be familiar with them.

- dsp^3 is a notation used to identify a hybrid atomic orbital which is formed from a combination of one d, one s, and three p atomic orbitals. The hybrid has a trigonal bipyramidal shape.
- d^2sp^3 is a notation used to identify a hybrid atomic orbital which is formed from a combination of two d, one s, and three p atomic orbitals. The hybrid has an octahedral shape.
- π is a symbol used to identify a molecular orbital which contributes to bonding. The electron density along the axis which connects the atoms being bonded is zero, but off the axis the orbital forms two "bridges" between the nuclei.
- π^* is a symbol used to identify a molecular orbital which is antibonding. The electron density along the axis which connects the atoms being bonded is zero, and no continuous "bridges" are formed between the nuclei.
- σ is a symbol used to identify a molecular orbital which contributes to bonding. The major electron density is along the axis which connects the atoms being bonded, and forms a continuous "bridge" between the nuclei.
- σ^* is a symbol used to identify a molecular orbital which is antibonding. The major electron density is along the axis which connects the atoms, but the orbital does not form a continuous "bridge" between the nuclei.
- sp is a notation used to identify a hybrid atomic orbital which is formed from a combination of one s and one p atomic orbitals. The hybrid has a linear shape.
- sp^2 is a notation used to identify a hybrid atomic orbital which is formed from a combination of one s and two p atomic orbitals. The hybrid has a triangular planar shape.

- sp^3 is a notation used to identify a hybrid atomic orbital which is formed from a combination of one s and three p atomic orbitals. The hybrid has a tetrahedral shape.
- s,p,d,f are notations used to represent electronic subshells in atoms. These subshells are the orbitals which are combined to form sp, sp^2, sp^3, dsp^3, sp^3d, d^2sp^3, sp^3d^2 hybrid orbitals.

II. A few additional symbols which are used to identify the elements have been used in this chapter. You should have memorized those in Table 9.1.

TABLE 9.1   Common Elements and Their Symbols

Symbol	Element	Symbol	Element
Al	aluminum	K	potassium
Ar	argon	Li	lithium
As	arsenic	Mg	magnesium
Au	gold	Mn	manganese
B	boron	N	nitrogen
Be	beryllium	Na	sodium
Br	bromine	Ne	neon
C	carbon	Ni	nickel
Ca	calcium	O	oxygen
Cl	chlorine	P	phosphorus
Cr	chromium	Pb	lead
Cu	copper	S	sulfur
F	fluorine	Sb	antimony
Fe	iron	Se	selenium
Ge	germanium	Sn	tin
H	hydrogen	Te	tellurium
He	helium	Tl	thallium
Hg	mercury	Xe	xenon
I	iodine	Zn	zinc

EXERCISES   I. Use the concept of electron pair repulsions, VSEPR, to predict the geometric configuration of a molecule of each of the following compounds:

1. XeF_4, xenon tetrafluoride
2. AsF_5, arsenic pentafluoride
3. SnF_4, tin tetrafluoride or tin(IV) fluoride
4. IF_2^+, the iodine difluoride cation
5. ClF_3, chlorine trifluoride
6. H_3O^+, the hydronium ion
7. I_3^-, the triiodide ion
8. $SnBr_6^{2-}$, the tin(IV) hexabromide anion
9. SF_4, sulfur tetrafluoride

II. Draw the most probable Lewis structure for each of the following compounds. Use VSEPR theory to predict the molecular shape.

1. PH_3
2. $SnCl_4$
3. $SnCl_2$
4. O_3
5. N_3^-

6. CO_2
7. SO_2
8. HCN
9. FNO
10. H_2CO

III. Predict the orbital hybridization of the central atom in each of the molecules described in Sections I and II.

IV. Write the notation for the electronic configuration of each of the following homopolar, diatomic molecules and ions. Determine the bond order of each and determine whether the molecule or ion is paramagnetic or diamagnetic. Also draw the Lewis structure of each.

1. NO 2. CN^- 3. O_2^+

V. Predict the geometric configuration of a molecule of each of the following compounds. Describe the bonding between all atoms in the molecule. Remember that the molecular shapes will only be approximations based on the theories with which you are now familiar.

1. H_2CO, formaldehyde
2. C_6H_6, a six-carbon ring compound, benzene
3. CCl_4, carbon tetrachloride
4. HCCH, acetylene
5. SO_2, sulfur dioxide
6. FNNF, dinitrogen difluoride
7. HF, hydrogen fluoride

VI. From the following list of molecules and atoms, described by their Lewis structures, choose the ones that answer the questions. A molecule or atom may be used to answer more than one question.

a. H—Ö—H

b. H—N̈—H
    |
    H

c. H—C̈l:

d.
   H H
   | |
H—C—C—H
   | |
   H H

e.
    :Cl:
     |
:C̈l—C—C̈l:
     |
    :Cl:

f.
   H   H
   | |
H—C—Ö—C—H
   | |
   H   H

g. Na·

h. :Ö—S̈=Ö

i. ·C̈·

j.
         H
         |
       :O:
        |
Ö=S—Ö—H
        ||
       :O:

k.
      :F:
  F· | ·F·
    S
  F· | ·F·
      :F:

l. :F:
    |
  :I—F:
    |
   :F:

m.
     :F:
     | /F·
     Xe
  ·F/ |
     :F:

_____ 1. Which molecule(s) contains one or more atoms that probably use sp^3 hybrid orbitals in bonding?

_____ 2. Which molecule(s) contains π bonds?

_____ 3. Which molecule(s) contains seven σ bonds?

_____ 4. Which molecule(s) or atom(s) contains unpaired electrons?

_____ 5. Which molecule(s) is planar?

_____ 6. Which molecule(s) does not contain nonbonding pairs of
electrons?

_____ 7. Which molecule(s) can exhibit $p\pi$-$d\pi$ overlap?

_____ 8. Which molecule(s) is linear?

_____ 9. Which molecule(s) could contain $d\pi$-$p\pi$ bonds?

ANSWERS TO
EXERCISES

I. Electron pair repulsions and molecular geometry.

1. square
planar
configu-
ration

To predict the geometry of a molecule from a consideration
of electron pair repulsions, the following method can be
used if the molecule contains no double bonds:

Electron-pair repulsion method:

(a) Determine the number of electrons in the valence
shell of the neutral central atom. If the molecule
is an ion, account for its charge by adding or
subtracting electrons in the valence shell of the
central atom.

(b) Determine the total number of electrons contributed
to the central atom's valence shell by the atoms
directly bonded to the central atom.

(c) Divide the sum of (a) and (b) by 2 to determine the
number of electron pairs.

(d) Distribute the electron pairs in the valence shell
of the central atom such that they are as far away
from each other as possible. Such orientations of
electron pairs are given in Table 9.2 of the study
guide.

TABLE 9.2   Orientations of Electron Pairs about the
               Center of the Central Atom of a Molecule

Number of Electron Pairs	Orientation
2	linear
3	triangular planar
4	tetrahedral
5	trigonal bipyramidal
6	octahedral

(e) Bond the peripheral atoms to the central atom such
    that minimum repulsion occurs. Molecular shapes
    predicted from a consideration of minimum electron
    pair repulsions are given in Table 9.3 of the study
    guide.

Using the electron-pair repulsion method to predict the
geometry of XeF_4:

(a) The total number of valence electrons in a neutral
    atom of xenon is 8.   Since XeF_4 is uncharged, no
    electrons must be subtracted from or added to the
    valence shell of xenon.
(b) One electron is donated to the valence shell of
    xenon by each fluorine atom.   The total number of
    electrons donated is therefore 4.
(c) 8 electrons + 4 electrons = 12 electrons
    12 electrons/2 = 6 electron pairs
(d) According to the information in Table 9.2 of the study
    guide, the electron pairs are predicted to have an
    octahedral orientation about the center of the xenon
    atom.
(e) According to the information in Table 9.3 of the study
    guide, the molecule is predicted to have a square planar
    shape.

TABLE 9.3   Number of Electron Pairs in the Valence Shell of the Central
             Atom and Molecular Shape

Number of Electron Pairs				
Total	Bonding	Nonbonding	Shape of Molecule or Ion	Examples
2	2	0	linear	$HgCl_2$, $CuCl_2^-$
3	3	0	triangular planar	BF_3, $HgCl_3^-$
3	2	1	angular	$SnCl_2$, NO_2
4	4	0	tetrahedral	CH_4, BF_4^-
4	3	1	trigonal pyramidal	NH_3, PF_3
4	2	2	angular	H_2O, ICl_2^+
5	5	0	trigonal bipyramidal	PCl_5, $SnCl_5^-$
5	4	1	irregular tetrahedral	$TeCl_4$, IF_4^+
5	3	2	T shaped	ClF_3
5	2	3	linear	XeF_2, ICl_2^-
6	6	0	octahedral	SF_6, PF_6^-
6	5	1	square pyramidal	IF_5, SbF_5^{2-}
6	4	2	square planar	BrF_4, XeF_4

2. trigonal bipyramidal

F—As with F atoms (two F above/below, F, F, F around)

Electron-pair repulsion method:

(a) Five valence electrons are in the valence shell of arsenic.
(b) Five electrons are contributed by the peripheral atoms, one by each fluorine atom.
(c) There is a total of five electron pairs.

$$(5 + 5)/2 = 5$$

(d) Electron pairs are distributed such that they have a trigonal bipyramidal orientation.
(e) The shape of the molecule is predicted to be a trigonal bipyramid.

3. tetrahedral

Electron-pair repulsion method:
(a) Four electrons are in the valence shell of tin.
(b) Four electrons are contributed by the peripheral atoms.
(c) There is a total of four electron pairs.
(d) Electron pairs are distributed such that they have a tetrahedral orientation.
(e) The shape of the molecule is predicted to be a tetrahedron.

4. angular

(a) There are seven electrons in the valence shell of a neutral iodine atom. Since the molecule has 1+ charge, one electron must be subtracted from the valence shell of the iodine atom. Thus, the valence shell of the iodine atom has six electrons.
(b) Two electrons are contributed by the peripheral atoms, one by each fluorine atom.
(c) There is a total of four electron pairs.
(d) Electron pairs are distributed such that they have a tetrahedral orientation.
(e) The molecule is predicted to have an angular shape.

5. T shaped

6. trigonal pyramidal

7. linear

8. octahedral

Br
Br   Br
   Sn
Br   Br
   Br

②⁻

9. irregular tetrahedron

F
   F
S
   F
F

II. VSEPR theory continued:

To predict molecular shapes by drawing Lewis structures use the following procedure:

(a) Draw the structure using the method described in Chapter 8.
(b) Count the number of lone pairs on the central atom.
(c) Count the number of peripheral atoms bonded to the central atom.
(d) Distribute the sum of (b) and (c) around the central atom so that they are as far away from each other as possible. See Table 9.2 in the study guide.
(e) Add the double bonds.

1. trigonal pyramidal

Applying these rules to PH$_3$:

(a) H—P̈—H
          |
          H

(b) One lone pair
(c) Three peripheral atoms
(d) Tetrahedral distribution

(e)

P
H   H
   H

2. tetrahedral

3. angular

Experimental evidence shows that the double bond does not exist. The correct angular shape is predicted by this approach anyway. The actual structure is

It would be a fair assumption to predict the possibility of some double bond character in the absence of evidence to the contrary. Your instructor will guide you by choosing molecules in which the experimental evidence and theory are consistent.

4. angular

Applying the rules outlined in problem 1 in this section of the study guide:

(a) Ö=Ö—Ö:
(b) One lone pair
(c) Two peripheral atoms
(d) Triangular planar distribution

(e)

5. linear

:N=N=N:⊖

(a) :N=N=N:⊖
(b) No lone pairs
(c) Two peripheral atoms
(d) Linear distribution

(e) N=N=N⊖

6. linear

Ö=C=Ö

7. angular

8. linear

H—C≡N:

9. angular

10. triangular
    planar

(a)   :O:
       ‖
   H—C—H

(b) No lone pairs
(c) Three peripheral atoms
(d) Triangular planar distribution
(e)   :O:
       ‖
       C
   H   H

III. Hybrid orbitals and molecular geometry

1. (I)
   d^2sp^3

2. (I)
   dsp^3

3. (I)
   sp^3

4. (I)
   sp^3

5. (I)
   dsp^3

6. (I)
   sp^3

Normally, experimental evidence is needed to predict the types of hybrid orbitals involved in bonding. In the absence of such evidence the hybrid orbitals can be determined from the assumed molecular shape and the number of nonbonded pairs of electrons. The total number of nonbonding electron pairs plus the number of peripheral atoms (atoms bonded to the central atom) always equals the total number of orbitals forming the hybrid. For instance, a d^2sp^3 orbital is a hybrid of two d orbitals plus one s orbital plus three p orbitals, for a total of six orbitals. Thus any combination of nonbonding pairs plus peripheral atoms which totals six might well indicate that the central atom is d^2sp^3 hybridized. Refer to Table 9.4 in the study guide to answer all questions in this section.

7. (I)$_3$
    dsp^3

8. (I)
    d^2sp^3

9. (I)
    sp^3

1. (II)
    sp^3

2. (II)
    sp^3

3. (II)
    sp^2

4. (II)
    sp^2

5. (II)
    sp

6. (II)
    sp

7. (II)
    sp^2

8. (II)
    sp

9. (II)
    sp^2

10. (II)
    sp^2

TABLE 9.4   Geometry and Hybridization of Atomic Orbitals

Number of Nonbonding Electron Pairs + Number of Atoms Bonded to the Central Atom	Type of Hybrid Orbitals Used by the Central Atom	Geometry of Hybrid Orbitals about the Center of the Central Atom
2	sp	linear
3	sp^2	triangular planar
4	sp^3	tetrahedral
5	dsp^3	trigonal pyramidal
6	d^2sp^3	octahedral

IV. Molecular orbitals

It is imperative that you read your class notes very meticulously to ascertain the depth to which your instructor wishes you to master this material.

1. In an NO molecule 15 electrons are distributed as described in your text:

$$(\sigma_{1s})^2(\sigma_{1s}^*)^2(\sigma_{2s})^2(\sigma_{2s}^*)^2(\sigma_{2p})^2(\pi_{2p})^4(\pi_{2p}^*)^1$$

The bond order is $2\frac{1}{2}$: bond order

$$= \frac{10 \text{ bonding electrons} - 5 \text{ antibonding electrons}}{2}$$

$$= 2\frac{1}{2}$$

A molecular orbital energy-level diagram for the higher-energy molecular orbitals of NO can be drawn:

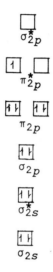

From this diagram we readily see that there is one unpaired electron in an NO molecule. Thus an NO molecule is paramagnetic. Lewis structures do not give an adequate description of the NO molecule:

$$\cdot\ddot{N}\!\!=\!\!\ddot{O}\cdot$$

2. In a CN⁻ ion there are 14 electrons to be distributed in molecular orbitals:

$$(\sigma_{1s})^2 (\sigma_{1s}^{\star})^2 (\sigma_{2s})^2 (\sigma_{2s}^{\star})^2 (\pi_{2p})^4 (\sigma_{2p})^2$$

The bond order is 3. A molecular orbital energy-level diagram shows that there are no unpaired electrons in a CN⁻ ion. A CN⁻ ion is therefore diamagnetic. The Lewis structure

$$:C\!\!\equiv\!\!N: {}^{\ominus}$$

gives an adequate description of the bond order of the molecule.

3. In an O_2^+ ion there are 15 electrons to be distributed in molecular orbitals:

$$(\sigma_{1s})^2 (\sigma_{1s}^*)^2 (\sigma_{2s})^2 (\sigma_{2s}^*)^2 (\sigma_{2p})^2 (\pi_{2p})^4 (\pi_{2p}^*)^1$$

The bond order is $2\frac{1}{2}$.

A molecular orbital diagram shows that there is one unpaired election in the O_2^+ ion.   An O_2^+ ion therefore should be paramagnetic.   Lewis structures do not give an adequate description of the bond order of the ion:

$$\ddot{\text{O}}\!=\!\dot{\text{O}}\!\cdot$$

V. Geometric configurations of molecules

1. triangular planar

The Lewis structure is

$$\begin{array}{c} :\text{O}: \\ \| \\ \text{H}\!-\!\text{C}\!-\!\text{H} \end{array}$$

According to the concept of electron pair repulsion, the σ bonding pairs of electrons are distributed about the center of the carbon atom in the configuration of a plane triangle:

The central carbon atom is said to be sp^2 hybridized. The $2s$ orbital and two of the $2p$ orbitals of carbon form the sp^2 hybrid orbitals of carbon while the remaining p orbital of carbon overlaps a p orbital of oxygen to form a π bond:

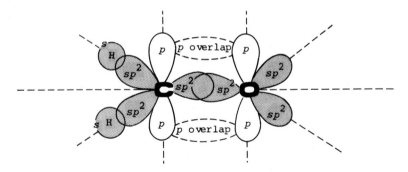

FIGURE 9.1

2. hexagonal
   planar

There are two resonance forms of benzene:

According to the concept of electron pair repulsion, the σ bonding pairs of electrons are distributed about the centers of the appropriate carbon atoms in the configuration of a plane triangle:

Each carbon is said to be sp^2 hybridized. The $2s$ orbital and two of the $2p$ orbitals of carbon form the sp^2 hybrid orbitals of carbon while the remaining p orbital of carbon enters into π bonding (see Figure 9.2).

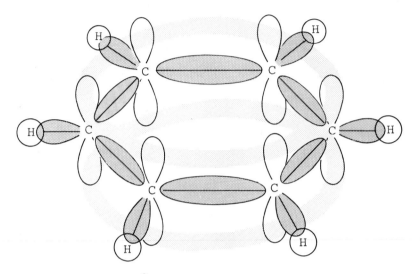

FIGURE 9.2 The sp^2 hybrid orbitals of the carbon atoms are shaded, and the unhybridized p orbitals are not. The p orbital overlap shown on a decreased scale is indicated by the more lightly shaded areas.

3. tetrahedral   The Lewis structure is

$$:\ddot{C}l:$$
$$|$$
$$:\ddot{C}l-C-\ddot{C}l:$$
$$|$$
$$:\ddot{C}l:$$

According to the concept of electron pair repulsion, a tetrahedral geometry is predicted:

$$
\begin{array}{c}
Cl \\
| \\
C \\
Cl \quad | \quad Cl \\
Cl
\end{array}
$$

The carbon is said to be sp^3 hybridized.

4. linear   The Lewis structure is

$$H-C\equiv C-H$$

Each carbon has two atoms bonded to it and no nonbonding pairs of electrons. According to the concept of electron pair repulsion, the molecule has a linear shape. Each carbon is said to be *sp* hybridized and forms two σ bonds.

 a σ bond formed by the overlap of an *sp* hybrid orbital of each carbon

a σ bond formed by the overlap of an orbital of hydrogen and an *sp* hybrid orbital of carbon

Each carbon also forms two π bonds. These bonds are formed by the overlap of the unhybridized *p* orbitals of each carbon.

$$H-C\equiv C-H$$ one σ bond and two π bonds

5. angular   The resonance forms of SO_2 are

$$:\overset{\ominus}{\underset{\cdot\cdot}{O}}-\overset{\oplus}{\underset{\cdot\cdot}{S}}=\overset{\cdot\cdot}{\underset{\cdot\cdot}{O}} \leftrightarrow \overset{\cdot\cdot}{\underset{\cdot\cdot}{O}}=\overset{\oplus}{\underset{\cdot\cdot}{S}}-\overset{\ominus}{\underset{\cdot\cdot}{O}}:$$

Sulfur is bonded to two atoms and has one nonbonding pair of electrons. According to electron pair repulsion, the σ and nonbonding electron pairs are distributed about the center of the sulfur atom in the configuration of a plane triangle:

It is difficult to predict the bonding of the oxygen atoms because the π bond is not localized between any two atoms:

6. planar
   bent

The Lewis structure is

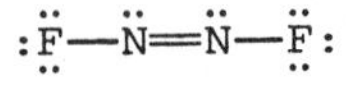

$$:\ddot{F}\!-\!\ddot{N}\!=\!\ddot{N}\!-\!\ddot{F}:$$

According to the concept of electron pair repulsion, the σ and nonbonding pairs of electrons are distributed about each nitrogen atom in a planar triangular configuration:

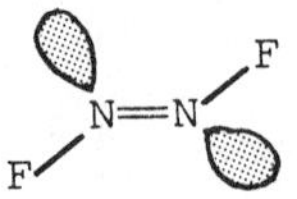

Each nitrogen is said to be sp^2 hybridized. A σ bond between the nitrogen atoms is formed from the overlap of these sp^2 hybrid orbitals. Each nitrogen also has one sp^2 orbital overlapping with a fluorine orbital to form an N-F σ bond. The remaining sp^2 nitrogen orbital contains a nonbonding pair of electrons. The unhybridized p orbitals overlap to form a π bond. A structure in which the fluorine atoms are on the same side of the molecule is also possible:

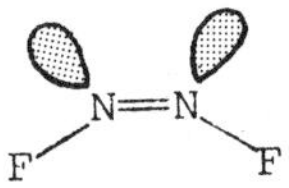

7. linear   The Lewis structure is

$$H : \overset{..}{\underset{..}{F}} :$$

The fluorine is said to be sp^3 hybridized. A tetrahedral arrangement of the bonding pair of electrons and the nonbonding pairs of electrons provides the most effective separation of electron pairs.

VI. Properties of atoms and molecules.

1. a, b, c,   See Section 9.3 of your text.
   d, e, f,j
   (a) The oxygen atom probably uses sp^3 hybrid orbitals.
   (b) The nitrogen atom probably uses sp^3 hybrid orbitals.
   (c) The chlorine atom *possibly* uses sp^3 hybrid orbitals.
   (d) Each carbon atom probably uses sp^3 hybrid orbitals.
   (e) The carbon atom probably uses sp^3 hybrid orbitals, and the chlorine atom *possibly* uses sp^3 hybrid orbitals.
   (f) The oxygen atom and the carbon atoms each probably uses sp^3 hybrid orbitals.
   (j) The oxygen atoms with single bonds probably use sp^3 hybrid orbitals.

2. h, j   See Sections 9.4 and 9.6 of your text.

3. d   See Section 9.4 of your text.

4. g, i   g: An atom of sodium has a single electron in the $2s$ subshell.
   i: An atom of carbon has two unpaired electrons in the $2p$ subshell.

5. a, c, h,   See Section 9.2 of your text.
   l, m

6. d

7. j   See Section 9.6 of your text.

8. c   See Section 9.2 of your text.

SELF-TEST   Complete the test in 25 minutes.

1. Complete the table:

| Formula | Number of Electron Pairs | | Shape of Molecule or Ion |
	Bonding	Nonbonding	
SCl_2	_____	_____	_____
XeF_4	_____	_____	_____
AlH_4	_____	_____	_____
$TeCl_4$	_____	_____	_____
SeF_5^-	_____	_____	_____

2. Crystalline PCl_5 consists of an ionic lattice of PCl_4^+ and PCl_6^- ions. In the vapor and liquid states, however, the compound exists as PCl_5 molecules. What type of hybrid orbitals does P employ in each of these species, and what is the geometry of the molecule or ion?

	Hybrid Orbitals	Geometric Shape
PCl_5	_____	_____
PCl_4^+	_____	_____
PCl_6^-	_____	_____

3. Complete the following table. The σ orbitals include only the $\sigma 2s$ and $\sigma 2p$ orbitals, and the σ^* orbitals include only the $\sigma^* 2s$ and $\sigma^* 2p$ orbitals.

| Molecule | Total Number of Electrons in | | | | Bond Order | Number of Unpaired Electrons |
	σ Orbitals	σ^* Orbitals	π Orbitals	π^* Orbitals		
Be_2	_____	_____	_____	_____	_____	_____
B_2	_____	_____	_____	_____	_____	_____
N_2	_____	_____	_____	_____	_____	_____
O_2	_____	_____	_____	_____	_____	_____
NO^+	_____	_____	_____	_____	_____	_____

GASES

CHAPTER 10

II. If all but one of the variables (pressure, P, volume, V, number of moles, n, and temperature, T) of the ideal gas law are given, you should be able to calculate the unknown variable.

III. If a gas under a defined set of conditions is changed to a new set of conditions, you should be able to use Boyle's, Charles', and Amonton's laws in calculations to determine the complete new set of conditions.

IV. You should understand vapor pressure and partial pressures and be able to perform calculations involving them.

V. You should be able to calculate relative rates of gaseous effusion.

VI. You should understand Gay-Lussac's law of combining volumes and be able to perform calculations involving chemical reactions of gases.

UNITS,
SYMBOLS,
MATHEMATICS

I. The following symbols are used in this chapter. You should be familiar with them.

- atm is the abbreviation for atmosphere, a unit of pressure frequently used by chemists. One atm equals 101,325 Pa.
- °C is the abbreviation for degrees Celsius.
- K is the abbreviation for Kelvin, the SI unit of temperature.
- M is the symbol used to represent molecular weight in equations such as $PV = \dfrac{g}{M} RT$.
- mL is the abbreviation for milliliters. One milliliter equals 10^{-3} liters.
- mol is the abbreviation for the mole. One mol equals 6.022×10^{23} atoms or molecules.
- n is the symbol used to represent the number of moles in equations such as $PV = nRT$.
- P is the symbol used to represent pressure in equations such as $PV = nRT$.
- Pa is the abbreviation for pascal, the SI unit of pressure.
- p_A is a symbol used to represent the partial pressure of any species in equations such as $P_{total} = X_A p_A + X_B p_B$.
- R is the symbol used to represent the gas constant. R equals 0.082056 L·atm/(K·mol).
- s is the abbreviation for second, the SI unit of time.
- STP is the abbreviation for standard temperature and pressure. The standard temperature is 273 K and the standard pressure is 1 atm in all problems in this chapter.

- T is a symbol used to represent temperature in Kelvin in equations such as $PV = nRT$.
- t is the symbol used to represent temperature in °C.
- torr is a unit of pressure equivalent to the pressure that will support a column of mercury to a height of 1 mm. 760 torr equals 1 atm.
- V is the symbol used to represent volume in equations such as $PV = nRT$.
- X_A is a symbol used to represent the mole fraction of A in equations such as $P_{total} = X_A p_A + X_B p_B$.

EXERCISES

I. Answer each of the following with *true* or *false*. If a statement is false, correct it.

_____ 1. One atmosphere equals 760 torr.

_____ 2. Boyle's law states that the volume of a gas varies inversely with the pressure under which it is measured.

_____ 3. Charles' law states that the volume of a gas varies inversely with the pressure under which it is measured.

_____ 4. STP stands for standard temperature, 273 K, and pressure, 1 atmosphere.

_____ 5. The value of R is 0.08206 L·atm/(K·mol).

_____ 6. $PV = nRT$.

_____ 7. Temperatures in °C are changed to the kelvin scale by adding 100: $T = t + 100$.

_____ 8. A mole of ideal gas contains 6.022×10^{23} molecules and occupies 22.414 liters at STP.

_____ 9. In a mixture of gases the total pressure is the sum of the partial pressures of the components.

_____ 10. According to Graham's law of effusion, gases with larger molecular weights effuse through small openings more rapidly than gases with smaller molecular weights.

_____ 11. The van der Waals equation accounts for the fact that real gas molecules have no volume and exert no attractive forces.

_____ 12. Below the critical temperature it is impossible to liquefy a gas regardless of pressure.

II. Complete the following statements with one of the
following:

increases
decreases
remains the same

1. If the temperature of a gas is increased and the pressure on the system is unchanged, the volume of the gas

_____.

2. If a gas is enclosed in a rigid container and the
container is heated, the pressure exerted by the gas

_____.

3. A vessel contains 2.5 mol of oxygen. If an additional
2.5 mol of oxygen is added to the vessel, the pressure

_____.

4. A gas is allowed to expand from 1 L to 22.4 L.

The number of moles of gas _____.

5. As a gas is heated, the average kinetic energy of the

molecules _____.

6. As a gas is compressed without changing the temperature,
the average kinetic energy of the molecules

_____.

7. As a gas is compressed at constant temperature, the mean

free path of a molecule _____.

8. A cylinder contains oxygen at a pressure of 1500 pounds
per square inch, and some gas is released from it.

The pressure _____. The number of

moles of gas in the cylinder _____.

If there is no temperature change, the average kinetic
energy of the molecules in the cylinder

_____ and the mean free path

_____.

9. A weather balloon is released from a station in Texas.

   As the balloon rises, its size _____
   due to decreased atmospheric pressure.

10. The combustion of hydrogen is represented by

    $$2H_2(g) + O_2(g) \rightarrow 2H_2O(g)$$

    If the temperature and volume do not change, the

    pressure _____ as the reaction
    proceeds in a closed container. If an appreciable
    temperature increase occurs during the reaction, the

    final pressure _____ due to the
    temperature change.

11. A gas in a 1-L container is heated from 0°C to
    100°C. Simultaneously the volume of the container
    increases to 2 L. The pressure

    _____ due to the temperature

    change ; the pressure _____ due to the
    volume change.

12. A naturally occurring mixture of $^{35}Cl_2$ and $^{37}Cl_2$ is
    enclosed in a cylinder. As a small leak allows gas
    to escape very slowly, a piston maintains the trapped
    sample at STP. The number of molecules in the cylinder

    _____. The density of the

    trapped gas _____. The molecular

    weight of the trapped gas _____.

13. A mole of helium at STP is enclosed in a rigid
    container. Argon is gradually added to the container.

    The pressure _____. The number of

    moles of helium _____. The

    partial pressure of helium _____.

    The partial pressure of argon _____.

14. A rigid 1-L vessel contains equimolar concentrations of He and Ar. As the temperature is increased, the

partial pressure of He, p_{He}, _____

and the partial pressure of Ar, p_{Ar}, _____.

The total pressure _____.

15. Gas is gradually escaping from a container. The number

of moles of gas, n, _____.
The volume and pressure are maintained constant. The

temperature of the gas _____.

III. Work the following problems:

1. A vessel containing argon at 0.10 atm is heated from
   0°C to 100°C. If the volume does not change, what is
   the final pressure?

2. A 75-mL sample of gas is heated at constant pressure
   from 0°C to 33°C. What is the volume at 33°C?

3. A gas in a 1.00-L container is allowed to expand to
   a volume of 5.00 L. If the initial pressure is 748
   torr, what is the final pressure if the temperature
   does not change?

4. What volume will 16 g of oxygen gas occupy at STP?

5. What is the density, in g/mL, of oxygen at STP?

6. How many moles of gas at 1.00 atm and 25°C are
   contained in a 5.00-L vessel?

7. A balloon containing 1.00 L of He at STP is
   purchased in an air-conditioned store, in which the
   temperature is 25°C, and carried outside where it
   heats to 45°C. If there is no pressure change, what
   is the final volume?

8. If the pressure on the balloon described in problem 7
   of this section does change from 745 torr to 757 torr
   during the temperature change, what is the final
   volume?

9. A gas sample is heated from -10°C to 87°C, and the volume is increased from 1.00 L to 3.30 L. If the initial pressure is 0.750 atm, what is the final pressure?

10. A 1.0-L flask contains 3.5 mol of gas at 27°C. What is the pressure of the gas?

11. An evacuated 1.00-L flask weighs 104.35 g. A quantity of nitrogen gas is added to the flask. After the nitrogen is added, the flask and contents weigh 105.68 g, and the pressure exerted by the gas is 2.00 atm. What is the temperature inside the flask?

12. When magnesium metal is burned in air, solid MgO forms:

$$2Mg(s) + O_2(g) \rightarrow 2MgO(s)$$

What volume of oxygen at STP is needed to react with 0.500 g of Mg?

13. The complete combustion of octane yields carbon dioxide and water:

$$2C_8H_{18}(g) + 25O_2(g) \rightarrow 16CO_2(g) + 18H_2O(g)$$

What volume of gas is produced from the complete combustion of 0.670 g of octane if the temperature is 400°C and the pressure is 1.2 atm?

14. Hydrogen gas can be generated by the action of hydrochloric acid on magnesium:

$$Mg(s) + 2HCl(aq) \rightarrow Mg^{+2}(aq) + 2Cl^-(aq) + H_2(g)$$

If all the hydrogen from the reaction of 2.00 g of Mg is collected over water at 1.00 atm and 25°C, what volume will the dried gas occupy at STP?

15. In a normal breath 2.00 L of gas can be inhaled. How many moles of gas at STP are in this volume? If 21 percent (by volume) of the inhaled gas molecules is oxygen, how many grams of oxygen are in this volume?

16. The atmospheric pressure on Mt. Everest is 0.330 atm with a temperature of -10°C. How many grams of oxygen are inhaled in a 2.00-L breath?

17. One mole of hemoglobin, which has a weight of 66,280 g, can bind 4 moles of oxygen gas. What volume of O_2 at STP can 100 mL of blood with a hemoglobin concentration of 160 g/L carry?

18. About 1.5×10^8 metric tons (1 metric ton = 1000 kg) of CO are released into the atmosphere each year. What is the volume of this quantity of CO at STP?

19. At STP 0.66 mol of H_2 and 0.33 mol of O_2 are mixed in an expandable container. After ignition the product of the reaction reaches a temperature of 1300°C with a pressure of 800 torr. What is the volume of the reaction product?

$$2H_2(g) + O_2(g) \rightarrow 2H_2O(g)$$

20. Atmospheric air is a gaseous mixture of water vapor, oxygen, carbon dioxide, nitrogen, and traces of other species. What is the mole fraction of nitrogen gas at STP if p_{O_2} = 159 torr, p_{CO_2} = 0.23 torr, and p_{H_2O} = 23.8 torr?

21. A gas, X, effuses 3.1 times faster than fluorine gas. What is the molecular weight of the gas X?

22. A storage tank at JFK Space Center can contain 900,000 gallons, which is approximately 3.4 million L, of liquid hydrogen, which has a density of 0.070 g/cc at -253°C. What volume in liters would this hydrogen occupy at STP if it were evaporated?

23. W. W. Ruby estimated the total mass of atmospheric oxygen to be 15×10^{20} g. If all this oxygen were at STP, what volume in liters would it occupy?

24. Uranium 235 and uranium 238 are separated by the effusion difference of the hexafluorides $^{235}UF_6$ and $^{238}UF_6$. What is the ratio of the effusion rates of the hexafluorides?

25. What is the density of phosgene gas, $COCl_2$, at STP?

26. A 100-mL flask contains 0.162 g of an unknown gas at 760 torr and 100°C. What is the molecular weight of the gas?

27. Acetylene is burned to form water and carbon dioxide:

$$2C_2H_2(g) + 5O_2(g) \rightarrow 2H_2O(g) + 4CO_2(g)$$

Answer the following:
(a) What volume of CO_2 at STP is formed from 3.00 g of C_2H_2?
(b) What volume of CO_2 at STP is formed from 2.00 L of C_2H_2 at STP?

28. The van der Waals constants for ammonia are given in Table 10.4 of your text. Use this data to determine the pressure at which 1 mole of NH_3 will occupy 22.4 L at 25°C.

29. Use the data in Table 10.4 of your text to determine which of the gases listed has the strongest inter-molecular interactions and which has the largest molecular volume.

30. One (1.00) mL of liquid water at 25°C is added to dry helium that is contained in a 10.0-liter cylinder at 25°C and 100 atm. What is the partial pressure of water vapor in the vessel? What is the partial pressure of He? What is the total pressure?

31. One (1.0) L of a gas collected over water at STP weighs 1.135 g before drying. What is the molecular weight of the unknown gas?

32. A gas mixture is known to contain only helium and nitrogen. What are the partial pressures of each gas if the density of the mixture is 0.475 g/L at STP?

33. In the atmosphere upon irradiation with ultraviolet light, oxygen is converted to ozone, O_3:

$$3O_2 \overset{h\nu}{\rightarrow} 2O_3$$

The symbol $h\nu$ above the arrow in the equation for the reaction indicates the quantum of radiation necessary to cause the reaction to proceed. This same reaction can be carried out under controlled laboratory conditions. If 1.47 L of O_2 in a sealed container at 1.01 atm is irradiated until 5.24 percent of the oxygen reacts, what is the final pressure inside the flask? Assume that the temperature also rises from 25°C to 54°C during the reaction.

ANSWERS TO EXERCISES

I. Properties of gases

1. True

2. True

3. False

Volume varies directly with temperature in K.

4. True

5. True

6. True

7. False $T = t + 273$

8. True

9. True

10. False   A molecule with a smaller molecular weight effuses more quickly.

11. False   In the van der Waals equation, a corrects for intermolecular interactions and b corrects for molecular volume.

12. False   Above the critical temperature liquefaction cannot occur.

II. Behavior of gases

1. increases

2. increases

3. increases

4. remains the same

5. increases

6. remains the same

7. decreases

8. decreases   Gas escapes.
   decreases
   remains the same
   increases   Molecules are less tightly packed.

9. increases

10. decreases   Three moles of reactants produce only 2 moles of product.
    increases

11. increases
    decreases

12. decreases
    increases

    increases

The ^{35}Cl isotope effuses more quickly than ^{37}Cl; therefore, the remaining gas is enriched in the heavier isotope. Atomic weights are weighted averages of isotopic masses.

13. increases
    remains the
    same
    remains the
    same
    increases

14. increases
    increases
    increases

15. decreases
    increases

III. Gas law calculations

1. 0.14 atm

An increase in temperature results in an increase in pressure. In the gas laws temperatures must be expressed in K, whereas pressures may be expressed in any pressure units. Thus,

$$? \text{ atm} = 0.10 \text{ atm} \begin{pmatrix} \text{temperature} \\ \text{correction} \\ \text{factor} \end{pmatrix}$$

Notice that the ratio of temperatures must be larger than 1 to reflect an increase in pressure:

$$? \text{ atm} = 0.10 \text{ atm} \left(\frac{373 \text{ K}}{273 \text{ K}}\right) = 0.14 \text{ atm}$$

Note that only two significant figures are reported since 0.10 atm has two significant figures. You may find it convenient to tabulate information for use in Boyle's and Charles' laws in the following way. Tabulations such as this should be helpful if you have a tendency to invert correction factors.

	Initial Conditions	Final Conditions
T	(0 + 273) K	(100 + 273) K
P	0.10 atm	?
V	constant	constant

Since the temperature increases and the volume is constant, the pressure must increase; therefore, the temperature correction factor must be larger than 1.

$$? \text{ atm} = 0.10 \text{ atm} \left(\frac{373 \text{ K}}{273 \text{ K}}\right) = 0.14 \text{ atm}$$

2. 84 mL

An increase in temperature is accompanied by an increase in volume. In the gas laws volume may be expressed in any volume units, whereas temperature must be expressed in K. Thus,

$$? \text{ mL} = 75 \text{ mL}\left(\frac{306 \text{ K}}{273 \text{ K}}\right) = 84 \text{ mL}$$

3. 150 torr

An increase in volume is accompanied by a decrease in pressure. (See Example 10.2 of your text.) Thus,

$$? \text{ atm} = 748 \text{ torr}\left(\frac{1.00 \text{ L}}{5.00 \text{ L}}\right) = 150 \text{ torr}$$

4. 11 L O_2

First calculate the number of moles of oxygen gas. Oxygen is diatomic; i.e., it exists as O_2 molecules, and the molecular weight is 32.0.

$$? \text{ mol } O_2 = \frac{16 \text{ g } O_2}{32.0 \text{ g } O_2/1 \text{ mol } O_2} = 0.50 \text{ mol } O_2$$

Each mole of ideal gas occupies 22.4 L at STP, 273 K and 1 atm. Therefore,

$$? \text{ L } O_2 = 0.50 \text{ mol } O_2\left(\frac{22.4 \text{ L } O_2}{1 \text{ mol } O_2}\right) = 11 \text{ L } O_2$$

The ideal gas law can also be used directly:

$$PV = nRT$$

$$V = \frac{nRT}{P}$$

$$= \frac{(0.50 \text{ mol})(0.0821 \text{ L}\cdot\text{atm}/(\text{K}\cdot\text{mol}))(273 \text{ K})}{1 \text{ atm}} = 11 \text{ L}$$

5. 1.43×10^{-3} g/mL

At STP a mole of oxygen occupies 22.4 L. Each mole weighs 32.0 g. Therefore,

$$? \text{ g/mL} = \left(\frac{1 \text{ mol } O_2}{22.4 \text{ L } O_2}\right)\left(\frac{32.0 \text{ g } O_2}{1 \text{ mol } O_2}\right)\left(\frac{1 \text{ L } O_2}{1000 \text{ mL } O_2}\right)$$

$$= 1.43 \times 10^{-3} \text{ g/ml}$$

6. 0.204 mol

Using the ideal gas law:

$$PV = nRT$$

$$n = \frac{(1.00 \text{ atm})(5.00 \text{ L})}{(0.0821 \text{ L}\cdot\text{atm}/(\text{K}\cdot\text{mol}))(298 \text{ K})}$$

If 0.0821 L·atm/(K·mol) is the value used for R, volume must be expressed in liters, temperature in K, and pressure in atmospheres.

7. 1.07 L

According to Charles' law, an increase in temperature is accompanied by an increase in volume. Therefore,

$$? L = (1.00\ L)\ \left(\frac{318\ K}{298\ K}\right) = 1.07\ L$$

8. 1.05 L

Charles' law predicts that an increase in temperature is accompanied by an increase in volume, and Boyle's law predicts that an increase in pressure is accompanied by a decrease in volume. Combining laws, it is found

$$? L = (1.00\ L)\ \left(\frac{318\ K}{298\ K}\right) \left(\frac{745\ torr}{757\ torr}\right)$$

$$= 1.05\ L$$

Note that the temperature correction factor must be larger than 1 to reflect an increase in volume. The pressure correction factor must be smaller than 1 to account for a decrease in volume due to an increase in pressure. The pressure change does not have a significant effect in this problem; however, this is not always the situation.

9. 0.311 atm

An increase in temperature causes an increase in pressure, and an increase in volume causes a decrease in pressure. Therefore,

$$? atm = 0.750\ atm\left(\frac{360\ K}{263\ K}\right)\left(\frac{1.00\ L}{3.30\ L}\right)$$

$$= 0.311\ atm$$

If you did not solve this problem correctly the first time, tabulate the data as shown in problem 1 of this section:

	Initial Conditions	Final Conditions
T	(-10 + 273) K	(87 + 273) K
P	0.750 atm	?
V	1.00 L	3.30 L

Use the preceding logic to solve the problem.

10. 86 atm

Using the ideal gas law:

$$PV = nRT$$

$$P = \frac{nRT}{V}$$

$$= \frac{(3.5)\ mol)\ (0.0821\ L \cdot atm/(K \cdot mol))\ (300\ K)}{1.0\ L}$$

$$= 86\ atm$$

If the pressure is to be expressed in torr, convert atmosphere to torr:

$$? \text{ torr} = 86 \text{ atom} \left(\frac{760 \text{ torr}}{1 \text{ atm}} \right) = 6.5 \times 10^4 \text{ torr}$$

11. 513 K,
    or 240°C

If a problem concerns an ideal gas and temperature, pressure, and volume are not changed, $PV = nRT$ can be used. In this problem temperature is to be calculated.

$$T = \frac{PV}{nR}$$

Since P, V, and R are known, the value of n must be calculated:

$$? \text{ mol } N_2 = \frac{(105.68 - 104.35) \text{ g } N_2}{(28.02 \text{ g } N_2/1 \text{ mol } N_2)} = 0.04747 \text{ mol } N_2$$

Substituting values into the ideal gas law, we find

$$T = \frac{(2.00 \text{ atm})(1.00 \text{ L})}{(0.04747 \text{ mol})(0.0821 \cdot L \text{ atm}/(K \cdot mol))}$$

$$= 513 \text{ K}$$

and

$$T = (513 - 273)(1°C) = 240°C$$

12. 0.230 L
    O_2

The balanced equation shows that 1 mol of O_2, which occupies 22.4 L at STP, combines with 2 mol of Mg, (2×24.31) g Mg. This relationship can be used to compute the volume of O_2 gas:

$$? \text{ L } O_2 = 0.500 \text{ g Mg } \left(\frac{22.4 \text{ L } O_2}{(2 \times 24.31) \text{ g Mg}} \right)$$

$$= 0.230 \text{ L } O_2$$

13. 4.6 L

Compute the volume of octane at STP:

$$= \left(\frac{0.670 \text{ g } C_8H_{18}}{114.2 \text{ g } C_8H_{18}/\text{mol}} \right) \left(\frac{22.4 \text{ L } C_8H_{18}}{\text{mol}} \right)$$

$$= 0.131 \text{ L}$$

The balanced equation shows that 16/2, or 8, times the volume of octane is the volume of CO_2 produced, and

18/2, or 9, times the volume of octane is the volume of water vapor produced. Therefore, the total volume of gas produced at STP would be

$$V_{total} = V_{CO_2} + V_{H_2O}$$

$$= (8 \times 0.131) \text{ L} + (9 \times 0.131) \text{ L}$$

$$= 2.23 \text{ L}$$

Changing from STP to the actual conditions:

$$? \text{ liters} = 2.23 \text{ L} \left(\frac{(673 \text{ K})(1.0 \text{ atm})}{(273 \text{ K})(1.2 \text{ atm})} \right)$$

$$= 4.6 \text{ L gas}$$

14. 1.84 L H_2

Since the water vapor is removed before the volume is measured, its effect need not be considered. Therefore,

$$? \text{ L H}_2 = 2.00 \text{ g Mg} \left(\frac{22.4 \text{ L H}_2}{24.30 \text{ g Mg}} \right)$$

$$= 1.84 \text{ L H}_2$$

15. 0.60 g O_2

First calculate the number of moles of gas at STP:

$$PV = nRT$$

$$n = \frac{PV}{RT}$$

$$= \frac{(1.00 \text{ atm})(2.00 \text{ L})}{(0.0821 \text{ L} \cdot \text{atm}/(\text{K} \cdot \text{mol}))(273 \text{ K})}$$

$$= 8.92 \times 10^{-2} \text{ mol}$$

Alternatively, since 1 mol of ideal gas at STP occupies 22.4 L,

$$? \text{ mol gas} = 2.00 \text{ L} \left(\frac{1 \text{ mol gas}}{22.4 \text{ L gas}} \right)$$

$$= 8.93 \times 10^{-2} \text{ mol gas}$$

Twenty-one percent of this quantity is O_2. Therefore,

$$? \text{ g O}_2 = 8.93 \times 10^{-2} \text{ mol gas} \left(\frac{0.21 \text{ mol O}_2}{1 \text{ mol gas}} \right) \left(\frac{32.0 \text{ g O}_2}{1 \text{ mol O}_2} \right)$$

$$= 0.60 \text{ g O}_2$$

16. 0.20 g O_2 On the top of Mt. Everest,

$$n = \frac{(0.330 \text{ atm}) (2.00 \text{ L})}{(0.0821 \text{ L} \cdot \text{atm}/(\text{K} \cdot \text{mol})) (263 \text{ K})}$$

$$= 3.06 \times 10^{-2} \text{ mol}$$

and

$$? \text{ g } O_2 = 3.06 \times 10^{-2} \text{ mol gas} \left(\frac{0.21 \text{ mol } O_2}{1 \text{ mol gas}}\right) \left(\frac{32.0 \text{ g } O_2}{1 \text{ mol } O_2}\right)$$

$$= 0.20 \text{ g } O_2$$

17. 21.6 mol O_2 This problem describes a balanced chemical equation. Note that 1 mol of hemoglobin combines with 4 moles of O_2. Therefore, if we let Hb stand for hemoglobin, the equation is

$$Hb + 4O_2 \rightarrow Hb \cdot 4O_2$$

in which $Hb \cdot 4O_2$ is the product. The problem can be solved simply by determining the number of grams of Hb available:

$$? \text{ L } O_2 = 16.0 \text{ g Hb} \left(\frac{(4 \times 22.4) \text{ L } O_2}{66,280 \text{ g Hb}}\right)$$

$$= 0.0216 \text{ L } O_2$$

or

$$? \text{ mL } O_2 = 0.0216 \text{ L } O_2 \left(\frac{1000 \text{ mL } O_2}{1 \text{ L } O_2}\right)$$

$$= 21.6 \text{ mL } O_2$$

18. 1.2×10^{14} L CO   Calculate the volume at STP

$? \text{ L CO}$

$$= 1.5 \times 10^8 \text{ metric tons CO} \left(\frac{10^6 \text{ g}}{1 \text{ metric ton}}\right) \left(\frac{22.4 \text{ L CO}}{28.0 \text{ g CO}}\right)$$

$$= 1.2 \times 10^{14} \text{ L CO}$$

The factors used are

$$10^3 \text{ kg} = 1 \text{ metric ton}$$

$$10^3 \text{ g} = 1 \text{ kg}$$

22.4 L CO = 1 mol CO = 28.0 g CO

19. 81 L H_2O

The oxygen and hydrogen are mixed in the proper molar ratio for complete reaction, 2 mol H_2 to 1 mol O_2. Therefore, either the number of moles of O_2 or H_2 can be used to begin the calculation. First calculate the volume of water vapor at STP:

? L H_2O

$$= 0.33 \text{ mol } O_2 \left(\frac{(2 \times 22.4) \text{ L } H_2O}{1 \text{ mol } O_2} \right)$$

$$= 14.8 \text{ L } H_2O$$

or

? L H_2O

$$= 0.66 \text{ mol } H_2 \left(\frac{(2 \times 22.4) \text{ L } H_2O)}{2 \text{ mol } H_2} \right)$$

$$= 14.8 \text{ L } H_2O$$

Then adjust the volume to the actual temperature and pressure conditions. An increase in temperature is accompanied by an increase in volume, and an increase in pressure is accompanied by a decrease in volume. Thus,

$$? \text{ L } H_2O = 14.8 \text{ L } H_2O \left(\frac{1573 \text{ K}}{273 \text{ K}} \right) \left(\frac{760 \text{ torr}}{800 \text{ torr}} \right)$$

$$= 81 \text{ L } H_2O$$

20. 0.759

The partial pressure of a component gas such as N_2 is directly proportional to the mole fraction of that component gas and the total pressure:

$$P_{N_2} = X_{N_2} P_{total}$$

The total pressure is the sum of the partial pressures of the components:

$$P_{total} = P_{N_2} + P_{O_2} + P_{CO_2} + P_{H_2O}$$

Therefore,

$$p_{N_2} = p_{total} - p_{O_2} - p_{CO_2} - p_{H_2O}$$

$$= 760 \text{ torr} - 159 \text{ torr} - 0.23 \text{ torr} - 23.8 \text{ torr}$$

$$= 577 \text{ torr}$$

and

$$X_{N_2} = \frac{p_{N_2}}{p_{total}} = \frac{577 \text{ torr}}{760 \text{ torr}} = 0.759$$

21. 4.0

Graham's law of effusion states

$$\frac{r_A}{r_B} = \sqrt{\frac{M_B}{M_A}}$$

Therefore,

$$\frac{3.1}{1} = \sqrt{\frac{38.0}{M_A}}$$

22. 2.6×10^9 L H_2

Calculate the number of moles of liquid hydrogen:

? mol H_2

$$= 3.4 \times 10^6 \text{ L } H_2 \left(\frac{10^3 \text{ cm}^3 \text{ } H_2}{1 \text{ L } H_2}\right)\left(\frac{0.070 \text{ g } H_2}{1 \text{ cm}^3 \text{ } H_2}\right)\left(\frac{1 \text{ mol } H_2}{2.02 \text{ g } H}\right)$$

$$= 1.18 \times 10^8 \text{ mol } H_2$$

Then calculate the volume of gas at STP:

$$? \text{ L } H_2 = 1.18 \times 10^8 \text{ mol } H_2 \left(\frac{22.4 \text{ L } H_2}{1 \text{ mol } H_2}\right)$$

$$= 2.6 \times 10^9 \text{ L } H_2$$

As the gas at STP, H_2 occupies approximately 1000 times the volume of the liquid.

23. 1.0×10^{21} L O_2

At STP each mol of O_2, i.e., 32.0 g O_2, occupies 22.4 L. Therefore,

$$? \text{ L } O_2 = 15 \times 10^{20} \text{ g } O_2 \left(\frac{22.4 \text{ L } O_2}{32.0 \text{ g } O_2} \right)$$

$$= 1.0 \times 10^{21} \text{ L } O_2$$

24. 0.996

According to Graham's law

$$\frac{r_A}{r_B} = \sqrt{\frac{M_B}{M_A}}$$

The molecular weights of the two gases must be calculated:

molecular weight of $^{235}UF_6$ = 235 + 6(19) = 349

molecular weight of $^{238}UF_6$ = 238 + 6(19) = 352

Substituting the molecular weights into the equation of Graham's law, we find

$$\frac{r_{238}UF_6}{r_{235}UF_6} = \sqrt{\frac{349}{352}} = \sqrt{0.992} = 0.996$$

25. 4.42 g/L

One mole of phosgene weighs 98.9 g, so at STP 98.9 g occupies 22.4 L. Therefore,

$$\text{density} = \frac{\text{weight}}{\text{volume}}$$

$$= \frac{98.9 \text{ g}}{22.4 \text{ L}} = 4.42 \text{ g/L}$$

26. 49.6 g/mol

The ideal gas law can be used. Since n, the number of moles, is the weight of material, g, divided by the molecular weight, M:

$$PV = nRT$$

$$PV = \frac{g}{M} RT$$

$$M = \frac{gRT}{PV}$$

$$= \frac{(0.162 \text{ g}) (0.0821 \text{ L·atm/(K·mol)}) (373 \text{ K})}{(1 \text{ atm}) (0.100 \text{ L})}$$

$$= 49.6 \text{ g/mol}$$

The preceding method is preferred. Any problem in which the temperature, pressure, and volume are not changed can be solved directly with the ideal gas law; problems that involve a change in any of these can be solved easily with a tabular method such as that suggested in problem 1 of this section. The molecular weight of an ideal gas is the mass of gas that would occupy 22.4 L at STP. Thus the problem could be solved in the following way:

First calculate the volume of 0.162 g of gas at STP:

$$? \text{ mL gas} = 100.0 \text{ mL gas} \left(\frac{273 \text{ K}}{373 \text{ K}}\right)$$

$$= 73.2 \text{ mL gas, or } 0.0732 \text{ L gas}$$

Then calculate the density of the gas at STP:

$$\text{density} = \frac{\text{weight}}{\text{volume}} = \left(\frac{0.162 \text{ g}}{0.0732 \text{ L}}\right) = 2.21 \text{ g/L}$$

Finally calculate the molecular weight:

$$M = 2.21 \text{ g/L} \left(\frac{22.4 \text{ L}}{1 \text{ mol}}\right) = 49.5 \text{ g/mol}$$

Another method can also be used:

$$? \text{ g/mol} = \left(\frac{0.162 \text{ g}}{100 \text{ ml}}\right)\left(\frac{373 \text{ K}}{273 \text{ K}}\right)\left(\frac{1000 \text{ ml}}{1 \text{ L}}\right)\left(\frac{22.4 \text{ L}}{1 \text{ mol}}\right)$$

$$= 49.6 \text{ g/mol}$$

Be careful with this method. Volume is in the denominator so the temperature correction factor must be inverted.

27.   (a) 5.17
     L CO_2

$$? \text{ L } CO_2 = 3.00 \text{ g } C_2H_2 \left(\frac{(4 \times 22.4) \text{ L } CO_2}{2(26.02) \text{ g } C_2H_2}\right)$$

$$= 5.17 \text{ L } CO_2$$

(b) 4.00
   L CO_2

Part (b) of this problem can be done by inspection. Gay-Lussac's law states that in a chemical reaction the ratio of moles is the same as the ratio of volumes measured at the same temperature and pressure.

28. 1.08 atm   The van der Waals equation is

$$P + \left(\frac{n^2 a}{v^2}\right)(V - nb) = nRT$$

Rearranging the equation, we find

$$p = \frac{nRT - \left(\frac{n^2 a}{V}\right) + \left(\frac{n^3 ab}{v^2}\right)}{V - nb}$$

Solving for P, we find

$$P = 1.08 \text{ atm}$$

From the ideal gas law an answer of 1.09 atm is obtained.

29. Cl_2   The value of a is largest for chlorine gas; thus, the intermolecular attraction of chlorine gas molecules is greater than that of the other gas molecules. Notice that the value of a is smallest for helium, the molecules of which show little intermolecular interaction. The value of a is large for molecules with a large dipole and a non-bonding electron pair, such as ammonia molecules. Chlorine gas is also the largest of the gases listed in Table 10.4 of your text; it has the largest value of b, a constant which is directly proportional to the molecular radius.

30. 100 atm   According to Table 10.3 in your text, the partial pressure of water vapor is 0.0313 atm, and the partial pressure of He is 100 atm. The total pressure is the sum of the two partial pressures:

$$P = p_{H_2 O} + p_{He}$$

$$= 0.0313 \text{ atm} + 100 \text{ atm}$$

$$= 100 \text{ atm}$$

31. 25.3 g/mol   Since the gas was weighed before drying, the mass of water vapor must be subtracted to obtain the mass of the unknown gas. Use the ideal gas law to find the mass of water vapor:

$$? \text{ g } H_2 O = \frac{MPV}{RT}$$

$$= \frac{(18.0 \text{ g/mol})(0.0060 \text{ atm})(1.00 \text{ L})}{(0.0821 \text{ L} \cdot \text{atm}/(\text{K} \cdot \text{mol}))(273 \text{ K})}$$

$$= 0.0048 \text{ g}$$

The mass of the unknown gas is

1.135 g - 0.005 g = 1.130 g

The molecular weight of the unknown gas is

$$? \text{ g/mol} = \left(\frac{1.130 \text{ g}}{1.00 \text{ L}}\right)\left(\frac{22.4 \text{ L}}{1 \text{ mol}}\right)$$

$$= 25.3 \text{ g/mol}$$

32. p_{He} = 549 torr and p_{N_2} = 210 torr

A mole of the gas mixture would weigh

(0.475 g/L)(22.4 L/mol) = 10.64 g/mol

Let x equal the mole fraction of He and $1 - x$ equal the mole fraction of N_2,

$$x(4.00 \text{ g/mol}) + (1 - x)(28.0 \text{ g/mol}) = 10.64 \text{ g/mol}$$
$$x = 0.723$$
$$1 - x = 0.277$$

The partial pressure of helium is the total pressure times the mole fraction of helium:

$$p_{He} = X_{He}P_{total}$$

$$= (0.723)(760 \text{ torr})$$

$$= 549 \text{ torr, or } 0.723 \text{ atm}$$

Similarly for nitrogen:

$$p_{N_2} = X_{N_2}P_{total}$$

$$= (0.277)(760 \text{ torr})$$

$$= 210 \text{ torr, or } 0.277 \text{ atm}$$

33. 1.09 atm

Since this problem contains so many variables, it is best to tabulate the information:

	Initial Conditions	Final Conditions
P	1.01 atm	?
V	1.47 L	1.47 L
T	(25 + 273) K	(54 + 273) K
n	? n_{O_2}	? $(n_{O_2} + n_{O_3})$

Use the ideal gas law, $PV = nRT$ to determine the initial number of moles of O_2:

$$n = \frac{PV}{RT} = 6.07 \times 10^{-2} \text{ mol } O_2$$

Then calculate the number of moles of O_2 that react (5.24% of the original amount);

$$n_{O_2} \text{ reacting} = (0.0524)(6.07 \times 10^{-2} \text{ mol } O_2)$$

$$= 3.18 \times 10^{-3} \text{ mol } O_2$$

and the number of moles of O_2 remaining:

$$n_{O_2} \text{ remaining} = (6.07 \times 10^{-2} \text{ mol}) - (3.18 \times 10^{-3} \text{ mol})$$

$$= 5.75 \times 10^{-2} \text{ mol } O_2$$

The number of moles of O_2 reacting and the balanced equation can be used to determine the number of moles of O_3 formed:

$$n_{O_3} = 3.18 \times 10^{-3} \text{ mol } O_2 \left(\frac{2 \text{ mol } O_3}{3 \text{ mol } O_2} \right)$$

$$= 2.12 \times 10^{-3} \text{ mol } O_3$$

The total amount of gas remaining after the reaction is

$$n_{total} = n_{O_2} + n_{O_3}$$

$$= (5.75 \times 10^{-2} \text{ mol } O_2) + (2.12 \times 10^{-3} \text{ mol } O_3)$$

$$= 5.96 \times 10^{-2} \text{ mol gas}$$

From $PV = nRT$ the final pressure is calculated:

$$PV = nRT$$

$$P = \frac{(5.96 \times 10^{-2} \text{ mol})(0.0821 \text{ L·atm/(K·mol)})(327K)}{1.47 \text{ L}}$$

$$P = 1.09 \text{ atm}$$

SELF-TEST   Complete the test in 60 minutes:

I. Work the following problems:

1. A bulb filled with a gas to a pressure of 1 atm weighs 116.3124 g. When the bulb is heated to 88°C at a pressure of 1 atm , 100 mL of gas is expelled and the bulb and remaining gas weigh 116.2584 g. What is the molecular weight of the gas?

2. A 1.168-g sample of an oxide, XO_2, reacts with exactly 500 mL of H_2 gas measured at STP. What is the molecular weight of XO_2?

$$XO_2(s) + 2H_2(g) \rightarrow X(s) + 2H_2O(g)$$

3. A 100-mL sample of gas is collected over water at temperature T_1, and the wet gas is found to exert a pressure of 0.987 atm at that temperature. The same sample of gas is found to occupy 97.1 mL at T_1 when under a pressure of 1 atm . Calculate the vapor pressure of water at T_1 from these data.

4. In a mixture of C_2H_6 and O_2 confined in a 1.00-L container, the partial pressure of C_2H_6 is 0.210 atm and the partial pressure of O_2 is 0.737 atm. The mixture is ignited and reacts according to the equation:

$$2C_2H_6(g) + 7O_2(g) \rightarrow 4CO_2(g) + 6H_2O(g)$$

The temperature is constant throughout the experiment and is high enough so that the H_2O is gaseous. What is the total pressure of the final mixture?

5. Calculate the density of a gas in g/L at STP if a given volume of the gas effuses through an apparatus in 5.00 minutes and the same volume of oxygen at the same temperature and pressure effuses through the apparatus in 6.30 minutes.

6. Assume that 10 L of $NH_3(g)$ and 10 L of $O_2(g)$ are mixed and react according to the equation:

$$4NH_3(g) + 5O_2(g) \rightarrow 4NO(g) + 6H_2O(g)$$

If the conditions under which the gases are measured are constant and such that all materials are gaseous, list the volumes of all materials at the conclusion of the reaction.

7. Calculate the molecular weight of a gas that has a density of 1.59 g/L at 50°C and a pressure of 0.960 atm.

II. Complete the statement or answer the question:

1. Of the two gases, H_2 (molecular weight = 2) and C_5H_{12} (molecular weight = 72), under the same conditions of temperature and pressure, _____ would effuse _____ times more rapidly through a given orifice.

2. If 1.00-L samples of $H_2(g)$ and $C_5H_{12}(g)$ are considered, both at STP, which sample has the larger number of molecules? _____. The molecules of which sample have the larger average kinetic energy?

3. The partial pressure of oxygen in a flask containing 32 g of O_2 and 32 g of H_2 is _____ of the total pressure. (Atomic weights: O = 16.0; H = 1.0.)

4. The behavior of real gases deviates from that described by the ideal gas law because the ideal gas law fails to take into account _____ and _____ .

III. Using the coordinates given, sketch the approximate shape of the curve for an ideal gas:

a. pressure vs. volume

c. absolute temperature vs. volume

b. pressure vs. product of pressure and volume

d. energy distribution of molecules

LIQUIDS AND SOLIDS

CHAPTER
11

OBJECTIVES

I. You should be able to demonstrate your knowledge of the following terms by defining them, describing them, and giving specific examples of them:

body-centered cubic [11.12]
boiling point [11.6]
Bragg equation [11.3]
Clausius-Clapeyron equation [11.7]
closest packed crystal [11.4]
coordination number [11.14, 11.15]
crystal lattice [11.12]
defect structures [11.6]
dipole-dipole forces [11.1]
equilibrium [11.5]
face-centered cubic unit cell [11.12]
freezing point [11.8]
hydrogen bond [11.2]
instantaneous dipoles [11.1]
intermolecular forces [11.1]
ionic crystals [11.11, 11.15]
London forces [11.1]
metallic crystals [11.11]
metastable [11.10]
molar enthalpy of condensation [11.7]
molar enthalpy of crystallization [11.8]

molar enthalpy of fusion [11.8]
molar enthalpy of sublimation [11.10]
molar enthalpy of vaporization [11.4]
molecular crystals [11.11]
network crystals [11.11]
normal boiling point [11.6]
normal freezing [11.8]
phase diagram [11.10]
simple cubic unit cell [11.12]
sublimation [11.10]
supercooling [11.8]
surface tension [11.3]
triple point [11.10]
unit cell [11.12]
vapor pressure [11.5]
viscosity [11.3]
X-ray diffraction [11.13]

II. You should be able to interpret phase diagrams.

III. You should be familiar with the various types of crystal lattices and be able to calculate densities from structural data and vice versa.

IV. You should be able to use the Bragg equation to determine the distance between diffraction planes.

V. You should be able to determine the heat of vaporization, ΔH_v, from the vapor pressure of a liquid at two temperatures and rearrange the Clausius-Clapeyron equation to determine either vapor pressure or temperature if given the other variables.

VI. You should be able to determine boiling point and melting point trends for a series of related compounds.

VII. You should be familiar with the properties and structures of crystalline solids.

UNITS,
SYMBOLS,
MATHEMATICS

I. The following symbols are used in this chapter. You should be familiar with them.

- atm is the abbreviation for atmosphere, a unit of pressure frequently used by chemists.
- °C is the abbreviation for degrees Celsius, a temperature unit on a scale on which water is defined as freezing at 0°C and boiling at 100°C under 1 atm pressure.
- D stands for dipole moment.

- ΔH_v is the symbol used to identify the enthalpy change associated with the vaporization of one mole of material.
- kJ is the abbreviation for kilojoules.
- λ stands for wavelength.
- p is the symbol for **vapor pressure**
- t_b stands for normal boiling point in °C.
- t_f stands for normal freezing point in °C.
- θ is the symbol used to show an angle.

II. Review all previous sections of "Units, Symbols, Mathematics."

EXERCISES   I. Answer each of the following with an entry from the list on the right:

1. Water can exist at 100°C and 1 atm as either gas or ___e___ .

    e

2. The phase in which molecular motion is most restricted is the ___f___ phase.

    f

3. As a liquid is heated, the vapor pressure ___b___ .

    b

4. ___h___ is a property of liquids that describes resistance to flow.

    h

5. In the ___e___ state matter assumes the shape of and completely fills the container into which it is placed.

6. A drop of liquid has a spherical shape due to the property of ___g___ .

    g

7. As pressure is decreased, boiling points ___a___ .

    a

8. As the temperature of a liquid increases, the rate of evaporation ___b___ .

    b

9. ___l___ is the condition in which the rates of two opposite tendencies are equal.

    L

a. decrease(s)

b. increase(s)

c. remains the same

d. gas

e. liquid

f. solid

g. surface tension

h. viscosity

i. melting point

j. normal boiling point

k. sublimation

l. equilibrium

m. vapor pressure

n. simple

o. body-centered

p. face-centered

10. The pressure of vapor in equi-
    librium with a liquid at a
    given temperature is called
    __m__.

11. At the __j__ the vapor pressure
    of a liquid equals 1 atmos-
    phere.

12. The __ι__ is the temperature at
    which liquid and solid are in
    equilibrium at 1 atmosphere.

13. __k__ is the process in which a
    solid passes directly to the
    vapor phase.

14. There is one atom per unit
    cell in the __f__ cubic
    crystal.

15. The __o__ cubic crystal is the
    most densely packed cubic
    crystal.

16. The __n__ cubic crystal contains
    four atoms per unit cell.

17. Each atom has twelve nearest
    neighbors in the __o__ cubic
    crystal.

18. The diagram

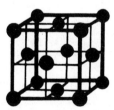

    is a representation of the
    __p__ cubic crystal.

19. For the series of elements
    F_2, Cl_2, Br_2, and I_2, the
    boiling point __b__ from
    F_2 to I_2.

20. For the series of gases O_2,
    NO, and CO, the boiling point
    __a__ from O_2 to CO.

II. Figure 11.1 shows the phase diagram for both monoclinic sulfur and carbon. Use this informatin to answer each of the following questions.

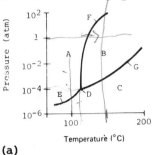
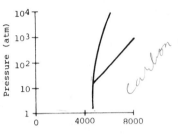

(a)   (b)

FIGURE 11.1   Phase diagrams for (a) monoclinic sulfur and (b) carbon.

m 1. In the phase diagram for sulfur   a. sulfur
      area A represents the _solid_ phase.
                                             b. carbon

l 2. In the phase diagram for sulfur
      area B represents the _liquid_ phase. c. sulfur and carbon

k 3. In the phase diagram for sulfur   d. gas-liquid
      area C represents the _gas_ phase.
                                             e. gas-solid

h 4. In the phase diagram for sulfur
      D represents the _triple point_   f. liquid-solid

h 5. In the phase diagram for sulfur
e   line E represents the _solid-gas_   g. liquid-gas
      equilibrium.
                                             h. solid-gas

   6. In the phase diagram for sulfur   i. solid-liquid
i or f line F represents the _solid liquid_
      equilibrium.   j. no

d 7. In the phase diagram for sulfur   k. gas
g   line G represents the _liquid gas_
      equilibrium.   l. liquid

a 8. Which element has the lower   m. solid
      melting point: <u>sulfur</u> or carbon?
                                             n. triple point

i 9. When sulfur is heated at one
      atmosphere from 100°C to 150°C,
      a _s→l_ phase change occurs.

j 10. When carbon is heated at one
       atmosphere from 100°C to 150°C,
       a _no_ phase change occurs.

(h) 11. As the pressure exerted on a
sample of sulfur maintained at
100°C is reduced from 1 atmosphere to 10^{-6} atmosphere, a
_____ phase change occurs.
solid - gas

12. As the pressure exerted on a
sample of sulfur maintained at
(g) 150°C is reduced from 1
atmosphere to 10^{-6} atmosphere,
a _g_ phase change occurs.
liquid gas.

(m) 13. Which carbon phase exists at
3000°C and 1 atmosphere? *solid*.

14. Which element, carbon or sulfur,
never exists as a liquid at 1
(b) atmosphere? *carbon*.

15. Which is more dense: liquid or
(m) solid sulfur? *solid*

III. Answer each of the following:

1. Argon crystallizes in a face-centered cubic structure
at -189°C. If the density of the solid is 1.7 g/cm^3,
calculate the length of an edge of a unit cell.

2. An oxide of zirconium forms a face-centered cubic
lattice of zirconium ions. In addition, each unit cell
contains eight oxide ions, O^{2-}, in the spaces between
the zirconium ions. What is the formula of the
zirconium oxide?

3. A sulfide of magnesium forms a face-centered cubic
lattice of magnesium ions. In addition, each unit
cell contains four sulfide ions, S^{2-}, in the spaces
between the magnesium ions. What is the formula of the
compound?

4. The mineral perovskite has a crystalline structure in
which the oxide ions occupy the face-centers and the
larger cations, Ca^{2+}, occupy the corners of the unit
cell. One Ti^{4+} is in the center of each unit cell.
What is the formula of perovskite?

5. Which of the following liquids should exert the
smaller vapor pressure at 25°C: CCl_4 or CH_4?

6. Which of the following liquids should have the smaller
heat of vaporization: NH_3 or BF_3?

7. In the diffraction studies of a gold crystal with X-rays of a wavelength equal to 154 pm, a first-order reflection is observed at an angle of 22°10'. What is the distance between the diffracted planes?

8. The vapor pressure of ethyl alcohol, C_2H_5OH, is 0.71 atm at 343 K. The normal boiling point of ethyl alcohol is 351 K. What is the enthalpy of vaporization of ethyl alcohol in this temperature range?

9. The heat of vaporization of mercury at 613 K is 65 kJ/mol and the vapor pressure of mercury at that temperature is 0.73 atm. Determine the vapor pressure of mercury at 623 K.

10. Aluminum crystallizes in a face-centered cubic unit cell with the length of an edge equal to 405 pm. Assume the atoms are hard spheres and each face-centered atom touches the four corner atoms of its face. Calculate the radius of a hard-sphere atom.

11. A pure metallic sample is studied by X-ray diffraction techniques. This data indicate that the metal forms a face-centered cubic lattice and that the length of the edge of a unit cell is 392 pm. The density of the metal is 21.5 g/cm^3. What is the molecular weight of the metal? What is the metal?

ANSWERS TO   I. Physical states of matter
EXERCISES

1. e., liquid   It is important to remember the terms that apply to phase
2. f., solid   changes. The following diagram may be of help:
3. b., increases
4. h., viscosity
5. d., gas
6. g., surface
        tension

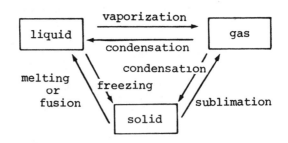

7. a., decreases
8. b., increases
9. l., equilibrium
10. m., vapor pressure
11. j., normal boiling
       point
12. i., melting point
13. k., sublimation
14. n., simple
15. p., face-centered
16. p., face-centered
17. p., face-centered   Hexagonal close-packed crystals also have twelve
18. p., face-centered   nearest neighbors per atom.
19. b., increases   London forces increase.
20. b., increases   Dipole-dipole forces increase.

II. Phase diagrams for monoclinic sulfur and carbon.

1. m., solid
2. l., liquid
3. k., gas
4. h., triple point
5. e., gas-solid or
       solid-gas
6. f., liquid-solid
       or i., solid-
       liquid
7. d., gas-liquid
       or g., liquid-
       gas
8. a., sulfur

Carbon does not form a liquid phase at
1 atmosphere; it sublimes when heated to 3652°C.

9. i., solid-liquid

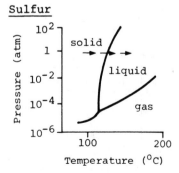

Sulfur

10. j., no   **Carbon**

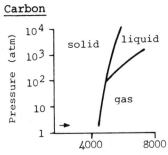

11. h., solid-gas   **Sulfur**

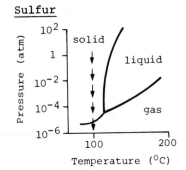

12. g., liquid-gas   **Sulfur**

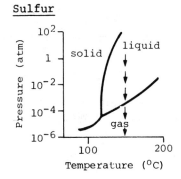

13. m., solid   **Carbon**

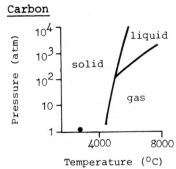

14. b., carbon   Carbon

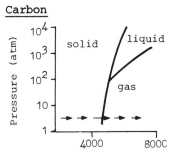

15. m., solid   As pressure is increased, the material becomes more dense,
                  and the solid form is produced:

                  Sulfur

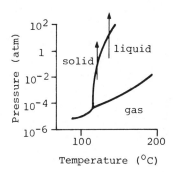

III. Molecular interactions in solids and liquids

1. 540 pm   First calculate the volume of a unit cell:

$$? \text{ cm}^3 = 4 \text{ atoms} \left(\frac{1 \text{ mol}}{6.02 \times 10^{23} \text{ atoms}}\right)\left(\frac{39.95 \text{ g}}{1 \text{ mol}}\right)\left(\frac{1 \text{ cm}^3}{1.7 \text{ g}}\right)$$

$$= 1.56 \times 10^{-22} \text{ cm}^3$$

                  Finally calculate the length of the edge of a unit cell:

$$\sqrt[3]{1.56 \times 10^{-22} \text{ cm}^3} = 5.4 \times 10^{-8} \text{ cm} = 540 \text{ pm}$$

2. ZrO_2 Since the zirconium ions form a face-centered cubic
                  lattice, there are four atoms per unit cell. The
                  problem states that there are eight oxide ions per unit
                  cell. The stoichiometry of a unit cell is Zr_4O_8 and
                  the simplest formula is ZrO_2.

3. MnS   Since the manganese ions form a face-centered cubic
                  lattice, there are four atoms per unit cell. The
                  problem states that there are four sulfide ions per unit
                  cell. The stoichiometry of a unit cell is Mn_4S_4 and the
                  simplest formula is MnS.

4. $CaTiO_3$ There are eight corner Ca^{2+} ions, each shared by eight
                  unit cells; thus, there is (1/8)(8), or one Ca^{2+} ion per
                  unit cell.

                  There are six face-centered positions containing an oxide
                  ion, each shared by another unit cell; thus, there are
                  (1/2)(6), or three oxide ions per unit cell.

                  The Ti^{4+} ion is in the center of the unit cell, so there
                  is one Ti^{4+} ion per unit cell.

                  The formula of perovskite is therefore $CaTiO_3$.

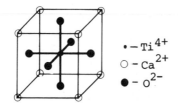

$\bullet - Ti^{4+}$
$\circ - Ca^{2+}$
$\bullet - O^{2-}$

5. CCl_4 Both CCl_4 and CH_4 are tetrahedral molecules which contain
                  no net dipole. Since CCl_4 is larger, London forces
                  between CCl_4 molecules would be larger than those between
                  CH_4 molecules. Thus CCl_4 molecules are less likely to
                  escape from solution and CCl_4 has the lower vapor pressure.

6. BF_3 In general, the lower the heat of vaporization, the weaker
                  the intermolecular forces of attraction. Boron trifluoride,
                  BF_3, is a nonpolar molecule with weaker attractive forces
                  than those of the polar ammonia molecule.

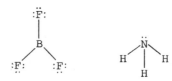

7. 204 pm

Substitute values into the Bragg equation:

$n = 2d \sin \theta$

$154 \text{ pm} = 2(d)(\sin 22°10')$

Then find the sin of 22°10' and substitute:

$154 \text{ pm} = 2(d)(0.377)$

Finally rearrange and solve:

$$d = \frac{154 \text{ pm}}{2(0.377)} = 204 \text{ pm}$$

8. 43 kJ/mol

The Clausius-Clapeyron equation is needed to solve this problem.

$$\log\left(\frac{p_2}{p_1}\right) = \left(\frac{H_v}{2.303R}\right)\left(\frac{T_2 - T_1}{T_1 T_2}\right)$$

Make a table of values to be sure to make the correct substitutions.

	condition 1	condition 2
p	0.71 atm	1.0 atm
T	343 K	351 K

$$\log\left(\frac{0.71 \text{ atm}}{1.0 \text{ atm}}\right) = \left(\frac{H_v}{(2.303)[8.314 \text{ J/(K·mol)}]}\right)\left(\frac{351 \text{ K} - 343 \text{ K}}{(343 \text{ K})(351 \text{ K})}\right)$$

$$\log 1.4 = \frac{H_v (8 \text{ K})}{2.3 \times 10^6 \text{ J·K/mol}}$$

$$0.15 = \frac{H_v (8)}{2.3 \times 10^6 \text{ J/mol}}$$

Rearranging and solving:

$H_v = 43 \text{ kJ/mol}$

9. 0.88 atm   The Clausius-Clapeyron equation is needed for solving this problem.

$$\log \left(\frac{p_2}{p_1} \right) = \left(\frac{H_v}{2.303\ R} \right) \left(\frac{T_2 - T_1}{T_1 T_2} \right)$$

Substituting:

$$\log \left(\frac{p_2}{0.73\ \text{atm}} \right) = \left(\frac{6.5 \times 10^4\, \text{J/mol}}{(2.303)\,[8.314\text{J}/(\text{K} \cdot \text{mol})]} \right) \left(\frac{10\text{K}}{(613\ \text{K})(623\ \text{K})} \right)$$

$$\log \left(\frac{p_2}{0.73\ \text{atm}} \right) = 0.089$$

$$\frac{p_2}{0.73\ \text{atm}} = \text{antilog}\ 0.089$$

$$\frac{p_2}{0.73\ \text{atm}} = 1.2$$

$$p_2 = 0.88\ \text{atm}$$

10. 143 pm   This is a problem of geometry. Envision the unit cell:

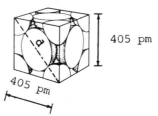

Using the Pythagorean theorem, we calculate the length of the hypotenuse, d:

$$d = \sqrt{2}\,(405\ \text{pm}) = 573\ \text{pm}$$

The length d corresponds to 4 radii; thus, $r = d/4 = $ 143 pm. The value of r is slightly larger than expected.

11. 195 u   The problem can easily be solved by using dimensional
    platinum   analysis to determine the mass of one mole of material:

First write the problem mathematically:

? g = 1 mol

Then use dimensional analysis:

$$? \ g = 1 \ \text{mol}\left(\frac{6.02 \times 10^{23} \ \text{atoms}}{1 \ \text{mol}}\right)\left(\frac{\text{cell}}{4 \ \text{atoms}}\right)\left(\frac{(392 \ \text{pm})^3}{\text{cell}}\right)\left(\frac{1 \ \text{cm}^3}{(10^{10} \ \text{pm})^3}\right)\left(\frac{21.5 \ \text{g}}{\text{cm}^3}\right)$$

$$= 195 \ \text{g}$$

Therefore the molecular weight is 195 u and the element is platinum.

SELF-TEST   I. Complete the test in 20 minutes:

1. Complete the phase diagram of H_2O. Label the triple point, the normal boiling point, the normal freezing point, and the solid, liquid, and vapor phases. Clearly show the approximate slopes of all lines.

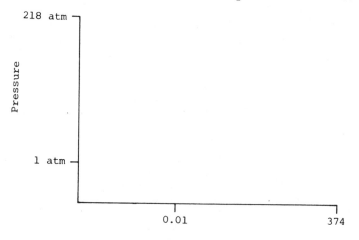

2. How many atoms are in the unit cell of
   (a) a simple cubic lattice
   (b) a body-centered cubic lattice

3. Which one of each pair has the higher boiling point?

   (a) Br_2 or I_2 (c) C or Ne

   (b) Na or K   (d) CO or N_2

                         (e) Si or Na

4. What is the atomic weight of an element that crystallizes as a face-centered cubic crystal, the density of which is 8.94 g/cm^3? The length of the diagonal through the center of the cube is 610 pm.

SOLUTIONS

CHAPTER
12

OBJECTIVES   I. You should be able to demonstrate your knowledge of the
following terms by defining them, describing them, or
giving specific examples of them:

azeotrope [12.10]
     maximum boiling azeotrope [12.10]
     minimum boiling azeotrope [12.10]
boiling point [12.8]
colligative properties [12.9]
concentration [12.6]
electrolyte [12.11]
enthalpy of hydration [12.4]
enthalpy of solution [12.4]
enthalpy of solvation [12.4]
fractional distillation [12.10]
freezing point [12.8]
Henry's law [12.5]
hydrated [12.3, 12.4]
ideal solution [12.7]
Le Chatelier's principle [12.5]
molality, m [12.6]
molarity, M [12.6]
mole fraction, X [12.6]
osmosis [12.9]
osmotic pressure [12.9]
percent by mass [12.6]
Raoult's law [12.7]
saturated solution [12.1]
simple distillation [12.10]
solute [12.1]
solvent [12.1]
supersaturated solution [12.1]
van't Hoff factor, i [12.12]

II. You should be able to determine the molarity, M, and the molality, m, of a solution. You should also be able to determine the percent by weight of a solute in a solution and the mole fraction, X, of a component of a solution.

III. You should be able to work problems involving the preparation of dilute solutions from weighed samples of more concentrated solutions.

IV. You should be able to determine the vapor pressure of a component of an ideal solution.

V. You should be able to determine molecular weights from data on freezing point depression, boiling point elevation, and osmotic pressure.

VI. You should understand the van't Hoff factor and be able to use it in freezing point depression and boiling point elevation calculations.

UNITS,
SYMBOLS,
MATHEMATICS

The following symbols are used in this chapter and you should be familiar with them.

- °C is the abbreviation for degrees Celsius.
- i is the symbol used to represent the van't Hoff factor in equations such as $\Delta t_f = i\, m\, k_f$; $t_b = i\, m\, k_b$; or $\pi = i\, M\, R\, T$.
- k_b is the symbol used to represent the molal boiling-point elevation constant in the equations $\Delta t_b = m\, k_b$ and $\Delta t_b = i\, m\, k_b$.
- k_f is the symbol used to represent the molal freezing point depression constant in the equations $\Delta t_f = m\, k_f$ and $\Delta t_f = i\, m\, k_f$.
- M is the symbol used to represent the molarity of a solution in units of moles of solute per one liter of solution.
- m is the symbol used to represent the molality of a solution in units of moles of solute per 1000 grams of solvent.
- n is the symbol used to represent the number of moles of a material.
- p_A is the symbol used to represent the partial pressure of species A.
- p_A° is the symbol used to represent the partial pressure of species A when it is pure, i.e., when it has nothing dissolved in it.
- π is the symbol used to represent osmotic pressure in equations such as $\pi V = nRT$.
- R is the symbol used to represent the gas constant. R equals 0.082056 liter·atm/(K·mol).
- T is the symbol used to represent the temperature in Kelvin: $T = 273 + t$.

- Δt_f is the symbol used to represent the change in freezing point in °C.
- Δt_b is the symbol used to represent the change in boiling point in °C.
- V is the symbol used to represent volume in liters.
- X_A is the symbol used to represent the mole fraction of A.

EXERCISES   I. Work the following problems

1. A total of 4.00 g of $AgNO_3$ is dissolved in a small quantity of water and diluted to 1.00 L. What is the molarity of the final $AgNO_3$ solution?

2. What volume of a 30.5% HCl solution, which has a density of 1.12 g/ml, is needed to prepare 500 mL of 0.200M HCl?

3. What volume of a standard riboflavin solution that has a concentration of $1.00 \times 10^{-2} M$ is necessary to prepare 100 mL of a solution that has a concentration of $1.00 \times 10^{-5} M$?

4. How many grams of $K_2Cr_2O_7$ are needed to prepare 1.00 L of 0.100M $K_2Cr_2O_7$?

5. A beverage contains 0.174 g of ethanol, C_2H_5OH, and 4.72 g of water, H_2O. What is the mole fraction of ethanol?

6. What is the molality of ethanol in the solution described in problem 5 above?

7. How many grams of $K_2Cr_2O_7$ are needed to prepare 60 mL of a 0.15M $K_2Cr_2O_7$ solution?

8. What volume of a 0.45M $KMnO_4$ solution is needed to prepare 1.0 L of 0.10M $KMnO_4$?

9. Is Henry's law valid for all solutions?

10. If NaCl costs $56.00 per 1.0 ton and $CaCl_2$ costs $88.00 per 1.0 ton, which compound would be more economical to prevent icy roads?

11. What is the molarity of an acetic acid solution if 25 mL of this solution are diluted to 100 mL to form a 0.75M solution?

12. How many mL of 0.065M riboflavin, vitamin B_2, are necessary to prepare 1.0 L of riboflavin solution that contains a solute concentration of 0.10 mg/mL? The molecular weight of riboflavin is 376.

13. Which of the following would be expected to have the greater hydration energy: $BeCl_2$ or $BaCl_2$?

14. At 25°C the osmotic pressure of 100 mL of β-lacto-globulin solution containing 1.49 g of that protein was found to be 0.0100 atm. Calculate the molecular weight of the protein.

15. The vapor pressure of pure water at 25°C is 3.12×10^{-2} atm. What is the vapor pressure of a 100-g sample of water in which 27.4 g of sucrose, $C_{12}H_{22}O_{11}$, is dissolved?

16. Assuming the aluminum compound $Al_2(SO_4)_3$ is a strong electrolyte, what is the vapor pressure above a solution prepared with 0.100 mol of $Al_2(SO_4)_3$ and 9.00 mol of H_2O at 30°C if the vapor pressure of pure H_2O at 30°C is 0.0395 atm? Hint: Assume each mol of $Al_2(SO_4)_3$ gives five moles of ions.

17. In which of the following liquids should the gas HF be most soluble: CF_4, Br_2, or H_2O?

18. Which of the following solutions would have the lowest freezing point? 1.00m $CaCl_2$, or 1.00m NaCl?

19. A 0.10M NaCl solution is heated and the volume increases by 10.0%. What is the molarity of the hot solution?

20. A 0.10m NaCl solution is heated and the volume increases by 10.0%. What is the molality of the hot solution?

II. Answer each of the following questions with either *higher than* (a larger more positive numerical value), *lower than* (a smaller more negative numerical value), or *remains the same* (has exactly the same numerical value).

1. The freezing point of a 0.10 *m* aqueous solution of NaCl is _____ the freezing point of a 0.010 *m* aqueous solution of NaCl.

2. The freezing point of a 0.10 *m* aqueous solution of NaCl is _____ the freezing point of a 0.10 *m* aqueous solution of K_2SO_4.

3. The freezing point of a 0.01 *m* aqueous solution of sugar (a non-electrolyte) is _____ the freezing point of a 0.01 *m* aqueous solution of NaCl.

4. The boiling point of a 0.10 *m* aqueous solution of Nacl is _____ the boiling point of a 0.10 *m* aqueous solution of K_2SO_4.

5. The boiling point of a 0.01 *m* aqueous solution of acetic acid (a weak electrolyte) is _____ the boiling point of a 0.01 *m* aqueous solution of NaCl.

6. The concentration of lead acetate in a supersaturated aqueous solution at 12°C is _____ the concentration of lead acetate in a saturated aqueous solution at 12°C.

7. The enthalpy of solution of KNO_2 is endothermic; therefore, the enthalpy of hydration is _____ the lattice energy of KNO_2.

8. The solubility of oxygen in water at 25°C is _____ the solubility at 50°C.

9. The solubility of KNO_2 in water at 25°C is _____ the solubility at 50°C. Remember that the enthalpy of solution of KNO_2 is endothermic.

10. A solution is prepared by dissolving 0.01 g of naphthalene, $C_{10}H_8$, in 60 mL of benzene, C_6H_6, at 25°C. When the solution is cooled at 10°C, the molality is _____ the molality of the original solution at 25°C.

11. Mixtures of carbon disulfide, CS_2, and acetone, C_3H_6O, form solutions which deviate negatively from Raoult's law. The intramolecular attractions between two molecules of carbon disulfide is _____ the intramolecular attractions between a carbon disulfide molecule and an acetone molecule.

12. Pentane, C_5H_{12}, and hexane, C_6H_{14}, form ideal solutions. The intramolecular attractions between two molecules of pentane are _____ the intramolecular attractions between a pentane molecule and a hexane molecule.

13. Pentane, C_5H_{12}, and hexane, C_6H_{14}, form ideal solutions. The equilibrium partial pressure of pentane above a solution where $X_{pentane}$ = 0.22 and X_{hexane} = 0.78 is _____ the equilibrium partial pressure of pentane above a solution where $X_{pentane}$ = 0.25 and X_{hexane} = 0.75.

14. In order to prevent cells from either swelling of shriveling, the solution into which the cells are placed must have an osmotic pressure _____ the osmotic pressure of the solution inside the cell itself.

15. Minimum boiling azeotropes have boiling points _____ either of the components of the solution.

ANSWERS TO EXERCISES

I. Problems

1. 0.0235M

Calculate the molarity:

$$? \; M \; AgNO_3 = \frac{\text{number of moles of } AgNO_3}{\text{number of liters of } AgNO_3 \text{ sol'n}}$$

$$= \frac{(4.00 \text{ g } AgNO_3) \left(\dfrac{1 \text{ mol } AgNO_3}{169.9 \text{ g } AgNO_3} \right)}{1 \text{ L } AgNO_3 \text{ sol'n}}$$

$$= 0.0235M \; AgNO_3$$

Alternatively, use the conversion factor method to calculate the number of moles of $AgNO_3$ in 1.00 L of solution:

$$? \text{ mol } AgNO_3 = 1.00 \text{ L sol'n } \left(\frac{4.00 \text{ g } AgNO_3}{1.00 \text{ L sol'n}}\right)\left(\frac{1 \text{ mol } AgNO_3}{169.9 \text{ g } AgNO_3}\right)$$

$$= 0.0235 \text{ mol } AgNO_3$$

2. 10.7 mL   Using the conversion factor method, the problem is easily solved:

$$? \text{ mL conc HCl} = 500 \text{ mL sol'n}$$

$$\times \left(\frac{0.200 \text{ mol HCl}}{1000 \text{ mL sol'n}}\right)\left(\frac{36.45 \text{ g HCl}}{1 \text{ mol HCl}}\right)\left(\frac{100 \text{ g conc HCl}}{30.5 \text{ g HCl}}\right)\left(\frac{1 \text{ mL conc HCl}}{1.12 \text{ g conc HCl}}\right)$$

$$= 10.7 \text{ mL conc HCl}$$

3. 0.100 mL   Remember that the number of moles of riboflavin in the concentrated solution $[V_1 M_1 = V_1(1.00 \times 10^{-2}M)]$ equals the number of moles of riboflavin in the dilute solution $[V_2 M_2 = (0.100 \text{ L})(1.00 \times 10^{-5}M)]$.   Therefore

$$V_1 M_1 = V_2 M_2$$

$$V_1(1.00 \times 10^{-2}M) = (0.100 \text{ L})(1.00 \times 10^{-5}M)$$

$$V_1 = 1.00 \times 10^{-4} \text{ L}$$

or

$$V_1 = 0.100 \text{ mL}$$

4. 29.4 g   Using the conversion factor method, calculate the number of grams of $K_2Cr_2O_7$ in 1.00 L of solution.

$$? \text{ g } K_2Cr_2O_7 = 1.00 \text{ L sol'n } \left(\frac{0.100 \text{ mol } K_2Cr_2O_7}{1.00 \text{ L sol'n}}\right)\left(\frac{294 \text{ g } K_2Cr_2O_7}{1 \text{ mol } K_2Cr_2O_7}\right)$$

$$= 29.4 \text{ g } K_2Cr_2O_7$$

5. 0.0142

Calculate the mole fraction of ethanol as follows:

$$? \text{ mol } C_2H_5OH = 0.174 \text{ g } C_2H_5OH\left(\frac{1 \text{ mol } C_2H_5OH}{46.07 \text{ g } C_2H_5OH}\right)$$

$$= 0.003776 \text{ mol } C_2H_5OH$$

$$? \text{ mol } H_2O = 4.72 \text{ g } H_2O\left(\frac{1 \text{ mol } H_2O}{18.02 \text{ g } H_2O}\right)$$

$$= 0.2619 \text{ mol } H_2O$$

$$X \; C_2H_5OH = \frac{\text{number of moles of } C_2H_5OH}{\text{total number of moles of } C_2H_5OH + H_2O}$$

$$= \frac{0.003776 \text{ mol}}{0.003776 \text{ mol} + 0.2619 \text{ mol}} = 0.0142$$

6. 0.800 *m*

Using the conversion factor method, calculate the number of moles of ethanol in 1000 g of solvent.

$$? \text{ mol } C_2H_5OH = 1000 \text{ g } H_2O\left(\frac{0.174 \text{ g } C_2H_5OH}{4.72 \text{ g } H_2O}\right)\left(\frac{1 \text{ mol } C_2H_5OH}{46.07 \text{ g } C_2H_5OH}\right)$$

$$= 0.800 \text{ mol } C_2H_5OH$$

Therefore the solution is 0.800 *m* C_2H_5OH

7. 2.6 g

Use the conversion factor method to determine the number of grams of $K_2Cr_2O_7$ in 60 mL of solution.

$$? \text{ g } K_2Cr_2O_7 = 0.060 \text{ L sol'n}\left(\frac{0.150 \text{ mol } K_2Cr_2O_7}{1.00 \text{ L sol'n}}\right)\left(\frac{294 \text{ g } K_2Cr_2O_7}{1 \text{ mol } K_2Cr_2O_7}\right)$$

$$= 2.6 \text{ g } K_2Cr_2O_7$$

8. 0.22 L

The number of moles of solute in the unknown volume of concentrated solution equals the number of moles of solute in the dilute solution:

$$V_1M_1 = V_2M_2$$

Rearranging the preceding equation and substituting values into it yields

$$V_1 = \frac{V_2M_2}{M_1} = \frac{(1.0 \text{ L})(0.10M)}{0.45M} = 0.22 \text{ L}$$

9. no   Henry's law, which states that the solubility of a gas in a solution is directly proportional to the partial pressure of that gas above the solution, is valid only for dilute solutions and low pressures. Also, extremely soluble gases usually react with the solvent and thus do not follow Henry's law.

10. NaCl   Calculate the cost of a single mole of ions for each compound because the number of ions is directly proportional to the freezing point depression.

For $CaCl_2$:

? \$/mol ions

$$= \left(\frac{\$88}{2000 \text{ lb } CaCl_2}\right)\left(\frac{1 \text{ lb } CaCl_2}{454 \text{ g } CaCl_2}\right)\left(\frac{111 \text{ g } CaCl_2}{1 \text{ mol } CaCl_2}\right)\left(\frac{1 \text{ mol } CaCl_2}{3 \text{ mol ions}}\right)$$

$$= 3.6 \times 10^{-3} \text{ \$/mol ions}$$

For NaCl:

? \$/mol ions

$$= \left(\frac{\$56}{2000 \text{ lb } NaCl}\right)\left(\frac{1 \text{ lb } NaCl}{454 \text{ g } NaCl}\right)\left(\frac{58.4 \text{ g } NaCl}{1 \text{ mol } NaCl}\right)\left(\frac{1 \text{ mol } NaCl}{2 \text{ mol ions}}\right)$$

$$= 1.8 \times 10^{-3} \text{ \$/mol ions}$$

Thus, it is cheaper to use NaCl than $CaCl_2$.

11. 3.0M Rearranging the equation

$$V_1 M_1 = V_2 M_2$$

and substituting values into it yields

$$M_1 = \frac{V_2 M_2}{V_1}$$

$$? \ M_1 = \frac{(100 \text{ mL})(0.75)}{(25 \text{ mL})}$$

$$= 3.0M$$

12. 4.1 mL   First determine the molarity of the 0.10 mg/ml solution of vitamin B_2:

$$? \ M \ B_2 = \left(\frac{0.10 \text{ mg } B_2}{1 \text{ mL sol'n}}\right)\left(\frac{1 \text{ g}}{1000 \text{ mg}}\right)\left(\frac{1000 \text{ mL}}{1 \text{ L}}\right)\left(\frac{1 \text{ mol } B_2}{376 \text{ g } B_2}\right)$$

$$= 2.66 \times 10^{-4} M \ B_2$$

Then rearrange the equation

$$V_1 M_1 = V_2 M_2$$

and substitute values into it:

$$V_1 = \frac{V_2 M_2}{M_1}$$

$$V_1 = \frac{(1.0 \text{ L})(2.66 \times 10^{-4} M \text{ B}_2)}{6.5 \times 10^{-2} M \text{ B}_2}$$

$$= 0.0041 \text{ L, or } 4.1 \text{ mL}$$

13. $BeCl_2$

An ion with a larger value of the ratio of ionic charge to ionic radius forms a more stable hydrated ion and thus has a larger enthalpy of hydration, $\Delta H_{\text{hydration}}$.

14. 3.64×10^4 g/mol

Rearranging the van't Hoff equation

$$\pi V = nRT$$

gives

$$n = \frac{\pi V}{RT}$$

Since

$$n = \text{number of moles} = \frac{\text{weight in grams}}{\text{molecular weight in grams}}$$

the van't Hoff equation becomes

$$\frac{\text{weight in grams}}{\text{molecular weight in grams}} = \frac{\pi V}{RT}$$

or

$$\text{molecular weight in grams} = \frac{(RT)(\text{weight in grams})}{\pi V}$$

Substituting values into the van't Hoff equation gives:

molecular weight in grams

$$= \frac{(0.08206 \text{ L} \cdot \text{atm}/(\text{K} \cdot \text{mol}))(298 \text{ K})(1.49 \text{ g})}{(1.00 \times 10^{-2} \text{ atm})(1.00 \times 10^2 \text{ mL})(1 \text{ L}/10^3 \text{ mL})}$$

$$= 3.64 \times 10^4 \text{ g/mol}$$

15. 3.08×10^{-2} atm

Using Raoult's law to solve this problem:

$$p_{H_2O} = X_{H_2O}p^o_{H_2O}$$

First determine the number of moles of water and sucrose:

$$? \text{ mol } H_2O = 100 \text{ g } H_2O \left(\frac{1 \text{ mol } H_2O}{18.02 \text{ g } H_2O} \right)$$

$$= 5.55 \text{ mol } H_2O$$

$$? \text{ mol } C_{12}H_{22}O_{11} = 27.4 \text{ g } C_{12}H_{22}O_{11} \left(\frac{1 \text{ mol } C_{12}H_{22}O_{11}}{342.3 \text{ g } C_{12}H_{22}O_{11}} \right)$$

$$= 8.00 \times 10^{-2} \text{ mol } C_{12}H_{22}O_{11}$$

The mole fraction of water is

$$X_{H_2O} = \frac{5.55 \text{ mol}}{5.55 \text{ mol} + 0.800 \text{ mol}}$$

$$= 0.986$$

Substituting values into Raoult's law

$$p_{H_2O} = X_{H_2O}p^o_{H_2O}$$

$$= 0.986(3.12 \times 10^{-2} \text{ atm})$$

$$= 3.08 \times 10^{-2} \text{ atm}$$

Since the solute is nonvolatile,

$$P_{total} = p_{H_2O} = 3.08 \times 10^{-2} \text{ atm}$$

16. 0.0374 atm

Since $Al_2(SO_4)_3$ is assumed to be a strong electrolyte, in solution a mole of it produces 5 mol of ions:

$$Al_2(SO_4)_3 \rightarrow 2Al^{3+} + 3SO_4^{2-}$$

Assume an ideal solution is formed and use Raoult's law to solve the problem:

$$p_{H_2O} = X_{H_2O}p^o_{H_2O}$$

$$= 0.0395 \text{ atm} \left(\frac{9.00 \text{ mol}}{9.00 \text{ mol} + 0.500 \text{ mol}} \right)$$

$$= 0.0374 \text{ atm}$$

Since the solute is nonvolatile,

P_{total} = 0.0374 atm

17. H_2O *Like dissolves like.* Hydrogen fluoride, HF, and water, H_2O, are highly polar and have hydrogen bonding ability. Tetrafluoromethane, CF_4, and bromine, Br_2, are nonpolar.

18. $CaCl_2$ Assuming each compound is a strong electrolyte, each would dissociate into the following number of ions:

$CaCl_2$: 3
NaCl: 2

Substitute values into the equation

$$\Delta T_f = iK_f m$$

The freezing point depression is largest for $CaCl_2$, for which i = 3.

19. 0.091 *M*   Assuming the initial volume of 1.00 L, the new volume would be 1.10 L and would contain the same number of moles of NaCl. Therefore

$$V_1 M_1 = V_2 M_2$$
$$(1.10 \text{ L}) M_1 = (1.00 \text{ L})(0.10M)$$
$$M_1 = 0.091M$$

20. 0.10 *m*   The molality is defined as the number of moles of solute per 1 kg of solvent. Neither the moles of solute, NaCl, nor the grams of solvent, H_2O, changes upon heating. Therefore, the molality does not change.

II. Answers to Higher Than/Lower Than/Same as Problems.

1. lower than   The freezing point of a 0.10 *m* aqueous solution of NaCl is lower than the freezing point of a 0.010 *m* aqueous solution of NaCl.

2. higher than   The freezing point of a 0.10 *m* aqueous solution of NaCl is higher than the freezing point of a 0.10 *m* aqueous solution of K_2SO_4.

3. higher than   The freezing point of a 0.01 *m* aqueous solution of sugar (a non-electrolyte) is higher than the freezing point of a 0.01 *m* aqueous solution of NaCl.

4. lower
   than

The boiling point of a 0.10 m aqueous soluton of NaCl
is lower than the boiling point of a 0.10 m aqueous
solution of K_3SO_4.

5. lower
   than

The boiling point of a 0.01 m aqueous solution of
acetic acid (a weak electrolyte) is lower than the
boiling point of a 0.01 m aqueous solution of NaCl.

6. higher
   than

The concentration of lead acetate in a supersaturated
aqueous solution at 12°C is higher than the
concentration of lead acetate in a saturated aqueous
solution at 12°C.

7. lower
   than

The enthalpy of solution of KNO_2 is endothermic;
therefore, the enthalpy of hydration is lower than the
lattice energy of KNO_2.

8. higher
   than

The solubility of oxygen in water at 25°C is higher
than the solubility at 50°C.

9. lower
   than

The solubility of KNO_2 in water at 25°C is lower than
the solubility at 50°C. Remember that the enthalpy of
solution of KNO_2 is endothermic.

10. the
    same as

A solution is prepared by dissolving 0.01 g of
napthalene, $C_{10}H_8$, in 60 mL of benzene, C_6H_6, at 25°C.
When the solution is cooled to 10°C, the molality is
the same as the molality of the original solution at
25°C.

11. lower
    than

Mixtures of carbon disulfide, CS_2, and acetone, C_3H_6O,
form solutions which deviate negatively from Raoult's
law. The intramolecular attractions between two
molecules of carbon disulfide is lower than the
intramolecular attractions between a carbon disulfide
molecule and an acetone molecule.

12. the
    same as

Pentane, C_5H_{12}, the hexane, C_6H_{14}, form ideal
solutions. The intramolecular attractions between two
molecules of pentane are the same as the intramolecular
attractions between a pentane molecule and a hexane
molecule.

13. lower
    than

Pentane, C_5H_{12}, and hexane, C_6H_{14}, form ideal
solutions. The equilibrium partial pressure of pentane
above a solution where $X_{pentane}$ = 0.22 and X_{hexane}
= 0.78 is lower than the equilibrium partial pressure
of pentane above a solution where $X_{pentane}$ = 0.25 and
X_{hexane} = 0.75.

14. the
    same as

In order to prevent cells from either swelling or shriveling, the solution into which the cells are placed must have an osmotic pressure the same as the osmotic pressure of the solution inside the cell itself.

15. lower
    than

Minimum boiling azeotropes have boiling points lower than either of the components of the solution.

SELF-TEST

I. Complete the test in 30 minutes:

1. A solution is 30.0% by mass HCl and has a density of 1.18 g/mL. What is the molality of the solution? Use a periodic table if necessary.

2. What is the molarity of the solution described in problem 1? Use a periodic table if necessary.

3. A sample of sodium is added to water to make 500 mL of solution. At STP 33.6 L of dry hydrogen gas are collected from the reaction of sodium metal with water:

$$2Na(s) + 2H_2O \rightarrow 2NaOH(aq) + H_2(g)$$

What is the molarity of the NaOH solution produced by the reaction?

4. What volume of a concentrated hydrochloric acid solution, which is 37% HCl and has a density of 1.189 g/mL, should be used to prepare 2000 mL of $0.100M$ HCl solution?

5. A solution containing 0.500 g of an unknown nonvolatile solute in 10.0 g of camphor has a freezing point of 159.4°C. What is the molecular weight of the solute if the normal freezing point of camphor is 179.0°C and the molal freezing-point depression constant for camphor is 49°C/m?

REACTIONS IN AQUEOUS SOLUTION

CHAPTER
13

OBJECTIVES   I. You should be able to demonstrate your knowledge of the following terms by defining them, or giving specific examples of them:

acid [13.4]
acidic oxide [13.5]
acid salt [13.4]
amphoterism [13.5]
Arrhenius acids and bases [13.4]
base [13.4]
basic oxide [13.5]
disproportionation [13.3]
equivalent weight [13.8]
half-reaction [13.3]
hydronium ion [13.4]
ion-electron method [13.3]
metathesis reactions [13.1]
neutralization [13.4]
normality, N [13.8]
oxidation [13.3]
oxidation number [13.2]
oxidation-reduction method [13.3]
oxidizing agent [13.3]
precipitate [13.1]
reducing agent [13.3]
reduction [13.3]
salt [13.4]
solubility [13.1]
standard solution [13.7]
titration [13.7]
volumetric analysis [13.7]

II. You should be able to write balanced metathesis equations in proper net ionic form. Know the information contained in Tables 11.1 and 11.2 in your text.

III. You should be able to predict common oxidation numbers of elements and determine oxidation numbers of these elements in polyatomic molecules and ions.

IV. You should be able to identify oxidizing agents and reducing agents.

V. You should be able to balance oxidation-reduction equations by using the oxidation-number method and/or the ion-electron method.

VI. You should be able to write chemical equations for reactions involving acids, bases, and salts.

VII. You should be able to name and write chemical formulas for oxyacids and their salts.

VIII. You should be able to calculate concentration and mass percent using titration data.

IX. You should be able to use normalities in computations involving oxidation-reduction and neutralization reactions.

UNITS,
SYMBOLS,
MATHEMATICS

The following symbols were used in this chapter. You should be familiar with them.

- (aq) is used after a chemical formula to indicate that the chemical exists dissolved in water under the physical conditions described.

- (g) is used after a chemical formula to indicate that the chemical exists in the gaseous state under the physical conditions specified.

- (l) is used after a chemical formula to indicate that the chemical exists in the liquid state under the physical conditions specified.

- M is the symbol used to stand for the solution concentration, molarity, which has units of moles per liter, mol/L.

- N is the symbol used to stand for the solution concentration, normality, which has units of equivalents per liter.

- (s) is used after a chemical formula to indicate that the chemical exists in the solid state under the physical conditions specified.

EXERCISES   I. Write balanced net ionic chemical equations for the
reactions that occur when aqueous solutions of the
following compounds are mixed. Some of them yield no
reaction; indicate this by writing NR.

✓ 1. $HCl + NaCl \rightarrow$ No Reaction

✓ 2. $HCl + NH_4Cl \rightarrow$ No Reaction

✓ 3. $H^+ + Cl^- + Ag^+ + NO_3$
$HCl + AgNO_3 \rightarrow AgCl + HNO_3$

✗ 4. $H^+ + Cl^- + Na^+ + CO_3^{2-}$
$HCl + Na_2CO_3 \rightarrow NaCl + HCO_3$ $(CO_2(g) + H_2O)$

✓ 5. $H^+ Cl^-$
$HCl + NaOH \rightarrow NaCl + H_2O$

✓ 6. $H^+ + Cl^- Na^+ + C_2H_3O_2^-$
$HCl + NaC_2H_3O_2 \rightarrow NaCl + HC_2H_3O_2$ ✓

✓ 7. $H^+ + Cl^- + H^+ + NO_3^-$
$HCl + HNO_3 \rightarrow$ No Reaction

✗ 8. $Na^+ + OH^- + H_3^+ + PO_4^-$
$NaOH + H_3PO_4 \rightarrow$

✓ 9. $Na^+ + OH^- + K^+ + SO_4^{--}$
$NaOH + K_2SO_4 \rightarrow$ No Reaction

10. $Na^+ + OH^- + NH_4^+ + SO_4^{-2}$
$NaOH + (NH_4)_2SO_4 \rightarrow NH_3 + H_2O$

✓ 11. $Na^+ + OH^- + Fe^{+2} + SO_4^{-2}$
$NaOH + FeSO_4 \rightarrow Fe(OH)_2$ ✗

✓ 12. $NH_4^+ + CO_3^{-2} + Na^+ + SO_4^{-2}$
$(NH_4)_2CO_3 + Na_2SO_4 \rightarrow$ No Reaction

II. What is the oxidation number of:

+4	1. B in BF_4	+4	7. Sn in $SnCl_4$
-3	2. N in NH_4^+	+6	8. Cr in $Cr_2O_7^{2-}$
+5	3. N in NO_3^-	+2	9. Cr in CrO_4^{2-}
+1	4. P in $H_2PO_2^-$	+4	10. Mn in MnO_2
+5	5. P in PO_4^{3-}	+3	11. Mn in MnO_4^-
+7	6. Cl in ClO_4^-	0	12. Cl in Cl_2

III. Work the following problems.

1. Hydrogen gas can be prepared commercially by each of
the following methods. Balance the equations:

(a) $\overset{0}{C}(s) + \overset{+1}{H_2}\overset{-2}{O}(g) \xrightarrow{1000°C} \overset{}{CO}(g) + \overset{0}{H_2}(g)$

(b) $2H_2O(1) \xrightarrow{\text{electrolysis}} 2H_2(g) + O_2(g)$

(c) $3Fe(s) + 4H_2O(g) \xrightarrow{400°C} Fe_3O_4(s) + 4H_2(g)$

(d) $Zn(s) + 2H^+(aq) \rightarrow Zn^{2+}(aq) + H_2(g)$

2. Determine the oxidation number of each atom in problem 1 of this section.

3. Identify the reducing agent in each reaction in problem 1 of this section.

4. Yellow phosphorus is very toxic and is used in some rat poisons. Since elemental phosphorus is normally not a constituent of biological material, the following nonquantitative screening test can be used to detect it:

Elemental phosphorus, P_4, is converted in the presence of water to hypophosphorous acid, H_3PO_2, and phosphine, PH_3 (reaction 1). The hypophosphorous acid converts silver ions to metallic silver, yielding phosphoric acid (reaction 2). The phosphine reacts with the silver ions of $AgNO_3$ to form silver phosphide, Ag_3P (reaction 3). Phosphine and silver phosphide each give a brown stain to filter paper impregnated with $AgNO_3$.

Do the following:

(a) Write a balanced equation for each reaction of the test.

(b) Identify the oxidizing and reducing agents in each redox reaction of the test.

5. Balance each of the following equations. All reactions occur in acid solution.

(a) $MnO_4^- + H_2C_2O_4 \rightarrow Mn^{2+} + CO_2$

(b) $Sn^{2+} + HgCl_2 \rightarrow Hg_2Cl_2 + Sn^{4+} + Cl^-$

(c) $MnO_4^- + Mo^{3+} \rightarrow MoO_4^{2-} + Mn^{2+}$

(d) $Ce^{4+} + H_3AsO_3 \rightarrow Ce^{3+} + H_3AsO_4$

6. Balance each of the following equations. All reactions occur in alkaline solution.

(a) $IO_4^- + I^- \rightarrow IO_3^- + I_2$

(b) $CO(NH_2)_2 + OBr^- \rightarrow CO_2 + N_2 + Br^-$

(c) $Al \rightarrow AlOH_4^- + H_2$

IV. Work the following problems.

1. A total of 4.00 g. of $AgNO_3$ is dissolved in a small quantity of water and diluted to 1.00 L.
This silver nitrate solution is to be used for titrating the chloride in natural ground water:

$$Ag^+(aq) + Cl^-(aq) \rightarrow AgCl(s)$$

What is the normality of the silver nitrate solution?

2. A solution is 0.625M $KMnO_4$. What is the normality of the solution if the permanganate is to be used in the reaction

$$MnO_4^- + Fe^{2+} \rightarrow Mn^{2+} + Fe^{3+}$$

3. Using Toepfer's reagent indicator, one can determine the concentration of free HCl present in gastric juice by reacting it with NaOH; usually concentrations are between 0 and 0.04N. If 5.0 mL of gastric juice requires 1.0 mL of 0.10N NaOH for complete neutralization, determine whether a patient's level of free HCl is within the normal limits.

4. How many mL of 0.150M HCl are required to completely neutralize 25.0 mL of 0.250M NaOH solution?

5. What is the molarity of 20.0 mL of acetic acid solution which requires 40.0 mL of 0.500M NaOH for neutralization?

6. A 0.376 g sample of soda ash (impure sodium carbonate) is dissolved and titrated with 14.5 mL of 0.105M HCl. What is the mass percent sodium carbonate in the soda ash?

V. Complete the following problems.

1. Write a balanced chemical equation in molecular form for the reaction of the following with water.

   (a) CaO

   (b) SO_3

   (c) CO_2

   (d) P_4O_6

   (e) Na_2O

   (f) Cl_2O

   (g) N_2O_5

2. Complete and balance the following equations in ionic form.

   (a) $Sc_2O_3(s) + H^+(aq) \rightarrow$

   (b) $SO_2(g) + OH^-(aq) \rightarrow$

   (c) $Ag_2O(s) + HCl(g) \rightarrow$

   (d) $Al_2O_3(s) + OH^-(aq) + H_2O \rightarrow$

   (e) $Al_2O_3(s) + H^+(aq) \rightarrow$

VI. Write the formula of each of the following compounds in the space provided.

_____ 1. potassium phosphate

_____ 2. hydrobromic acid

_____ 3. hypobromous acid

_____ 4. bromous acid

_____ 5. bromic acid

_____ 6. perbromic acid

_____ 7. potassium perbromate

_____   8. barium sulfate

_____   9. copper (II) dihydrogen phosphate

_____   10. sodium monohydrogen phosphite

_____   11. sodium bisulfate

_____   12. barium arsenate

_____   13. potassium arsenite

_____   14. boric acid

_____   15. calcium nitrite

_____   16. sulfuric acid

_____   17. sulfurous acid

_____   18. perchloric acid

_____   19. nitric acid

_____   20. ammonium nitrate

VII. Write the name of each of the following compounds in the space provided.

_____   1. $NaClO$

_____   2. $NaClO_2$

_____   3. $NaClO_3$

_____   4. $NaClO_4$

_____   5. $NaIO_4$

_____   6. $ZnHPO_4$

_____   7. $Sn_3(PO_4)_2$

_____   8. $Cu(NO_3)_2$

_____   9. H_3PO_4

_____   10. HNO_2

ANSWERS TO
EXERCISES

I. Metathesis Reactions

1. NR

2. NR

3. $Cl^-(aq) + Ag^+(aq) \rightarrow AgCl(s)$

4. $2H^+(aq) + CO_3^{2-}(aq) \rightarrow CO_2(g) + H_2O$

5. $H^+(aq) + OH^-(aq) \rightarrow H_2O$

6. $H^+(aq) + C_2H_3O_2^-(aq) \rightarrow HC_2H_3O_2(aq)$

7. NR

8. $3OH^-(aq) + H_3PO_4(aq) \rightarrow 3H_2O + PO_4^{3-}(aq)$

9. NR

10. $OH^-(aq) + NH_4^+(aq) \rightarrow NH_3(g) + H_2O$

11. $2OH^-(aq) + Fe^{2+}(aq) \rightarrow Fe(OH)_2(s)$

12. NR

II. Oxidation numbers

1. 3+

The sum of the oxidation numbers of the atoms in a polyatomic ion always equals the charge of that ion. For
BF_4^- the sum of the oxidation numbers is 1-. Boron's
oxidation number must therefore be 3+:

4 F atoms (1- per F atom) = 4-
1 B atom (3+ per B atom) = 3+
sum = 1-

2. 3-

For NH_4^+

4 H atoms (1+ per H atom) = 4+
1 N atom (3- per N atom) = 3-
sum = 1+

and the charge of the ion is 1+.

3. 5+

For NO_3^-

3 O atoms (2- per O atom) = 6-
1 N atom (5+ per N atom) = 5+
sum = 1-

and the charge of the ion is 1-.

4. 1+

For $H_2PO_2^-$

2 O atoms (2- per O atom) = 4-
2 H atoms (1+ per H atom) = 2+
1 P atom (1+ per P atom) = 1+
                    sum = 1-

and the charge of the ion is 1-.

5. 5+

For PO_4^{3-}

4 O atoms (2- per O atom) = 8-
1 P atom (5+ per P atom) = 5+
                    sum = 3-

and the charge of the ion is 3-.

6. 7+

For ClO_4^-

4 O atoms (2- per O atom) = 8-
1 Cl atom (7+ per Cl atom) = 7+
                    sum = 1-

and the charge of the ion is 1-.

7. 4+

For $SnCl_4$ the sum of the oxidation numbers is 0.

4 Cl atoms (1- per Cl atom) = 4-
1 Sn atom (4+ per Sn atom) = 4+
                    sum = 0

8. 6+

For $Cr_2O_7^{2-}$

7 O atoms (2- per O atom)   = 14-
2 Cr atoms (6+ per Cr atom) = 12+
                    sum = 2-

and the charge of the ion is 2-.

9. 6+

For CrO_4^{2-}

4 O atoms (2- per O atom)   = 8-
1 Cr atom (6+ per Cr atom) = 6+
                    sum = 2-

and the charge of the ion is 2-.

10. 4+

For MnO_2 the sum of the oxidation numbers is 0.

2 O atoms (2- per O atom)   = 4-
1 Mn atom (4+ per Mn atom) = 4+
                    sum = 0

11. 7+ For MnO_4^-

4 O atoms (2- per O atom) = 8-
1 Mn atom (7+ per Mn atom) = 7+
$$\text{sum} = \overline{1-}$$

and the charge of the ion is -1.

12. 0 Any atom in a molecule of an element has an oxidation number of 0.

III. Oxidation-reduction

1., 2., 3. Oxidation numbers are given below the balanced equations, and reducing agents are identified by the abbreviation *Red* below the equation:

(a) $C(s) + H_2O(g) \xrightarrow{1000°C} CO(g) + H_2(g)$

0   1+ 2-   2+ 2-   0
Red.

(b) $2H_2O(l) \xrightarrow{\text{electrolysis}} 2H_2(g) + O_2(g)$

1+ 2-   0   0
Red.

(c) $3Fe(s) + 4H_2O \xrightarrow{400°C} Fe_3O_4(s) + 4H_2(g)$

0   1+ 2-   8/3+ 2-   0
Red.

(d) $Zn(s) + 2H^+(aq) \rightarrow Zn^{2+}(aq) + H_2(g)$

0   1+   2+   0
Red.

4. (a) Reaction 1,

$$P_4 + 6H_2O \rightarrow 3H_3PO_2 + PH_3$$

is a disproportionation reaction:

$$3[P_4 + 8H_2O \rightarrow 4H_3PO_2 + 4H^+ + 4e^-]$$
$$P_4 + 12H^+ + 12e^- \rightarrow 4PH_3$$

$$4P_4 + 24H_2O \rightarrow 12H_3PO_2 + 4PH_3$$

or

$$P_4 + 6H_2O \rightarrow 3H_2PO_2 + PH_3$$

Reaction 2,

$$H_3PO_2 + 2H_2O + 4AgNO_3 \rightarrow 4HNO_3 + H_3PO_4 + 4Ag$$

is a redox reaction:

$$H_3PO_2 + 2H_2O \rightarrow H_3PO_4 + 4H^+ + 4e^-$$

$$4[e^- + Ag^+ + NO_3^- \rightarrow Ag + NO_3^-]$$

$$\overline{H_3PO_2 + 2H_2O + 4Ag+ + 4NO_3^- \rightarrow 4H^+ + H_3PO_4 + 4Ag + 4NO_3^-}$$

or

$$H_3PO_2 + 2H_2O + 4AgNO_3 \rightarrow 4HNO_3 + H_3PO_4 + 4Ag$$

Reaction 3,

$$PH_3 + 3AgNO_3 \rightarrow 3HNO_3 + Ag_3P$$

is a metathesis reaction.

When you balance any equation by the ion-electron method, you can check your answer by answering each of the following questions:

(a) Is the same number of atoms of each element on both sides of the equation?

(b) Is the total charge the same on each side of the equation?

(c) Do particular ions or molecules appear only once in the equation?

(d) Are the coefficients in the equation the lowest whole numbers?

If the answer to each question is *yes*, you have balanced the equation correctly.

(b) P_4 ox agent   Reaction (1) is a disproportionation reaction since P_4 is
P_4 red agent   both the oxidizing and reducing agent and undergoes oxidation and reduction.

Ag^+ ox agent

H_3PO_4
red agent

Not a redox
reaction

In reaction (2) H_3PO_2 is oxidized and is therefore the reducing agent. The silver ion, Ag^+, is reduced and is therefore the oxidizing agent.

Reaction (3) is a metathesis reaction.

5.

(a) $2MnO_4^- + 5H_2C_2O_4 + 6H^+ \rightarrow 2Mn^{2+} + 10CO_2 + 8H_2O$

(b) $Sn^{2+} + 2HgCl_2 \rightarrow Hg_2Cl_2 + Sn^{4+} + 2Cl^-$

(c) $8H_2O + 3MnO_4^- + 5Mo^{3+} \rightarrow 5MoO_4^{2-} + 3Mn^{2+} + 16H^+$

(d) $2Ce^{4+} + H_3AsO_3 + H_2O \rightarrow 2Ce^{3+} + H_3AsO_4 + 2H^+$

6.

(a) $IO_4^- + 2I^- + H_2O \rightarrow IO_3^- + I_2 + 2OH^-$

(b) $CO(NH_2)_2 + 3OBr^- \rightarrow CO_2 + N_2 + 3Br^- + 2H_2O$

(c) $2OH^- + 2Al + 6H_2O \rightarrow 2Al(OH)_4^- + 3H_2$

IV. Volumetric analysis

1. 0.0235N

For the reaction

$$Ag^+(aq) + Cl^-(aq) \rightarrow AgCl(s)$$

the equivalent weight of $AgNO_3$ is equal to the molecular weight of $AgNO_3$. Therefore,

$$? \ N \ AgNO_3 = \frac{\text{number of equiv of } AgNO_3}{\text{number of liters of } AgNO_3 \ \text{sol'n}}$$

$$= \frac{(4.00 \ g \ AgNO_3)\left(\dfrac{1 \ \text{equiv } AgNO_3}{169.9 \ g \ AgNO_3}\right)}{1 \ L \ AgNO_3 \ \text{sol'n}}$$

$$= 0.0235N \ AgNO_3$$

2. 3.12N

For the reaction

$$5e^- + 8H^+ + \underset{7+}{MnO_4^-} \rightarrow \underset{2+}{Mn} + 4H_2O$$

the equivalent weight of $KMnO_4$ is 1/5 of the molecular weight of $KMnO_4$. Thus, the normality is 5 times the molarity since 1 mol $KMnO_4$ is 5 equivalents:

$$? \ N \ KMnO_4 = \left(\frac{0.625 \ mol \ KMnO_4}{1 \ L \ KMnO_4 \ sol'n} \right) \left(\frac{5 \ equivalents \ KMnO_4}{1 \ mol \ KMnO_4} \right)$$

$$= \frac{3.12 \ equivalents \ KMnO_4}{1 \ L \ KMnO_4 \ sol'n}$$

$$= 3.12N \ KMnO_4$$

3. yes

    0.020N

Substitute values into the equation

$$V_1 N_1 = V_2 N_2$$

and solve to find the unknown normality

$$N_1 = \frac{(1.0 \times 10^{-3} \ 1)(0.10N)}{(5.0 \times 10^{-3} \ L)}$$

$$= 0.020N$$

4. 41.7 mL

First determine the number of moles of NaOH used

$$? \ mol \ NaOH = 25.0 \times 10^{-3} \ L \left(\frac{0.250 \ mol \ NaOH}{L \ sol'n} \right)$$

$$= 6.25 \times 10^{-3} \ mol \ NaOH$$

Since one mole of NaOH reacts with one mole of HCl the problem can easily be solved using that relationship.

$$? \ 1 \ HCl = 6.25 \times 10^{-3} \ mol \ NaOH \left(\frac{1 \ mol \ HCl}{1 \ mol \ NaOH} \right) \left(\frac{1 \ L \ sol'n}{0.150 \ mol \ HCl} \right)$$

$$= 0.0417 \ L \ or \ 41.7 \ mL$$

5. 1.00M

Determine the number of moles of NaOH used.

$$? \ mol \ NaOH = 0.0400 \ L \ \frac{0.500 \ mol \ NaOH}{L \ sol'n}$$

$$= 0.0200 \ mol \ NaOH$$

Using the 1:1 stoichiometry for the reaction, determine the number of moles of acetic acid in a liter of sol'n.

$$? \ mol \ acetic \ acid = 1 \ L \ sol'n \ \frac{0.0200 \ mol \ acetic \ acid}{0.0200 \ L \ sol'n}$$

$$= 1.00 \ mol \ acetic \ acid \ in \ 1 \ L \ sol'n$$

or the solution is 1.00M

6. 21.5%   First determine the number of moles of HCl used to titrate
the soda ash:

$$? \text{ mol HCl} = 0.0145 \text{ L sol'n} \left(\frac{0.105 \text{ mol HCl}}{1 \text{ L sol'n}} \right)$$

$$= 1.52 \times 10^{-3} \text{ mol HCl}$$

Since the stoichiometry is such that it takes two moles of
HCl to neutralize one mole of Na_2CO_3, the relationship

$$1 \text{ mol } Na_2CO_3 = 2 \text{ mol HCl}$$

must be used in the calculation. Now calculate the number
of grams of Na_2CO_3 in the sample

$$? \text{ g } Na_2CO_3 = 1.52 \times 10^{-3} \text{ mol HCl} \left(\frac{1 \text{ mol } Na_2CO_3}{2 \text{ mol HCl}} \right) \left(\frac{106 \text{ g } Na_2CO_3}{1 \text{ mol } Na_2CO_3} \right)$$

$$= 8.06 \times 10^{-2} \text{ g } Na_2CO_3$$

Determine the percent Na_2CO_3 in the sample

$$\left(\frac{8.06 \times 10^{-2} \text{ g } Na_2CO_3}{0.376 \text{ g sample}} \right) 100\% = 21.4\%$$

V. Reactions of Oxides

1. (a) $CaO + H_2O \rightarrow Ca(OH)_2$

(b) $SO_3 + H_2O \rightarrow H_2SO_4$

(c) $CO_2 + H_2O \rightarrow H_2CO_3$

(d) $P_4O_6 + 6H_2O \rightarrow 4H_3PO_3$

(e) $Na_2O + H_2O \rightarrow 2NaOH$

(f) $Cl_2O + H_2O \rightarrow 2HOCl$

(g) $N_2O_5 + H_2O \rightarrow 2HNO_3$

2. (a) $Sc_2O_3(s) + 6H^+(aq) \rightarrow 2Sc^{3+}(aq) + 3H_2O$

(b) $SO_2(g) + 2OH^-(aq) \rightarrow SO_3^{2-}(aq) + H_2O$

(c) $Ag_2O(s) + 2HCl(g) \rightarrow 2AgCl(s) + H_2O$

(d) $Al_2O_3(s) + 2OH^-(aq) + 3H_2O \rightarrow 2Al(OH)_4^-(aq)$

$$(e) \ Al_2O_3(s) + 6H^+(aq) \rightarrow 2Al^{3+}(aq) + 3H_2O$$

VI. Formulas and nomenclature

1. K_3PO_4
2. HBr
3. $HBrO$ These are the formulas of oxybromine acids.
4. $HBrO_2$ Formulas for oxyacids of other elements are given
5. $HBrO_3$ in Table 13.1 of the study guide.
6. $HBrO_4$
7. $KBrO_4$ The perbromate ion has a charge of 1-.
8. $BaSO_4$ The sulfate ion has a charge of 2-.
9. $Cu(H_2PO_4)_2$
10. Na_2HPO_3
11. $NaHSO_4$
12. $Ba_3(AsO_4)_2$
13. K_3AsO_3
14. H_3BO_3
15. $Ca(NO_2)_2$
16. H_2SO_4
17. H_2SO_3
18. $HClO_4$
19. HNO_3
20. NH_4NO_3

TABLE 13.1   Some Common Oxyacids

Name	Formula	Name	Formula
sulfurous acid	H_2SO_3	nitrous acid	HNO_2
sulfuric acid	H_2SO_4	nitric acid	HNO_3
hypophosphorous acid	H_3PO_2	hypoiodous acid	HIO
phosphorous acid	H_3PO_3	iodic acid	HIO_3
phosphoric acid	H_3PO_4	periodic acid	HIO_4

VII. Nomenclature and formulas

1. sodium hypochlorite   The names of anions of other
2. sodium chlorite   common oxyacids are given in
3. sodium chlorate   Table 13.2 of the study guide.
4. sodium perchlorate
5. sodium periodate
6. zinc(II) hydrogen phosphate
7. tin(II) phosphate
8. copper(II) nitrate or cupric nitrate
9. phosphoric acid
10. nitrous acid

TABLE 13.2   Anions of Some Common Oxyacids

Name	Formula	Name	Formula
sulfate ion	SO_4^{2-}	nitrite ion	NO_2^-
sulfite ion	SO_3^{2-}	nitrate ion	NO_3^-
hypophosphite ion	PO_2^{3-}	hypoiodite ion	IO^-
phosphite ion	PO_3^{3-}	iodate ion	IO_3^-
phosphate ion	PO_4^{3-}	periodate ion	IO_4^-

SELF-TEST

Complete this test in 40 minutes. You may use the periodic chart of the elements as needed.

1. Write balanced equations for the reactions that take place when the following are mixed. Write all reactants and products in proper form by indicating the physical state.

   (a) $SO_2(g) + OH^-(aq) \rightarrow$

   (b) $ZnSO_4(s) + Ba(OH)_2(aq) \rightarrow$

   (c) $Fe^{2+}(aq) + MnO_4^-(aq) + H^+(aq) \rightarrow$ (in acid solution)

   (d) $CO_2(g) + H_2O(l) \rightarrow$

   (e) $Na^+(aq) + SO_4^{2-}(aq) + Cu^{2+}(aq) + Cl^-(aq) \rightarrow$

   (f) $K_2O(s) + H_2O \rightarrow$

2. A 20.00-mL portion of commercial vinegar, the density of which is 1.01 g/mL, is titrated with 28.75 mL of 0.503M NaOH. What is the weight percentage of acid, $HC_2H_3O_2$, in vinegar?

3. A 0.2070-g sample of an unknown organic acid is dissolved in 100 mL of 0.02560N sodium hydroxide. The resulting solution is titrated to the end point with 36.50 mL of 0.0240N HCl. Calculate the equivalent weight of the unknown acid.

4. Balance the following equations by the ion-electron or oxidation-number method.

   (a) $ClO_2 \rightarrow ClO_2^- + ClO_3^-$ (basic solution)

   (b) $IO_4^- + H_2AsO_3^- \rightarrow IO_3^- + H_2AsO_4^-$ (acid solution)

   (c) $Sb + NO_3^- \rightarrow Sb_4O_6 + NO$ (acid solution)

   (d) $O_3 + I^- \rightarrow H_2O + I_2$ (acid solution)

   (e) $NO_3^- + H_2S \rightarrow NO + S$ (acid solution)

   (f) $Cr_2O_7^{2-} + Cl^- \rightarrow Cr^{3+} + Cl_2$ (acid solution)

5. What is the oxidation number of each element written in the equations in problem number 4?

6. A solution is 0.625M $KMnO_4$. What is the normality of the solution if the permanganate is to be used in the reaction

$$MnO_4^- + Fe^{2+} \rightarrow Mn^{2+} + Fe^{3+} \text{ (acid solution)}$$

CHEMICAL KINETICS

CHAPTER

14

OBJECTIVES

I. You should be able to demonstrate your knowledge of the following terms by defining them, describing them, or giving specific examples of them:

activated complex [14.4]
Arrhenius equation [14.7]
bimolecular [14.6]
catalyst [14.8]
chain mechanism [14.6]
chemisorption [14.8]
collision theory [14.4]
energy of activation [14.4]
enzyme [14.8]
half-life [14.3]
heterogeneous catalyst [14.8]
homogeneous catalyst [14.8]
initial rate [14.1]
order [14.2], 14.3]
    zero-order reactions [14.2, 14.3]
    first-order reactions [14.2, 14.3]
    second-order reactions [14.2, 14.3]
    third-order reactions [14.2]
rate constant [14.2]
rate equation [14.2]
rate-determining step [14.5]
reaction intermediates [14.4]
reaction mechanism [14.5]
reaction rate [14.1]
single step [14.3]
termolecular [14.5]
transition state [14.4]
unimolecular [14.5]

II. Given experimental data, you should be able to formulate the rate equation for a reaction.

III. You should be able to determine the half-life of a chemical reaction.

IV. You should be able to propose a reaction mechanism based on experimental data.

V. You should be able to use the Arrhenius equation to calculate E_a or k from appropriate experimental data.

VI. You should understand the effect that a change in conditions such as temperature and concentration will have on the reaction rate, as well as the effect that the presence of a catalyst will have.

UNITS,
SYMBOLS,
MATHEMATICS

I. The following symbols were used in this chapter. You should be familiar with them.

- $\xrightarrow{A}$ is the symbolic way of indicating that species A is a catalyst for the reaction.
- A is the symbol used to represent the frequency factor in the Arrhenius equation, $k = Ae^{-E_a/RT}$.
- $\Delta[A]$ is the symbol used to indicate a change in the concentration of species A.
- $E_{a,f}$ is the symbol used to designate the potential energy difference between the reactants and the highest energy activated complex in a chemical reaction.
- $E_{a,r}$ is the symbol used to designate the potential energy difference between the products and the highest energy activated complex in a chemical reaction.
- ΔH is the symbol used to identify an enthalpy change.
- Δt is the symbol used to indicate a small interval of time.
- k is the symbol used to represent the rate constant in a rate equation.
- s is the abbreviation for second, the SI unit of time.
- t is the symbol used to represent time in equations.
- ln is the symbol used to represent a natural logarithm, i.e., a logarithm to   base e.
- log is the symbol used to represent a logarithm to base 10.
- R is the symbol used to represent the molar gas constant, 8.3143 J/(K·mol).
- $t_{1/2}$ is used to express the half-life of a chemical reaction. That is, the time it takes for one-half of the reaction to proceed.

II. Graphing of data becomes extremely important in chemical
   kinetics. At a minimum, you need to be familiar with the
   concepts outlined here.

   Figure 14.1 shows a series of straight lines and the
   corresponding formulas for the lines. The general formula
   for a straight line that goes through the intercept--the
   point on the graph where both x and y values equal zero--is
   $y = mx$. Note that m represents the slope of the line.
   Larger values of m signify greater slopes.

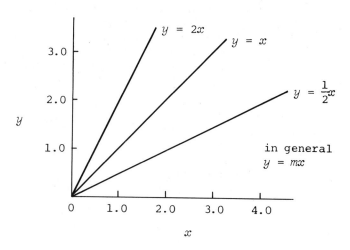

FIGURE 14.1

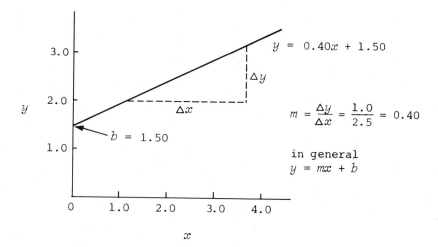

FIGURE 14.2

When the straight line does not pass through the intercept as in Figure 14.2, the general formula of the line is expressed as $y = mx + b$. Here b represents the value of y when $x = 0$, the intercept of the line with the y axis. As before, the slope of the line is m.

Determine the slope of a straight line by subtracting a smaller value of y from a larger value of y. Then subtract the value of x that corresponds to the smaller value of y from the value of x that corresponds to the larger value of y. The difference of the y's divided by the difference of the x's is the slope, which is commonly expressed as $m = \Delta y / \Delta x$. Note that positive slopes indicate that the value of y increases as x increases (as shown in Figure 14.1 and 14.2), and negative slopes indicate that the y value decreases as x increases (as shown in Figure 14.3).

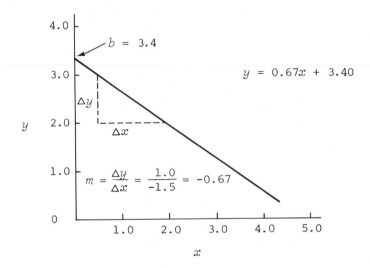

FIGURE 14.3

III. The common logarithm--the logarithm to the base 10--of a number is the power to which 10 must be raised in order to obtain the number.

For instance, if

$$a = 10^n$$

then

$$\log(a) = n.$$

It follows that

since $10 = 10^1$ then $\log(10) = 1$

since $100 = 10^2$ then $\log(100) = 2$

since $0.01 = 10^{-2}$ then $\log(0.01) = -2$

Using your calculator you will find that the log of 3.54 equals 0.549, or since $3.54 = 10^{0.549}$ then $\log(3.54) = 0.549$.

Logarithms are frequently used when values to be graphed extend over a very large range. Consider the following sets of data where small incremental changes in x cause large changes in the value of y:

x	y	$\log(y)$
1	1000000	6
2	100000	5
3	10000	4
4	1000	3
5	100	2

Attempts to plot these values on a standard rectilinear sheet of graph paper are shown in Figure 14.4. If we take the log of y and plot it versus x, the graph (Figure 14.5) becomes more meaningful since the values can be determined clearly from the graph. The slope of this new plot of $\log(y)$ vs. x is then $m = \Delta(\log(y))/\Delta x$, and since the graph is now linear the intercept b equals $\log(y)$.

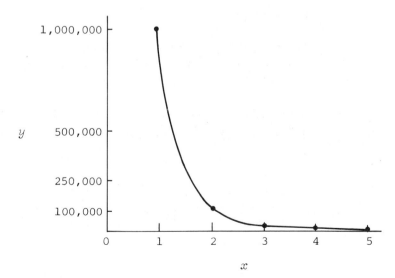

FIGURE 14.4

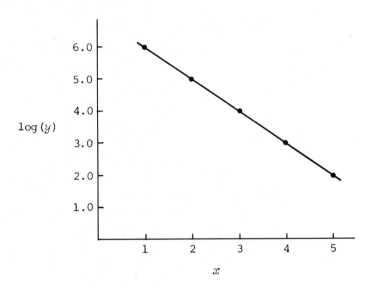

FIGURE 14.5

Once the log of y is known, the antilogarithm will return the value of y. For instance, if

$$\log(y) = 7$$

then the original number is 10 raised to the seventh power, i.e., 10^7.

Likewise if

$$\log(y) = 5 \qquad \text{then } y = 10^5$$
$$\log(y) = -2 \qquad \text{then } y = 10^{-2}$$
$$\log(y) = 3.74 \qquad \text{then } y = 10^{3.74} = 5,500$$

Since common logarithms are exponents of 10 and since $10^m \times 10^n = 10^{m+n}$, we can write

$$\log(10^m \times 10^n) = \log(10^{m+n})$$

$$= m + n$$

$$= \log(10^m) + \log(10^n)$$

or

Thus, the log of a product of two numbers is equal to the sum of the individual logs of those numbers. In general,

$\log(ab) = \log(a) + \log(b)$

A similar derivation for division shows that

$\log(a/b) = \log(a) - \log(b)$.

Manipulating logarithms in this fashion will become increasingly important to you as you continue in chemistry.

Logarithms to bases other than 10 are used. In this chapter natural logarithms (logarithms to the base 2.71828...) are also used. Convert natural logarithms--for which we use the symbol "ln"--to common logarithms--for which we use the symbol "log"--by using the following equations.

$\ln(a) = 2.303\log(a)$.

or

$$\log(a) = \frac{\ln(a)}{2.303}$$

EXERCISES I. Answer each of the following with true or false. If the answer is false, correct it.

_____ 1. The reaction represented in the energy diagram of Figure 14.6 of the study guide is exothermic.

_____ 2. In the energy diagram of Figure 14.6 of the study guide, E_1 is the activation energy for the forward reaction.

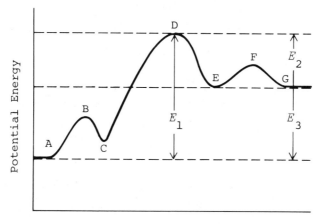

FIGURE 14.6

_____   3. In the energy diagram of Figure 14.6 of the study guide, E_2 is ΔE for the reaction.

_____   4. Point A in Figure 14.6 represents the potential energy of all reactants.

_____   5. Point B in Figure 14.6 represents the potential energy of a reaction intermediate.

_____   6. The energy diagram, Figure 14.6, could be only for a catalyzed reaction.

_____   7. If the specific rate constant for a reaction is large, the reaction will proceed rapidly.

_____   8. Rate constants are independent of temperature.

_____   9. Reaction intermediates appear in the net equation of a reaction.

_____   10. A catalyst changes the mechanism of a reaction.

II. Answer each of the following questions.

1. The rate equation for the decomposition of N_2O_5 to nitrogen dioxide and oxygen is found to be first order to N_2O_5. Write a rate equation for the decomposition.

2. The following data were obtained from a study of the decomposition reaction

$$2N_2O \rightarrow 2N_2 + O_2$$

Write a rate equation for the decomposition.

Experiment	[N_2O]	Rate of Formation of O_2
1	0.10 mol/L	1.0×10^{-5} mol/(L·s)
2	0.20 mol/L	1.0×10^{-5} mol/(L·s)
3	0.60 mol/L	1.0×10^{-5} mol/(L·s)

What is the value of the rate constant, k?

3. For the reaction

A(g) + B(g) → C(g)

the following data were obtained from three experiments:

Experiment	[A]	[B]	Rate (Formation of C)
1	0.20 mol/L	0.20 mol/L	3.0×10^{-4} mol/(L·s)
2	0.60 mol/L	0.60 mol/L	81.0×10^{-4} mol/(L·s)
3	0.60 mol/L	0.20 mol/L	9.0×10^{-4} mol/(L·s)

Use the data to answer the following:

(a) What is the rate equation for the reaction?
(b) What is the order of the reaction?
(c) What is the numerical value of the rate constant?
(d) If the volume of the reaction vessel were halved, what would be the effect on the reaction rate?
(e) Is the reaction as written a one-step process?

4. Assume that the rate-determining step of a reaction is

2A(g) + B(g) → C(g)

and that 2 mol of A and 1 mol of B are mixed in a liter container. Compare the following with the initial reaction rate of the mixture:
(a) rate when half of A has been consumed
(b) rate when two-thirds of A has been consumed
(c) initial reaction rate of a mixture of 2 mol of A and 2 mol of B in a liter container
(d) initial reaction rate of a mixture of 4 mol of A and 2 mol of B in a liter container

5. For the reaction

H_2(g) + I_2(g) → 2HI(g)

the slope of the plot of log k vs. $1/T$ is -8880 K. What is the energy of activation for the reaction?

6. For the reaction

2NO(g) + O_2(g) → 2NO_2(g)

the rate constant, k, equals 7.0×10^3 $L^2/(mol^2 \cdot s)$ and the rate equation is rate of appearance of $NO_2 = k[NO]^2[O_2]$

In a system, the initial oxygen concentration is 5.0×10^{-2} mol/L and the initial rate of appearance of NO_2 is 5.0×10^{-3} mol/(L·s).

(a) What is the initial concentration of NO?
(b) What is the concentration of oxygen when half of the NO has been consumed?
(c) What is the rate of appearance of NO_2 when half the initial amount of NO has been consumed?
(d) What is the concentration of oxygen when 9/10 of the NO has been consumed?
(e) What is the rate of appearance of NO_2 when 9/10 of the initial NO has been consumed?

7. For the reaction and the conditions described in problem number 6, calculate the concentration of NO when 1.0×10^{-3} mol/L O_2 has been consumed.

8. For a reaction

$$2A \rightarrow B + C$$

the rate equation is

rate of appearance of $C = k[A]^2$

At 300 K the rate constant is 4.8×10^{-9} L/(mol·s) and at 450 K the rate constant is 3.7×10^{-3} L/(mol·s). What is the energy of activation, E_a, for the reaction?

9. What is the energy of activation, E_a, for a reaction if a temperature increase from 350 K to 400 K results in a 100-fold increase in the rate of the reaction?

10. A suspected catalyst is found which causes the reaction described in problem 9 to proceed 500 times faster with the same temperature change. What is the energy of activation, E_a, for the reaction? Is the reaction catalyzed?

III. Refer to the following data which describe a hypothetical experiment in which a gas decomposed in a closed vessel, and the pressure of the gas was monitored for 100 minutes. Pressures can be used instead of concentrations in this problem since pressures are proportional to concentrations.

Time (minutes)	Pressure (torr)
0	350
10	200
20	114
30	65
40	37
50	21
60	12
70	7
80	4
90	2
100	1

1. Plot a graph of the pressure of the gas as a function of time.

2. Plot a graph that will allow you to determine the rate constant of the reaction from the slope. What is the rate constant?

3. What is the half-life of the reaction?

4. Is the half-life of the reaction dependent on the original pressure of the gas?

ANSWERS TO EXERCISES

I. Principles of chemical kinetics

1. False

The reaction is endothermic. The energy of the products is higher than that of the reactants; thus, the addition of energy is necessary for the reaction to occur.

2. True

3. False

Since E_2 is the difference between the energy of the products and that of the activated complex, it is the activation energy for the backward reaction. The ΔE for the forward reaction is E_3, the difference between the energy of the products and that of the reactants.

4. True

5. False   This is the potential energy of an activated complex.
                     Intermediates are at points C and E in Figure 14.6 of
                     the study guide.

6. False   You cannot tell from a single diagram whether a catalyst
                     is present or not.

7. True

8. False   The rate constant, k, varies with temperature, T, in a
                     manner described by the Arrhenius equation

                     $$k = Ae^{-E_a/RT}$$

9. False   Reaction intermediates are not final products; thus,
                     they would not appear in the net equation.

10. True

II. Problems

1. rate of   The reaction order is the exponent in the rate equation.
   disappear-   In this case the reaction is first order in N_2O_5; there-
   ance of   fore, the exponent in the rate equation is 1.
   N_2O_5
   $= k[N_2O_5]$

2. rate of   The data show that changing the concentration of N_2O does
   formation   not change the rate of formation of O_2. The reaction is
   of $O_2 = k$ therefore zero order in N_2O. The rate constant is
                     1.0×10^5 mol/(L·s).

3. (a) rate   Comparing experiments 1 and 3, we find
   $= k[A][B]^2$
                     [B] = constant
                     [A] changes
                     rate = $k[A]^x[B]^y$

                     Since k and, in this case, $[B]^y$ are constants, the rate
                     is proportional to $[A]^x$. In experiment 3, [A] is 3 times
                     greater than in experiment 1. The rate in experiment 3
                     is also 3 times greater than that in experiment 1. The
                     rate is directly proportional to the change in [A]:

                     $x = 1$

                     and

                     rate = $k[A][B]^y$

Comparing experiments 2 and 3 we find

[A] = constant
[B] changes
rate = k[A][B]^y

Since k and, in this case, [A] are constants, the rate is proportional to [B]^y. In experiment 2, [B] is 3 times greater than that in experiment 3. The rate in experiment 2 is 9 times greater than that in experiment 3. The rate is directly proportional to the square of the change in [B]; thus,

$y = 2$

and

rate = k[A][B]^2

(b) 3   The reaction order is the sum of the exponents of the concentrations appearing in the equation. Thus,

reaction order = sum of exponents
$$= 1 + 2$$
$$= 3$$

(c) 0.038 mol^2/L^2·s)   Use any set of data in the equation

rate = k[A][B]^2

Using the data from experiment 1:

$$k = \frac{rate}{[A][B]^2}$$

$$= \frac{3.0 \times 10^{-4} \text{ /s}}{(0.20 \text{ mol/L})(0.20 \text{ mol/L})^2}$$

$$= 3.8 \times 10^{-2} \text{ mol}^2/(\text{L}^2 \cdot \text{s})$$

(d) The rate would be 8 times greater than that of the original.   The rate equation is

rate = k[A][B]^2

If the volume of the reaction vessel were halved, all concentrations would be doubled. For example, if the concentrations were originally 1 mol/L, they would be

1 mol/0.5 L, or 2 mol/L if the volume were halved. Thus, the rate of the reaction if the volume were halved would be 8 times greater than that of the original:

rate = k [A][B]^2

= k(2 mol/L)(2 mol/L)^2

= k8(mol/L)^3

(e) no

If this were a one-step process, the reaction as written would be the rate-determining step and the exponents of the concentrations of the reactants in the rate equation would be the same as the coefficients in the chemical reaction:

rate =k [A]^1[B]^1

4. (a) one-eighth of the original rate

The rate-determining step is

2A(g) + B(g) → C(g)

Thus,

rate = k[A]^2[B]

and initially:

rate = k[A]^2[B]

= k(2 mol/L)^{2}1(mol/L)

= k4(mol/L)^3

When half of each reactant has been consumed,

rate = k[A]^2[B]

= k(1 mol/L)^2(0.5 mol/L)

= k(0.5 mol/L^3) = k4(mol/L)^3 $\left(\dfrac{1}{8}\right)$

Thus, the rate when half of each reactant has been consumed is one-eighth of the initial rate.

(b) one-twenty-seventh of the initial rate

When two-thirds of A has been consumed, one-third of A remains $(1/3) \times 2 = (2/3)$. Likewise, one-third of B remains:

$$rate = k[A]^2[B]$$
$$= k(2/3 \text{ mol/L})^2(1/3 \text{ mol/L})$$
$$= k(4/9 \text{ mol/L}^2)(1/3 \text{ mol/L})$$
$$= k(4/27 \text{ mol/L}^3)$$
$$= k4(\text{mol/L})^3(1/27)$$

Thus, the rate when two-thirds of A has been consumed is one-twenty-seventh of the initial rate.

(c) The initial rate of the mixture of 2 mol of A and 2 mol of B is twice as great as that of the mixture of 2 mol of A and 1 mol of B.

The initial rate of a mixture of 2 mol of A and 2 mol of B in a liter container is

$$rate = k[A]^2[B]$$
$$= k(2 \text{ mol/L})^2(2 \text{ mol/L})$$
$$= k\ 8 \text{ mol/L}^3)$$
$$= k4(\text{mol L})^3(2)$$

Thus, the initial rate of a mixture of 2 mol of A and 2 mol of B in a liter container is 2 times as great as that of a mixture of 2 mol of A and 1 mol of B in a liter container

(d) The initial rate of the mixture of 4 mol of A and 2 mol of B is 8 times as great as that of the mixture of 2 mol of A and 1 mol of B.

The initial rate of a mixture of 4 mol of A and 2 mol of B in a liter container is

$$rate = k[A]^2[B]$$
$$= k(4 \text{ mol/L})^2(2 \text{ mol/L})$$
$$= k\ 32\ (\text{mol/L})^3$$
$$= k4(\text{mol L})^3(8)$$

Thus, the initial rate of a mixture of 4 mol of A and 2 mol of B in a liter container is 8 times as great as the initial rate of a mixture of 2 mol of A and 1 mol of B in a liter container.

5. 169.9 kJ/mol   The Arrhenius equation is

$$k = Ae^{-E_a/RT}$$

If we take the natural log of the Arrhenius equation, we find

$$\ln k = \ln A - \left(\frac{E_a}{RT}\right)$$

If we change the preceding equation to common logs, we find

$$2.303 \log k = 2.303 \log A - \left(\frac{E_a}{RT}\right)$$

$$\log k = \log A - \frac{E_a}{2.303RT}$$

$$\log k = \frac{-E_a}{2.303RT} + \log A$$

The preceding equation is the equation of a straight line, $y = mx + b$. The slope of a plot of $\log k$ vs. $1/T$ is $-E_a/2.303R$. Thus,

$$m = -\frac{E_a}{2.303R}$$

$$E_a = -2.303Rm$$

$$= (-2.303)[8.314 \text{ J}/(K \cdot mol)](-8880 \text{ K})$$

$$= 169.9 \text{ kJ}$$

6. (a) 3.8×10^{-3}
   mol/liter   Calculate the initial NO concentration by substituting into its rate equation:

rate of appearance of $NO_2 = k[NO]^2[O_2]$

$$5.0 \times 10^{-3} \text{ mol}/(L \cdot s) = 7.0 \times 10^3 L^2/(mol^2 \cdot s) [NO]^2 [0.050 \text{ mol/L}]$$

and rearranging

$$[NO] = \sqrt{\frac{5.0 \times 10^{-3} \text{ mol}/(L \cdot s)}{[7.0 \times 10^3 \text{ } L^2/(mol^2 \cdot s)] \cdot [0.050 \text{ mol/L}]}}$$

$$[NO] = 3.8 \times 10^{-3} \text{ mol/L}$$

(b) 4.9×10^{-2} mol/L The amount of NO consumed can easily be calculated.

$\frac{1}{2}(3.8 \times 10^{-3}$ mol/L$) = 1.9 \times 10^{-3}$ mol/L

The reaction stoichiometry

$$2NO(g) + O_2(g) \rightarrow 2NO_2(g)$$

shows that the amount of oxygen used is half of the amount of NO used. Therefore the amount of oxygen used can be calculated

$$\text{oxygen used} = \frac{1}{2}(1.9 \times 10^{-3} \text{ mol/L})$$

$$= 0.95 \times 10^{-3} \text{ mol/L}$$

The amount of oxygen remaining is the new concentration of oxygen

$$[O_2] = (5.0 \times 10^{-2} \text{ mol/L}) - (0.95 \times 10^{-3} \text{ mol/L})$$

$$= 4.9 \times 10^{-2} \text{ mol/L}$$

There is a very small change in the concentration of oxygen.

(c) rate of appearance of NO_2 = 1.2×10^{-3} mol/(L·s) Use the rate equation

rate of appearance of NO_2

$$= k[NO]^2[O_2]$$

$$= 7.0 \times 10^3 \text{ L}^2/(\text{mol}^2 \cdot \text{s}) [1.9 \times 10^{-3} \text{ mol/L}]^2 [4.9 \times 10^{-2} \text{ mol/L}]$$

$$= 1.2 \times 10^{-3} \text{ mol/(L} \cdot \text{s)}$$

(d) 4.8×10^{-2} mol/L The amount of oxygen used is $\frac{1}{2}$ of the amount of NO

$$\text{oxygen used} = \frac{1}{2}(9/10)(3.8 \times 10^{-3} \text{ mol/L})$$

$$= 1.7 \times 10^{-3} \text{ mol/L}$$

or the concentration of oxygen is

$$[O_2] = 5.0 \times 10^{-2} \text{ mol/L} - 1.7 \times 10^{-3} \text{ mol/L}$$

$$= 4.8 \times 10^{-2} \text{ mol/L}$$

Again there is a very small change in oxygen concentration because oxygen is very much in excess.

(e) 4.9×10^{-5} mol/(L·s)   Use the rate equation

rate of appearance of NO_2

$$= 7.0 \times 10^3 \ L^2/(mol^2 \cdot s) \ [3.8 \times 10^{-4} \ mol/L]^2 [4.8 \times 10^{-2} \ mol/L]$$

$$= 4.9 \times 10^{-5} \ mol/(L \cdot s)$$

The rate of the reaction is approximately one hundred times slower than the original rate. Since oxygen is in great excess, only the change in the concentration of NO greatly affects the reaction rate.

7. 1.8×10^{-3} mol/L   If 1.0×10^{-3} mol/L O_2 is consumed, twice as much NO is consumed or 2.0×10^{-3} mol/L. Thus

$$[NO] = (3.8 \times 10^{-3}) - (2.0 \times 10^{-3}) = 1.8 \times 10^{-3} \ mol/liter$$

8. 1.0×10^5 J/mol   Using the relationship described by Arrhenius and substituting the problem is easily solved.

$$E_a = 2.303R \left(\frac{T_1 T_2}{T_2 - T_1}\right) \log\left(\frac{k_2}{k_1}\right)$$

$$= 2.303 \, (8.31 \ J/(mol \cdot K)) \left(\frac{(300 \ K)(450 \ K)}{450 \ K - 300 \ K}\right) \times$$

$$\log\left(\frac{3.7 \times 10^{-3} \ L/(mol \cdot s)}{4.8 \times 10^{-9} \ L/(mol \cdot s)}\right)$$

$$= 19.1 \ J/(mol \cdot K) \, (900 \ K)(\log 7.7 \times 10^5)$$

$$= (1.7 \times 10^4 \ J/mol)(5.89) = 1.0 \times 10^5 \ J/mol \ \text{or} \ 100 \ kJ/mol$$

9. 1.1×10^5 J/mol   Use the equation to find energy of activation and let x equal the original rate.

$$E_a = 2.303R \left(\frac{T_1 T_2}{T_2 - T_1}\right) \log\left(\frac{k_2}{k_1}\right)$$

$$= 2.303 \, (8.31 \ J/(mol \cdot K)) \left(\frac{(350 \ K)(400 \ K)}{400 \ K - 350 \ K}\right) \log\left(\frac{100 \ x}{x}\right)$$

$$= (19.1 \ J/(mol \cdot K)) \, (2800 \ K) \log 100$$

$$= (5.3 \times 10^4 \ J/mol)(2)$$

$$= 1.1 \times 10^5 \ J/mol$$

10. 1.4×10^5 J/mol

Solving in the same manner as problem 9 of this section:

$$E_a = 2.303R \left(\frac{T_1 T_2}{T_2 - T_1}\right) \log \left(\frac{k_2}{k_1}\right)$$

$$= 19.1 \ J/(mol \cdot K) \ (2800 \ K) \ \log 500$$

$$= (5.3 \times 10^4 \ J/mol) \ (2.70)$$

$$= 1.4 \times 10^5 \ J/mol$$

no

The energy of activation is higher: The reaction is not catalyzed.

III. Working with Experimental Data

1. see Figure 14.7

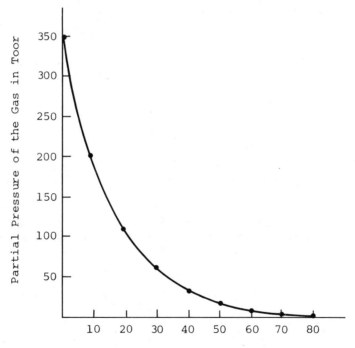

Time of the Reaction Proceeding in Minutes

FIGURE 14.7

2.   see Figure 14.8

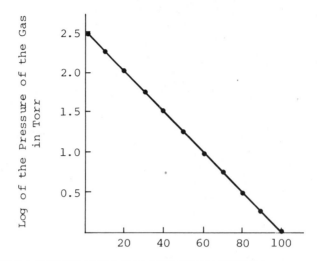

Time of the Reaction
Proceeding in Minutes

FIGURE 14.8

Attempts were made to obtain a straight-line plot by
first noting that a plot of the partial of the gas
versus time (Figure 14.7) is not linear.   Thus the
reaction is not zero-order.   Next a plot of the
logarithm of the pressure of the gas versus time
(Figure 14.8) is first-order plotted.   Since this gives
a straight-line graph, the reaction is first-order.

The slope of the line for a first order reaction is
related to the rate constant by the equation:

$m = -k/2.303$

From the graph the slope can be determined:

$$m = \frac{\Delta \log(y)}{\Delta x} = \frac{1.2}{-50 \text{ min}} = -0.024/\text{min}$$

Therefore

rate constant = $-0.024 = -k/2.303$

or

$k = 0.055/\text{min}.$

The half-life for the reaction is given by the equation

3. $t_{1/2} = 13$ min

$t_{1/2} = 0.693/k$

$= 0.693/(0.055/\text{min})$

$= 12.6$ min

4. no

SELF-TEST

I. Complete the test in 20 minutes. Answer each of the following:

1. The chemical equation for the reaction between oxalate ion and mercuric chloride in aqueous solution is

$$C_2O_4^{2-} \text{ (aq)} + 2HgCl_2 \text{ (aq)} \rightarrow 2CO_2 \text{ (g)} + Hg_2Cl_2 \text{ (s)} + 2Cl^- \text{ (aq)}$$

At a constant temperature, the following kinetic data were obtained:

Experiment	$[HgCl_2]$	$[C_2O_4^{2-}]$	Rate (Formation of CO_2)
1	0.080 M	0.200 M	1.5×10^{-5} M/min
2	0.080 M	0.400 M	6.0×10^{-5} M/min
3	0.040 M	0.400 M	3.0×10^{-5} M/min

Calculate the reaction order for $HgCl_2$, the reaction order for $C_2O_4^{2-}$, and the experimental rate constant.

2. The bromination of acetone in aqueous solution is given by the equation

$$H_3C-\overset{\overset{\displaystyle O}{\|}}{C}-CH_3(aq) + Br_2(aq) \rightarrow H_3C-\overset{\overset{\displaystyle O}{\|}}{C}-CH_2Br(aq) + HBr(aq)$$
acetone

and has an experimental rate law

rate = $k[\text{acetone}]^1$

Answer each of the following with *increases, decreases, or remains the same*:

(a) As the concentration of acetone is increased, the experimental rate constant _____.

(b) As the concentration of acetone is increased, the rate _____.

(c) As the concentration of Br_2 is increased, the rate _____.

(d) If a suitable positive catalyst is introduced, the overall activation energy for the reaction _____.

(e) If the temperature is increased, the rate constant for the rate-determining step _____.

3. For the reaction:

$$NO(g) + N_2O(g) \longrightarrow NO_2(g) + N_2(g)$$

which is first order in each of the reactants, the energy of activation is 2.09×10^5 J/mol. What temperature increase from an initial temperature of 25°C would cause the rate of the reaction to increase 100-fold?

CHEMICAL EQUILIBRIUM

CHAPTER 15

OBJECTIVES

I. You should be able to demonstrate your knowledge of the following terms by defining them, describing them, or giving specific examples of them:

chemical equilibrium [15.1]
endothermic [15.4]
equilibrium constant [15.1, 15.2, 15.3]
exothermic [15.4]
heterogeneous equilibrium [15.2]
homogeneous equilibrium [15.2]
LeChatelier's principle [15.4]
reaction quotient [15.2]
reversible reaction [15.2]

II. You should be able to use equilibrium constants and balanced chemical equations to determine equilibrium concentrations of all species participating in the chemical reaction.

III. You should be able to use Le Chatelier's principle to determine how an equilibrium is shifted due to a specific stress such as a changing pressure, temperature, or concentration.

UNITS,
SYMBOLS,
MATHEMATICS

I. The following symbols were used in this chapter; you should be familiar with them.

- K is the symbol for the equilibrium constant.
- K_p is the symbol for the equilibrium constant when pressures are used rather than concentrations.
- k_f is the symbol used to represent the rate constant of a reaction taking place in the forward direction, left to right.
- k_r is the symbol used to represent the rate constant of a reaction taking place in the reverse direction, right to left.
- kJ is the abbreviation for kilojoules.
- Δn is the symbol used to represent the difference between the number of moles of product gases and the number of moles of reactant gases in a balanced chemical equation.
- Q is the symbol for the reaction quotient which is used to predict the direction of a reaction.
- R is the symbol used to represent the gas constant in equations such as $K_p = K(RT)^{\Delta n}$.
- T is the symbol used to represent temperature in Kelvin in mathematical equations.

EXERCISES

I. Choose the best answer for each of the problems in this section. Avoid making any calculations.

1. Assume one mole of each of the reactants is placed in a 1 liter container and equilibrium is established. Which reaction would result in the lowest concentration of reactants?

   a. $NO_2(g) + NO_2(g) \rightleftharpoons N_2O_4(g)$ $K = 2.1 \times 10^2$ L/mol

   b. $PCl_3(g) + Cl_2(g) \rightleftharpoons PCl_5(g)$ $K = 11$ L/mol

   c. $Br(g) + Br(g) \rightleftharpoons Br_2(g)$ $K = 2.3 \times 10^6$ L/mol

   d. $CO(g) + Cl_2(g) \rightleftharpoons COCl_2(g)$ $K = 4.6 \times 10^9$ L/mol

2. Which of the molecules, O_2, N_2, Cl_2, or Br_2, is most dissociated at 1000 K? All equilibrium constants are given at 1000 K.

a. $O_2(g) \rightleftharpoons 2O(g)$ $K_p = 3.3 \times 10^{-20}$ atm

b. $N_2(g) \rightleftharpoons 2N(g)$ $K_p = 1.2 \times 10^{-31}$ atm

c. $Cl_2(g) \rightleftharpoons 2Cl(g)$ $K_p = 2.5 \times 10^{-7}$ atm

d. $Br_2(g) \rightleftharpoons 2Br(g)$ $K_p = 3.5 \times 10^{-5}$ atm

3. At 400 K the K is 7.0 for the reaction

   $$Br_2(g) + Cl_2(g) \rightleftharpoons 2BrCl(g)$$

   Initially 3.0 moles of each gas are mixed in a 1.0 L
   container at 400 K.   The reaction will proceed
   a. to the right
   b. to the left
   c. in neither direction (at equilibrium)

4. Which of the following equilibria will not be shifted
   when the pressure is increased by decreasing the
   volume of the container?

   a. $H_2(g) + CO_2(g) \rightleftharpoons H_2O(g) + CO(g)$
   b. $2C(s) + O_2(g) \rightleftharpoons 2CO(g)$
   c. $N_2O(g) + 4H_2(g) \rightleftharpoons 2NH_3(g) + H_2O(g)$
   d. $LaCl_3(s) + H_2O(g) \rightleftharpoons LaOCl(s) + 2HCl(g)$

5. Which of the equilibria described in problem 4 would
   be shifted to the right by adding $H_2O(g)$?

6. Assume one mole of each of the reactants is placed in
   a container sized so that the total initial pressure
   is one atm and that no temperature change occurs as
   the system goes to equilibrium. Which system would
   have a higher total pressure at equilibrium?

   a. $N_2O(g) + 4H_2O(g) \rightleftharpoons 2NH_3(g) + H_2O(g)$
   b. $2SO_2(g) + O_2(g) \rightleftharpoons 2SO_3(g)$
   c. $3Fe(s) + 4H_2O(g) \rightleftharpoons Fe_3O_4(s) + 4H_2(g)$
   d. $C(s) + CO_2(g) \rightleftharpoons 2CO(g)$

7. Which system described in problem 6 would not give any
   pressure change?

8. Which system described in problem 6 would have the same numerical value of K and K_p?

9. For the reaction

$$H_2(g) + I_2(g) \rightleftharpoons 2HI(g) + heat$$

the equilibrium constant at 700 K is 54.8.
The reaction producing HI is exothermic. Which of the following changes applied to the system will not shift the equilibrium?

a. increase the pressure
b. decrease the temperature
c. remove I_2
d. increase the temperature

10. For the same system described in problem 9, which change would cause more HI(g) to be formed?

11. For the system described in problem 9, which change would cause the equilibrium constant, K, to increase.

II. Answer each of the following questions.

1. Write the expression for the equilibrium constant, K, for each of the following reactions:

(a) $H_2(g) + I_2(g) \rightleftharpoons 2HI(g)$

(b) $PCl_5(g) \rightleftharpoons PCl_3(g) + Cl_2(g)$

(c) $N_2(g) + 3H_2(g) \rightleftharpoons 2NH_3(g)$

(d) $C(s) + CO_2(g) \rightleftharpoons 2CO(g)$

(e) $2NOBr(g) \rightleftharpoons 2NO(g) + Br_2(g)$

(f) $2Cl_2(g) + 2H_2O(g) \rightleftharpoons 4HCl(g) + O_2(g)$

(g) $2Hg(g) + O_2(g) \rightleftharpoons 2HgO(s)$

2. A 1.00-liter vessel initially contains 2.01 mol of N_2 and 2.08 mol of H_2 at 500°C. After the system reaches equilibrium, 0.50 mol of NH_3 has formed. Answer the following:

(a) What are the equilibrium concentrations of N_2 and H_2?

(b) What is the equilibrium constant, K, at 500°C?

(c) What is K_p at 500°C?

3. For the equilibrium

$$A(g) + B(g) \rightleftharpoons C(g)$$

K is 1.00×10^{-3} liter/mol at 100°C and 1.00×10^{-7} L/mol at 200°C. Is the reaction producing C endothermic or exothermic?

4. At temperature T_1 and a total pressure of 1.00 atm, N_2O_4 is 20.0 percent dissociated:

$$N_2O_4(g) \rightleftharpoons 2NO_2(g)$$

Assume that 1.00 mol of N_2O_4 is present initially and answer the following:

(a) How many moles of $N_2O_4(g)$ and $NO_2(g)$ are present at equilibrium?

(b) What is the total number of moles of gas present at equilibrium?

(c) What are the equilibrium pressures of $N_2O_4(g)$ and $NO_2(g)$?

(d) What is the value of K_p at temperature T_1?

5. At 100°C the equilibrium constant, K, for the reaction

$$CO(g) + Cl_2(g) \rightleftharpoons COCl_2(g)$$

is 4.57×10^9 L/mol. If 1.00 mol of $COCl_2$ is placed in a 1.00-liter container, what is the concentration of CO after equilibrium has been established?

6. Exactly one mole of $CO(g)$ and two moles of $Cl_2(g)$ are mixed and placed in a 1.00-liter container at 100°C. The reaction is

$$CO(g) + Cl_2(g) \rightleftharpoons COCl_2(g)$$

and the equilibrium constant at 100°C is 4.57×10^9 liter/mol. What is the concentration of $COCl_2$ at equilibrium? What is the concentration of CO and Cl_2?

7. At 400°C equal amounts of H_2 and I_2 are mixed in a 1.00-liter container. The reaction

$$H_2(g) + I_2(g) \rightleftharpoons 2HI(g)$$

proceeds until 95% of the H_2 is consumed. What is the equilibrium constant for the system? How much does the pressure change during the reaction if no temperature change occurs?

8. Pure SO_3 is placed in a 1.00-liter container and heated to 800°C. Gas-law calculations indicate that the

pressure of the SO_3 should be 6.00 atm, but the following equilibrium is established

$$2SO_3(g) \rightleftharpoons 2SO_2(g) + O_2(g)$$

and the actual equilibrium pressure of SO_3 is only 2.70 atm. What are the pressures of SO_2 and O_2 at equilibrium? What is the value of K_p? What is the total pressure at equilibrium if only SO_3, SO_2, and O_2 are present?

9. For the reaction described in problem 8, determine the value of K. What are the concentrations of SO_3, SO_2, and O_2 in mol/L?

10. For the equilibrium

$$C(s) + CO_2(g) \rightleftharpoons 2CO(g)$$

at 1000°C, the value of K_p is 167.5 atm. In a system the total pressure is 4.20 atm and the pressure of $CO_2(g)$ is 0.100 atm. What is the equilibrium pressure of CO?

11. For the equilibrium

$$PCl_5(g) \rightleftharpoons PCl_3(g) + Cl_2(g)$$

K_p is 2.25 atm. If PCl_3 and Cl_2 are added to an empty container at initial pressures of 0.50 atm each, what are the equilibrium pressures of PCl_5, PCl_3, and Cl_2?

12. If the volume of the container were doubled after the equilibrium described in problem 11 was established, what would be the new equilibrium pressures of PCl_3, PCl_5, and Cl_2?

13. What is the concentration of HI at equilibrium at 357°C if 6.22 mol and H_2 and 5.71 mol I_2 are in a 1.00-liter container and K for the reaction

$$H_2(g) + I_2(g) \rightleftharpoons 2HI(g)$$

is 71.3?

14. The equilibrium constant, K_p, for the reaction

$$C_4H_{10}(g) \rightleftharpoons 2H_2(g) + C_4H_6(g)$$

is 1.0×10^{-6} atm at 600°C. If 2.0 mol of H_2 and 1.0 mol of C_4H_{10} are put into a 1.00-L flask at 600°C, how many moles of C_4H_6 will be formed?

15. Consider the equilibrium

$$2SO_2(g) + O_2(g) \rightleftharpoons 2SO_3(g) + \text{heat}$$

What effect would each of the following stresses have on the reaction?
(a) addition of SO_2 at constant V and T
(b) reduction of volume at constant temperature
(c) increase in temperature

16. At 400°C equilibrium for the reaction

$$27 \text{ kcal} + 2Cl_2(g) + 2H_2O(g) \rightleftharpoons 4HCl(g) + O_2(g)$$

is established. What would be the effect on the number
of moles of $Cl_2(g)$ present if each of the following
changes were made?
(a) increase of temperature to 600°C
(b) addition of $O_2(g)$
(c) removal of $H_2O(g)$
(d) reduction of the volume of the container
(e) introduction of a catalyst

| ANSWERS TO EXERCISES | I. Multiple-choice problems relating to equilibria. |

1. (d)   Each answer contains two reactants and one product. Thus
           the equilibrium constant expressions are of the same form

           $$A + B \rightleftharpoons C$$

           $$K = \frac{[C]}{[A][B]}$$

           The equilibrium which has the largest equilibrium constant
           will have the largest numerator and the smallest denominator,
           i.e., smallest reactant concentrations.

2. (d)   Again, the largest equilibrium constant indicates the
           largest shift to the right, i.e., the most dissociation.

3. (a) Q is $(3.0)^2/[(3.0)(3.0)]$ or 1.0. Since $Q < K$, the
           reaction proceeds to the right.

4. (a)   This is the only equilibrium with the same number of moles
           of gaseous products as gaseous reactants.

5. (d)   Adding H_2O to the left would shift the equilibrium to the
           right.

6. (d)   This is the only equilibrium with more moles of gaseous
           products than gaseous reactants.

7. (c)   The number of moles of gaseous products equals the
           number of moles of gaseous reactants.

8. (c)   Since Δn is zero

$$K_p = K(RT)^{\Delta n}$$
$$K_p = K(RT)^0$$

and since any number raised to the zeroth power is one:

$$K_p = K$$

9. (a)

10. (b)   Exothermic reactions are shifted to the right by decreasing temperatures.

11. (b)   Larger equilibrium constants yield more product.

II. Answers to problems

1.

(a) $K = \dfrac{[HI]^2}{[H_2][I_2]}$ The exponent of each species is equal to the coefficient of that species in the balanced equation of the chemical reaction.

(b) $K = \dfrac{[PCl_3][Cl_2]}{[PCl_5]}$

(c) $K = \dfrac{[NH_3]^2}{[N_2][H_2]^3}$

(d) $K = \dfrac{[CO]^2}{[CO_2]}$ The concentrations of solids at constant temperature are constant and are included in the value of K; therefore, they are not written in the expression for the equilibrium constant.

(e) $K = \dfrac{[NO]^2[Br_2]}{[NOBr)^2}$

(f) $K = \dfrac{[HCl]^4[O_2]}{[Cl_2]^2[H_2O]^2}$

(g) $K = \dfrac{1}{[Hg]^2[O_2]}$

2. (a) $[N_2]$ = 1.76 mol/L

    $[H_2]$ = 1.33 mol/L

At equilibrium

$[NH_3]$ = 0.50 mol/1.00 L = 0.50 mol/L

The balanced equation

$$N_2(g) + 3H_2(g) \rightleftharpoons 2NH_3(g)$$

indicates that for every 2 mol of NH_3 formed, 1 mol of N_2 is lost. Thus, at equilibrium

$$[N_2] = \frac{2.01 \text{ mol } N_2 - (0.50 \text{ mol } NH_3 \text{ formed})\left(\dfrac{1 \text{ mol } N_2 \text{ lost}}{2 \text{ mol } NH_3 \text{ formed}}\right)}{1.00 \text{ L sol'n}}$$

$$= 1.76 \text{ mol/L}$$

Similarly, for every 2 mol of NH_3 formed, 3 mol of H_2 is lost:

$$[H_2] = \frac{2.08 \text{ mol } H_2 - (0.50 \text{ mol } NH_3 \text{ formed})\left(\dfrac{3 \text{ mol } H_2 \text{ lost}}{2 \text{ mol } NH_3 \text{ formed}}\right)}{1.00 \text{ L sol'n}}$$

$$= 1.33 \text{ mol/L}$$

(b) K = 6.0×10^2 L^2/mol^2

Substitute concentrations into the expression for the equilibrium constant:

$$K = \frac{[NH_3]^2}{[N_2][H_2]^3}$$

$$= \frac{(0.50)^2}{(1.76 \text{ mol/liter})(1.33 \text{ mol/liter})^3}$$

$$= 6.0 \times 10^{-2} \text{ } L^2/mol^2$$

(c) K_p = 1.5×10^{-5} $/atm^2$

The relationship between K_p and K is

$$K_p = K(RT)^{\Delta n}$$

For the reaction,

$$\Delta n = -2$$

and

$$K_p = K(RT)^{\Delta n}$$

$$= (6.0 \times 10^{-2} \ \text{L}^2/\text{mol}^2)$$

$$\times \ [(0.0821 \ \text{L} \cdot \text{atm}/(\text{K} \cdot \text{mol}))(773 \ \text{K})]^{-2}$$

$$= 1.5 \times 10^{-5}/\text{atm}^2$$

3. exothermic The expression for the equilibrium constant is

$$K = \frac{[C]}{[A][B]}$$

The equilibrium constant, K, decreases with increasing temperature, a fact that implies that the numerator of the equilibrium expression is becoming smaller. Thus, the equilibrium is shifting to the left. Since an increase in temperature shifts the equilibrium to the left, heat must appear on the right side; heat is released, and the reaction is exothermic.

4. (a) 0.80 mol N_2O_4

0.40 mol NO_2

The N_2O_4 is 20.0 percent dissociated; therefore, 0.80 mol of N_2O_4 remains:

$$? \ \text{mol} \ N_2O_4 \ \text{remaining} = 1.00 \ \text{mol} \ N_2O_4$$

$$- \ 0.200(1.00) \ \text{mol} \ N_2O_4$$

$$= 0.80 \ \text{mol} \ N_2O_4$$

Since 0.20 mol of N_2O_4 is dissociated, we know from the balanced equation that 0.40 mol of NO_2 is formed:

$$? \ \text{mol} \ NO_2 \ \text{formed} = 0.20 \ \text{mol} \ N_2O_4 \ \text{lost} \left(\frac{2 \ \text{mol} \ NO_2 \ \text{formed}}{1 \ \text{mol} \ N_2O_4 \ \text{lost}} \right)$$

$$= 0.40 \ \text{mol} \ NO_2$$

(b) 1.20 mol
gas

Calculate the total number of moles of gas present at equilibrium:

$$\text{moles of gas} = 0.80 \text{ mol } N_2O_4 + 0.40 \text{ mol } NO_2$$
$$= 1.20 \text{ mol gas}$$

(c) $p_{NO_2} = 0.33$ atm

Determine the partial pressure of each gas:

$$p_{N_2O_4} = \left(\frac{n_{N_2O_4}}{n_{N_2O_4} + n_{NO_2}}\right)p$$

$$= \left(\frac{0.80 \text{ mol}}{0.80 \text{ mol} + 0.40 \text{ mol}}\right)1.00 \text{ atm}$$

$$= 0.67 \text{ atm}$$

$$p_{NO_2} = \left(\frac{n_{NO_2}}{n_{N_2O_4} + n_{NO_2}}\right)p$$

$$= \left(\frac{0.40 \text{ atm}}{0.80 \text{ atm} + 0.40 \text{ atm}}\right)1.00 \text{ atm}$$

$$= 0.33 \text{ atm}$$

(d) 1.6×10^{-1} atm

Substituting partial pressures into the expression for K_p

$$K_p = \frac{(p_{NO_2})^2}{p_{N_2O_4}}$$

$$= \frac{(0.333 \text{ atm})^2}{(0.67 \text{ atm})}$$

$$= 0.16 \text{ atm}$$

5. [CO] = 1.48×10^{-5} mol/L

At 100°C the equilibrium constant is

$$K = 4.57 \times 10^9 \text{ L/mol} = \frac{[COCl_2]}{[CO][Cl_2]}$$

We make the following assumptions:

Species	Original Concentration	Equilibrium Concentration
CO	0	(x) mol/L
Cl_2	0	(x) mol/L
$COCl_2$	1.00	$(1.00 - x = 1.00)$ mol/L

Substitute concentrations into the expression for the equilibrium constant:

$$4.57 \times 10^9 \text{ L/mol} = \frac{[COCl_2]}{[CO][Cl_2]}$$

$$4.57 \times 10^9 \text{ L/mol} = \frac{1.00}{x^2}$$

$$x^2 = 2.19 \times 10^{-10} \text{ mol}^2/\text{L}^2$$

$$x = 1.48 \times 10^{-5} \text{ mol/L}$$

Note that 1.48×10^{-5} mol/L is negligible compared with 1.00 mol/L.

6. $[COCl_2] = 1.00 \text{ mol/L}$

$[CO] = 2.19 \times 10^{-10} \text{ mol/L}$

$[Cl_2] = 1.00 \text{ mol/L}$

For the equilibrium

$$CO(g) + Cl_2(g) \rightleftharpoons COCl_2(g)$$

the equilibrium constant is very large, 4.57×10^9 L/mol. Therefore it is safe to assume that almost no CO remains after equilibrium is established. Let x equal the small amount of CO which does not react:

Species	Original Concentration	Equilibrium Concentration
CO	1.00 mol/L	x mol/L
Cl_2	2.00 mol/L	$(1.00 + x)$ mol/L
$COCl_2$	0	$(1.00 - x)$ mol/L

Write the equilibrium expression

$$K = \frac{[COCl_2]}{[CO][Cl_2]}$$

then substitute values for K and all equilibrium concentrations:

$$4.57 \times 10^9 \text{ L/mol} = \frac{(1.00 - x) \text{ mol/L}}{(x \text{ mol/liter})(1.00 + x) \text{ mol/L}}$$

Since x is very small, it can be dropped wherever it is added or subtracted from 1.00. Units also cancel to give:

$$4.57 \times 10^9 = \frac{1.00}{x}$$

$$x = 2.19 \times 10^{-10} \text{ mol/L}$$

7. $K =$
   1.4×10^3
   no change
   in p_{total}

For the reaction

$$H_2(g) + I_2(g) \rightleftharpoons 2HI(g)$$

determine the equilibrium concentrations of all species:

Species	Original Concentration	Equilibrium Concentration
H_2	x mol/liter	$(x - 0.95x)$ mol/L
I_2	x mol/liter	$(x - 0.95x)$ mol/L
HI	0	$2(0.95x)$ mol/L

Notice that the same 95% decrease in the concentration of I_2 occurs because of the 1:1 stoichiometry, but since each H_2 consumed produces 2 HI molecules, the corresponding increase in HI is twice the decrease in H_2 and I_2. Substituting values into the equilibrium constant expression gives:

$$K = \frac{[HI]^2}{[H_2][I_2]} = \frac{(1.90x)^2}{(0.05x)(0.05x)}$$

$$= 1.4 \times 10^3$$

There is no pressure change because the number of moles of gaseous product equals the number of moles of gaseous reactant.

8. $p_{SO_2} =$
   3.30 atm

   $p_{O_2} =$
   1.65 atm

   $K_p =$
   2.46 atm

   $p_{total} =$
   7.65 atm

For the reaction

$$2SO_3(g) \rightleftharpoons 2SO_2(g) + O_2(g)$$

each mole of SO_3 which dissociates gives one mole of SO_2 and $\frac{1}{2}$ mole of O_2. All other conditions being equal, the number of moles of each gas is proportional to the partial pressure of the gas, $PV=nRT$. The equilibrium pressures can be determined:

Species	Original Pressure	Equilibrium Pressure
SO_3	6.00 atm	2.70 atm
SO_2	0	$(6.00-2.70)$ atm = 3.30 atm
O_2	0	$\frac{1}{2}(6.00-2.70)$ atm = 1.65 atm

The moles of SO_2 produced equals the moles of SO_3 used; thus, the pressure of SO_2 equals the decrease in pressure of SO_3. The partial pressure of O_2 is half that of SO_2 because half as much is produced.

The value of K_p can be calculated from the equilibrium constant expression:

$$K_p = \frac{(p_{SO_2})^2 (p_{O_2})}{(p_{SO_3})^2}$$

$$= \frac{(3.30\ atm)^2 (1.65\ atm)}{(2.70\ atm)^2}$$

$$= 2.46\ atm$$

The total pressure is the sum of the partial pressures of each gas:

$$p_{total} = p_{SO_3} + p_{SO_2} + p_{O_2}$$

$$= 2.70\ atm + 3.30\ atm + 1.65\ atm$$

$$= 7.65\ atm$$

9. $K = 0.0279$ mol/L

Remember that K and K_p are related by the equation derived in section 15.3 of your text:

$$K_p = K(RT)^{\Delta n}$$

and that Δn equals the total number of moles of gaseous products minus the total number of moles of gaseous reactants as deduced from the coefficients in the properly balanced equation. Here Δn equals 3 minus 2, or 1. Thus

$$2.46\ atm = K[0.0821\ L\cdot atm/(K\cdot mol)](1073\ K)$$

or

$$K = \frac{2.46\ atm}{88.1\ L\cdot atm/mol}$$

$$= 0.0279\ mol/L$$

$[SO_3]$ = 0.0306 mol/L

$[SO_2]$ = 0.0375 mol/L

$[O_2]$ = 0.0187 mol/L

The concentration of each species can be determined by using the ideal-gas law. For SO_3 calculate the number of moles in 1 liter:

$$PV = nRT$$

$$(2.70 \text{ atm})(1 \text{ L}) = n[0.0821 \text{ L·atm/(K·mol)}](1073 \text{ K})$$

$$n = 0.0306 \text{ mol}$$

or since the volume is one liter

$$[SO_3] = 0.0306 \text{ mol/L}$$

Similar calculations give the other concentrations.

10.

p_{CO_2} = 4.09 atm

Substitute values into the equilibrium constant expression:

$$K_p = \frac{(p_{CO})^2}{p_{CO_2}}$$

$$167.5 \text{ atm} = \frac{(p_{CO})^2}{0.100 \text{ atm}}$$

$$p_{CO} = 4.09 \text{ atm}$$

11.

p_{PCl_5} = 0.079 atm

p_{PCl_3} = 0.42 atm

p_{Cl_2} = 0.42 atm

Assume that x represents the partial pressure of the PCl_5 which is formed, then the partial pressure of both PCl_3 and Cl_2 are reduced by the same amount:

Species	Original Pressure	Equilibrium Pressure
PCl_5	0 atm	x atm
PCl_3	0.50 atm	$(0.50 - x)$ atm
Cl_2	0.50 atm	$(0.50 - x)$ atm

Substitute into the equilibrium constant expression:

$$K_p = \frac{(p_{PCl_3})(p_{Cl_2})}{(p_{PCl_5})}$$

$$2.25 \text{ atm} = \frac{(0.50 - x) \text{ atm} \, (0.50 - x) \text{ atm}}{x \text{ atm}}$$

Rearranging

$$x^2 - 3.25x + 0.25 = 0$$

and solving by using the quadratic formula

$$x = \frac{-b \pm \sqrt{b^2 - 4ac}}{2a}$$

$x = +0.079$ atm or 3.17 atm.

Since a maximum of 0.50 atm of PCl_5 can be produced, the answer 3.17 is not reasonable. Thus,

$$p_{PCl_5} = x = 0.079 \text{ atm}$$

$$p_{PCl_3} = (0.50 - x) \text{ atm} = 0.42 \text{ atm}$$

$$p_{Cl_2} = (0.50 - x) \text{ atm} = 0.42 \text{ atm}$$

12.

$p_{PCl_5} = 0.023$ atm

$p_{PCl_3} = 0.22$ atm

$p_{Cl_2} = 0.22$ atm

If the container volume were doubled, the new initial pressures, i.e., those before any shift in the equilibrium occurs, would be half the values computed in problem 11. Since the equilibrium shifts to the right, assume the pressure of Cl_2 increases by x. Then the pressure of PCl_5 decreases by x:

Species	Initial Pressure	Equilibrium Pressure
PCl_5	$\dfrac{0.079 \text{ atm}}{2}$	$(0.0395 - x)$ atm
PCl_3	$\dfrac{0.42 \text{ atm}}{2}$	$(0.21 + x)$ atm
Cl_2	$\dfrac{0.42 \text{ atm}}{2}$	$(0.21 + x)$ atm

Solve by substituting into the equilibrium constant expression

$$2.25 \text{ atm} = \frac{(0.21 + x) \text{ atm} (0.21 + x) \text{ atm}}{(0.0395 - x) \text{ atm}}$$

Use the quadratic formula to solve for x and find

$x = 0.017$ or -2.68

Only the 0.017 value is possible. Use it to determine all partial pressures.

13. [HI] = 9.55 mol/L

The expression for the equilibrium constant is

$$K = \frac{[HI]^2}{[H_2][I_2]}$$

From the balanced equation we know that for every 2 mol of HI formed, 1 mol of H_2 and 1 mol of I_2 are lost, or for every 1 mol of HI formed, $\frac{1}{2}$ mol of H_2 and $\frac{1}{2}$ mol of I_2 are lost. Therefore, we make the following assumptions:

Species	Original Concentration	Equilibrium Concentration
HI	0	(x) mol/L
H_2	6.22 mol/L	(6.22 - $x/2$) mol/L
I_2	5.71 mol/L	(5.71 - $x/2$) mol/L

Substitute concentrations into the expression for the equilibrium constant and use the quadratic equation

$$x = \frac{-b \pm \sqrt{b^2 - 4ac}}{2a}$$

to solve for x:

$$K = 71.3 = \frac{[HI]^2}{[H_2][I_2]}$$

$$= \frac{x^2}{(6.22 - x/2)(5.71 - x/2)}$$

$$= \frac{x^2}{35.5 - 2.86x - 3.11x + x^2/4}$$

$$x^2 = 2531 - 426x + 17.8x^2$$

$$0 = 16.8x^2 - 426x + 2531$$

$$x = \frac{-(-426) \pm \sqrt{(-426)^2 - 4(16.8)(2531)}}{2(16.8)}$$

$$= 9.55 \text{ or } 15.8$$

Since the concentration of H_2 and that of I_2 originally present are smaller than $x/2$ when $x = 15.8$, the solution is $x = 9.55$. Therefore

HI = 9.55 mol/L

14. 4.8×10^{-11} mol C_4H_6

The change in the number of moles, Δn, is +2. Therefore,

$$K = K_p (RT)^{-\Delta n}$$
$$= (1.0 \times 10^6 \text{ atm}^2)\{[0.0821 \text{ L·atm}/(K \cdot mol)](873 \text{ K})\}^{-2}$$
$$= 1.9 \times 10^{-10} \text{ mol}^2/L^2$$

The equilibrium expression is

$$K = \frac{[H_2]^2 [C_4H_6]}{[C_4H_{10}]}$$

From the reaction equation we know that for every 1 mol of C_4H_6 formed, 2 mol of H_2 are formed. Also, for every 1 mol of C_4H_6 formed, 1 mol of C_4H_{10} is lost. Therefore, we make the following assumptions:

Species	Original Concentration	Equilibrium Concentration
C_4H_6	0	(x) mol/L
H_2	2.0 mol/L	$(2.0 + 2x = 2.0)$ mol/L
C_4H_{10}	1.0 mol/L	$(1.0 - x = 1.0)$ mol/L

Substitute concentrations into the expression for the equilibrium constant:

$$K = 1.9 \times 10^{-10} \text{ mol}^2/L^2 = \frac{(2.0 \text{ mol/L})^2 (x)}{(1.0 \text{ mol/L})}$$

$$x = \frac{1.9 \times 10^{-10} \text{ mol/L}}{4.0 \text{ mol/L}}$$

$$x = 4.8 \times 10^{-11} \text{ mol/L}$$

15. (a) shift to   A shift to the right occurs since the stress is on the
        right   left.

    (b) shift to   A shift to the right occurs since the stress is on the
        right   left. The 2 moles of gas on the right would occupy less
                     volume than the 3 moles of gas on the left.

    (c) shift to   A shift to the left occurs since the stress is on the right.
        left   The reaction is exothermic; thus, the shift is toward the
                     left when the temperature is increased.

16. (a) decrease   The number of moles of Cl_2 would decrease. The reaction
                     is endothermic, and the increase in temperature would shift
                     the reaction to the side without heat.

    (b) increase   The number of moles of Cl_2 would increase. An increase in
                     the number of moles of O_2 would shift the reaction away
                     from the side with O_2.

    (c) increase   The number of moles of Cl_2 would increase. The removal of
                     H_2O would cause a shift to the left.

    (d) increase   The number of moles of Cl_2 would increase. There are
                     fewer moles of gas on the left side than on the right.
                     Thus, to alleviate the stress of volume reduction, the
                     equilibrium would shift to the side that contains fewer
                     molecules since such molecules collectively would occupy
                     less volume.

    (e) no change   There would be no change in the number of moles of Cl_2
                     since a catalyst would have no effect on the position of
                     equilibrium.

SELF-TEST   Complete this test in 30 minutes:

                 1. The reaction of gaseous chlorine and water

                    $$2Cl_2(g) + 2H_2O(g) \rightleftharpoons 4HCl(g) + O_2(g)$$

                    is endothermic. The four reactant gases are mixed in
                    a reaction vessel and allowed to attain an equilibrium
                    state. Indicate the effect (i.e., *increase, decrease,*
                    or *no change*) of the following operations described in
                    the left column on the equilibrium value of the quantity
                    in the right column. Each operation is to be considered
                    separately. Temperature and volume are constant except
                    when the contrary is indicated.

	Operation	Quantity
_____	(a) increase in volume of container	number of moles of HCl
_____	(b) increase in volume of container	K_p
_____	(c) addition of O_2 (g)	K_p
_____	(d) addition of O_2 (g)	number of moles of O_2
_____	(e) increase in temperature	K_p

2. The decomposition of $CaSO_3$ (s)

$$CaSO_3 (s) \rightleftharpoons CaO (s) + SO_2 (g)$$

is endothermic. At equilibrium there are significant quantities of all three compounds in the reaction vessel. Suggest two ways in which the yield of SO_2 (g) might be appreciably increased.

3. At 269.9 K nitric oxide and bromine are mixed in a vessel. Initially the partial pressures are

$$p_{NO} = 0.129 \text{ atm}$$

$$p_{Br_2} = 0.0543 \text{ atm}$$

The following equilibrium is established:

$$2NO (g) + Br_2 (g) \rightleftharpoons 2NOBr (g)$$

The total pressure at equilibrium is 0.145 atm. Calculate K_p.

4. At temperature T 1.00 mol of NOCl is sealed in a 1.00-liter container. At equilibrium 20.0 percent of the NOCl is dissociated:

$$2NOCl (g) \rightleftharpoons 2NO (g) + Cl_2 (g)$$

What is the value of the equilibrium constant? Use molar concentrations.

THEORIES OF ACIDS AND BASES

CHAPTER

16

OBJECTIVES

I. You should be able to demonstrate your knowledge of the following terms by defining them, describing them, or giving specific examples of them:

acid anhydride [16.1]
amphiprotic [16.2]
Arrhenius concept [16.1]
base anhydride [16.1]
Bronsted-Lowry concept [16.2]
conjugate pair [16.2]
electrophilic [16.5]
hydride [16.4]
leveling effect [16.3]
Lewis concept [16.5]
neutralization [16.1, 16.6]
nucleophilic [16.5]
oxyacid [16.4]
solvent systems [16.5]

II. You should be able to identify acids and bases in given reactions and deduce relative strengths of each from displacement of equilibrium and/or from a qualitative knowledge of factors that influence the strength of each.

EXERCISES

I. Fill in each of the statements with the appropriate entry or entries from the following list. An entry may be used more than once.

(a) ammonium ion
(b) chloride ion
(c) hydrofluoric acid
(d) hydroiodic acid

(e) sodium acetate
(f) water
(g) zinc(II) ion

1. In glacial acetic acid, i.e., pure acetic acid, _____ would be a strong base.

2. The strongest acid that can exist in liquid ammonia is _____ .

3. _____ is the weakest hydro acid of the elements of group VII A.

4. _____ are species that can act as Brønsted acids in water.

5. _____ are species that can act as Brønsted bases in water.

6. _____ is a species that can act as a Lewis acid.

7. _____ are species that can act as Lewis bases.

8. The strongest electrophilic species is _____ .

9. The conjugate acid of ammonia is _____ .

10. _____ is an amphiprotic material.

11. If 0.1 mol of _____ were added to 1 liter of water, a basic solution would be formed. Assume that the species are added as chloride salts of cations and sodium salts of anions.

12. If 0.1 mol of _____ were added to 1 liter of water, a neutral solution would be formed. Assume that the species are added as chloride salts of cations and sodium salts of anions.

II. Answer each of the following. Assume that each equilibrium is displaced to the right.

1. The strongest Lewis acid in the reaction

$$HCl + H_2O \rightleftharpoons Cl^- + H_3O^+$$

is _____ .

2. The strongest Brønsted base in the reaction

$$H_2SO_4 + F^- \rightleftharpoons HSO_4^- + HF$$

is _____ .

3. The strongest Brønsted acid in the reaction

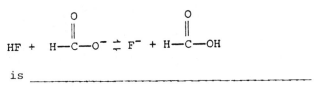

is _____

III. Use qualitative knowledge of factors that influence the strength of an acid to answer the following:

For each pair of compounds, determine which compound is the stronger acid:

1. HI and HF

2.
$$\begin{array}{ccccc} & F & O & & \\ & | & \| & & \\ F-&C-&C-&OH & \\ & | & & & \\ & F & & & \end{array} \quad \text{and} \quad \begin{array}{ccccc} & H & O & & \\ & | & \| & & \\ H-&C-&C-&OH & \\ & | & & & \\ & H & & & \end{array}$$

3.
$$\begin{array}{ccccc} & Cl & O & & \\ & | & \| & & \\ H-&C-&C-&OH & \\ & | & & & \\ & Cl & & & \end{array} \quad \text{and} \quad \begin{array}{ccccc} & Cl & O & & \\ & | & \| & & \\ Cl-&C-&C-&OH & \\ & | & & & \\ & Cl & & & \end{array}$$

4. $HClO_4$ and $HClO_3$

5. HOI and HOCl

IV. Use qualitative knowledge of factors that influence the strength of an acid or a base to answer the following. Assume that each equilibrium is displaced to the right.

1. List all Brønsted acids in the following reactions in order of decreasing strength:
   (a) $HOBr + CN^- \rightleftharpoons OBr^- + HCN$
   (b) $HCl + H_2O \rightleftharpoons H_3O^+ + Cl^-$
   (c) $HI + BrO_3^- \rightleftharpoons HBrO_3 + I^-$
   (d) $OH^- + HOBr \rightleftharpoons H_2O + OBr^-$
   (e) $H_3O^+ + IO_3^- \rightleftharpoons HIO_3 + H_2O$

2. Which acids in problem 1 above are strong acids?

3. List the Brønsted bases in problem 1 above in order of decreasing strength.

4. Identify the Lewis acid in the following reaction:

   $PCl_3 + Cl_2 \rightarrow PCl_5$

I. Acid-base systems

1. (e), sodium acetate

The acetate ion is the strongest base that can exist in acetic acid. The basicity of strongly basic materials in acetic acid is leveled to that of the acetate ion.

2. (a), ammonium ion

3. (c), hydrofluoric acid

There is an increase in acid strength of the hydro acids, HX, of the elements in any group of the periodic table with increasing atomic size of the electronegative element, X.

4.

A Brønsted acid is a species that can donate a proton:

(a), ammonium ion

$$NH_4^+ \quad + \quad OH^- \quad \rightleftharpoons \quad NH_3 \quad + \quad H_2O$$

| Brønsted acid | Brønsted base | Brønsted base | Brønsted acid |

(c), hydrofluoric acid

$$HF \quad + \quad H_2O \quad \rightleftharpoons \quad F^- \quad + \quad H_3O^+$$

| Brønsted acid | Brønsted base | Brønsted base | Brønsted acid |

(d), hydroiodic acid

$$HI \quad + \quad H_2O \quad \rightleftharpoons \quad I^- \quad + \quad H_3O^+$$

| Brønsted acid | Brønsted base | Brønsted base | Brønsted acid |

(f), water

$$H_2O \quad + \quad H^- \quad \rightleftharpoons \quad H_2 \quad + \quad OH^-$$

| Brønsted acid | Brønsted base | Brønsted base | Brønsted acid |

5.   A Brønsted base is a species that can accept a proton:

(e), sodium acetate

$$C_2H_5O_2^- \quad + \quad HCl \quad \rightleftharpoons \quad C_2H_5O_2H \quad + \quad Cl^-$$

| Brønsted base | Brønsted acid | Brønsted acid | Brønsted base |

(f), water

$$H_2O \quad + \quad HBr \quad \rightleftharpoons \quad H_3O^+ \quad + \quad Br^-$$

| Brønsted base | Brønsted acid | Brønsted acid | Brønsted base |

6. (g), Zn(II) ion   A Lewis acid *accepts* an electron pair to form a covalent bond.

7. (b), chloride ion   A Lewis base *donates* an electron pair to form a covalent bond.

8. (g), Zn(II) ion   The word *electrophilic* is used to describe the ability of a species to attract an electron pair.

9. (a), ammonium ion

$$H_2O \quad + \quad NH_3 \quad \rightleftharpoons \quad NH_4^+ \quad + \quad OH^-$$

$$acid_1 \qquad base_1 \rightleftharpoons acid_2 \qquad base_2$$

10. (f), water   Water can function as an acid and as a base:

(g), Zn^{2+}

$$H_2O + NH_2^- \rightleftharpoons NH_3 + OH^-$$

$$H_2O + HBr \rightleftharpoons Br^- + H_3O^+$$

11. (e), sodium acetate   Acetate ion hydrolyzes:

$$H_2O + C_2H_3O_2^- \rightleftharpoons HC_2H_3O_2 + OH^-$$

12. (b), chloride ion
    (f), water

II. Strengths of acids and bases determined from equilibrium displacement

1. HCl   HCl is the electron acceptor.

2. F⁻   The fluoride ion, F^-, is the proton acceptor.

3. HF   Hydrogen fluoride, HF, is the proton donor.

III. Strengths of acids and bases determined from qualitative knowledge of factors that influence strength.

1. HI   The proton is more easily removed from the hydride with the larger central atom.

2.   The electronegative F atoms have an electron withdrawing effect and the proton is more easily removed.

$$\begin{array}{c} \quad\ \ F\ \ \ O \\ \quad\ \ |\ \ \ \ || \\ F-C-C-OH \\ \quad\ \ | \\ \quad\ \ F \end{array}$$

3.   More electronegative chlorines result in increased acidity.

$$\begin{array}{c} \quad\ \ Cl\ \ O \\ \quad\ \ |\ \ \ || \\ Cl-C-C-OH \\ \quad\ \ | \\ \quad\ \ Cl \end{array}$$

4. $HClO_4$ The more O atoms bonded to an electronegative nonmetal, the more electron density is withdrawn from the H-O bond, and therefore the higher the acidity.

5. HOCl   The more electronegative Cl withdraws electron density from the H-O bond and the proton is allowed to dissociate more readily.

IV. Relative strengths of acids and bases determined from information on equilibrium displacement and qualitative knowledge of factors that influence strength

1. HI
   HCl

   The order of acid strength of the hydrogen halides is
   HI > HBr > HCl > HF

   H_3O^+

   Hydrochloric acid, HCl, is stronger than the hydronium ion, H_3O^+ (see reaction b).

   $HBrO_3$
   HIO_3

   Iodic acid, HIO_3, is weaker than the hydronium ion, H_3O^+ (see reaction e). Bromic acid, $HBrO_3$, is expected to be more acidic than iodic acid, HIO_3, and less acidic than hydroiodic acid, HI (see reaction c).

   HOBr

   Hydrobromous acid, HOBr, is stronger than HCN (see reaction a) and H_2O (see reaction d).

   HCN
   H_2O

2. HCl and HI

   Hydrochloric acid and hydrobromic acid are leveled by water:

   $$HCl + H_2O \rightleftharpoons Cl^- + H_3O^+$$
   $$HI \; + H_2O \rightleftharpoons I^- \; + H_3O^+$$

3. OH^-
   CN^-
   OBr^-

   IO_3^-
   BrO_3^-
   H_2O
   Cl^-
   I^-

   Increasing acid strength parallels decreasing base strength of the conjugate base. Thus, the conjugate base of the strongest acid is the weakest base. Bases weaker than water are not normally considered to be basic species in aqueous solution.

4. PCl_3

   Trichlorophosphine, PCl_3, is a Lewis acid and accepts an electron pair from each chlorine atom of Cl_2 to form PCl_5.

SELF-TEST

_____

Complete the test in 10 minutes:

1. The reaction of ammonia and the hydride ion goes to completion:

   $$NH_3 + H^- \rightarrow H_2 + NH_2^-$$

What is the strongest Brønsted acid in the reaction?
What is the strongest base?

2. What is the conjugate acid of HS^-?

3. The following reaction may be interpreted as an acid-base reaction in a SO_2 solvent system:

$$Cs_2SO_3 + SOCl_2 \rightarrow 2CsCl + 2SO_2$$

What is the acid?

4. What is the Lewis base in the following reaction?

$$BF_3 + F^- \rightarrow BF_4^-$$

5. What is the Lewis acid in the following reaction?

$$[:\ddot{\underset{..}{S}}:]^{2-} + \ddot{\underset{..}{S}}: \rightarrow S_2^{2-}$$

6. The following equilibrium is displaced to the right:

$$HC_2H_3O_2 + HS^- \rightleftharpoons H_2S + C_2H_3O_2^-$$

What is the strongest Brønsted base in the reaction?
What is the strongest Brønsted acid?

7. What is the conjugate base of $H_2PO_4^-$?

IONIC EQUILIBRIUM, PART I

CHAPTER

17

OBJECTIVES   I. You should be able to demonstrate your knowledge of the
                 following terms by defining them, describing them, or
                 giving specific examples of them:

acid-base titration [17.9]
acid dissociation constant, K_a [17.2]

base dissociation constant, K_b [17.1]

buffer [17.6]
common-ion effect [17.5]
degree of dissociation [17.1]
endpoint [17.9]
equivalence point [17.9]
Henderson-Hasselbalch equation [17.6]
hydrolysis [17.8]
hydronium ion [17.1]
indicator [17.4]
ionization constant [17.1]
percent ionization [17.1]
pH [17.3]
pK [17.6]
pOH [17.3]
polyprotic acid [17.7]
quadratic formula [17.1]
strong electrolyte [17.1]
titration curve [17.9]
water dissociation constant, K [17.2]
weak electrolyte [17.1]

II. You should be able to calculate the pH and the concentration of all species in a solution containing either strong acids or bases, weak acids or bases, buffers, polyprotic acids, cations of weak bases, or anions of weak acids.

III. You should be able to identify the type of problem you are being asked to solve:

If there is just an acid or a base in solution, it is a simple ionic equilibrium problem;

If there is an acid or base with its salt present, then it is a buffer problem;

If there is a salt in solution, then it is a hydrolysis problem.

IV. You should be able to predict the shape of titration curves.

UNITS,
SYMBOLS,
MATHEMATICS

I. There are several symbols you should be familiar with:

- α is the symbol for degree of dissociation.

- K_a is the symbol for the dissociation constant of a weak acid.

- K_b is the symbol for the dissociation constant of a weak base.

- K_w is the symbol for the dissociation constant of water.

- pH is equal to -log of the hydrogen ion concentration.

- pK is equal to -log K.

- pOH is equal to -log of the hydroxide ion concentration.

II. It is imperative that you be able to solve the quadratic equation. For an equation that contains the unknown to the second power as well as the unknown to the first power, rewrite it in the form:

$$ax^2 + bx + c = 0$$

Here the values of a, b, and c are constants that can be either positive, negative, or zero, and x represents the unknown. The values of x, which are proper solutions of the equation, are found by solving

$$x = \frac{-b \pm \sqrt{b^2 - 4ac}}{2a}$$

See example 17.2 of your text. You should review the use of logarithms in the appendix of your text.

III. Remember that pH values have no units and since pH is a logarithm, the number of significant figures is determined by the mantissa, that is the portion of the number after the decimal. The characteristic, that is the portion of the number before the decimal, indicates the position of the decimal in the $[H^+]$. All of the following pH values have two significant figures: 1.57, 8.93, 6.00, 7.05, 12.40, 0.10.

EXERCISES   I. Answer each of the following questions without performing
              any calculations. This exercise will increase your quali-
              tative understanding of equilibria. You may refer to
              equilibrium constant data if necessary.

_____ 1. The pH of pure water at 25°C is closest to
              (a) 6.0   (d) 7.5
              (b) 6.5   (e) 8.0
              (c) 7.0

_____ 2. The pH of 10^{-6} M HCl is closest to
              (a) 6.0   (d) 7.5
              (b) 6.5   (e) 8.0
              (c) 7.0

_____ 3. The pH of a 10^{-8} M HCl solution is closest to
              (a) 6.0   (d) 7.5
              (b) 6.5   (e) 8.0
              (c) 7.0

_____ 4. The pH of a 10^{-3} M NaOH solution is closest to
              (a) 3.0   (c) 9.0
              (b) 5.0   (d) 11.0

_____ 5. The pH of a 10^{-3} M NaCl solution is closest to
              (a) 3.0   (d) 8.0
              (b) 6.0   (e) 11.0
              (c) 7.0

_____ 6. The pH of 10^{-3} M acetic acid $(K_a = 1.8 \times 10^{-5})$ is
              closest to
              (a) 1.0   (d) 10.0
              (b) 4.0   (e) 13.0
              (c) 7.0

_____ 7. The species in a 10^{-3} M acetic acid solution that has
              the highest concentration is
              (a) H^+ (d) $C_2H_3O_2^-$
              (b) OH^- (e) $HC_2H_3O_2$
              (c) Na^+

_____ 8. The species in a 10^{-5} M HCl solution that has the
              highest concentration is
              (a) H^+ (d) Cl^-
              (b) OH^- (e) HCl
              (c) Na^+

9. Five solutions are prepared. Each solution is 10^{-2} M. Which solution is the most acidic?

(a) $HClO_2$, $K_a = 1.1 \times 10^{-2}$
(b) $HC_2H_3O_2$, $K_a = 1.8 \times 10^{-5}$
(c) HCN, $K_a = 4.0 \times 10^{-10}$
(d) HF, $K_a = 6.7 \times 10^{-4}$
(e) NH_3, $K_a = 1.8 \times 10^{-5}$

10. A solution that is 10^{-2} M in acetic acid and 10^{-2} M in sodium acetate is prepared. The pK_a of acetic acid is 4.7. The pH of the solution is closest to
(a) 2.0   (d) 5.7
(b) 4.7   (e) 7.0
(c) 5.0

11. Which of the following indicators would be best for the titration of 0.1 M HNO_2 with 0.1 M NaOH?
(a) methyl orange, which has a pH range for color change of 3.1 to 4.5
(b) methyl red, which has a pH range for color change of 4.2 to 6.3
(c) litmus, which has a pH range for color change of 5.0 to 8.0
(d) thymol blue, which has a pH range for color change of 8.0 to 9.6
(e) all of the preceding

12. The pH of a solution is 4.50. What is the pOH of the solution?
(a) 4.50   (d) 8.50
(b) 5.50   (e) 9.50
(c) 6.50

13. How many mL of 0.05 M NaOH are needed to titrate 50 mL of 0.1 M acetic acid?
(a) 25   (d) 200
(b) 50   (e) 500
(c) 100

14. The pH of 10^{-3} M $NH_4C_2H_3O_2$ is closest to
(a) 3   (d) 9
(b) 5   (e) 11
(c) 7

15. What is the relative pH of a solution of NH_4NO_2?
    (a) greater than 7
    (b) less than 7
    (c) equal to 7

II. Do the following:

1. The parietal cells of the human stomach secrete 0.155 M HCl. What is the pH of the secretion?

2. What is the pH of a solution with a $[OH^-]$ of 4.8×10^{-4} M?

3. A certain enzyme loses its ability to catalyze a reaction below a pH of 5.70. To what $[H^+]$ does the pH correspond?

4. What are the pH and pOH of a 0.020 M solution of KOH?

5. Formic acid, HCOOH, has an ionization constant of 1.77×10^{-4}. What is the pH of 0.10 M formic acid?

6. What are the concentrations of H^+, ClO^-, and HClO in 0.10 M hypochlorous acid? The ionization constant of hypochlorous acid is 3.2×10^{-8}.

7. How many moles of nitrous acid, HNO_2, must be added to water to prepare 1.0 L of solution with a pH of 3.0? The ionization constant of nitrous acid is 4.5×10^{-4}.

8. What is the concentration of a solution in which HX is 1.0 percent dissociated? The ionization constant of HX is 5.0×10^{-4}.

9. In metabolic acidosis the HCO_3^-/CO_2 ratio in the blood may be 16/1, a ratio that corresponds to a pH of 7.3. What is the pK_a of carbonic acid?

10. In metabolic alkalosis the HCO_3^-/CO_2 ratio in the blood may be 40/1. What is the pH of blood under such conditions? Use the value of the ionization constant of carbonic acid calculated in problem 9 above.

11. The indicator bromothymol blue has an ionization constant of 5.0×10^{-8}. The acid color is yellow, and the alkaline color is blue. The acid color is visible when the ratio of yellow to blue is 20 to 1, and the blue is visible when the ratio of blue to yellow is 2 to 1. What is the pH range of color change of bromothymol blue?

12. A solution is prepared by adding 0.040 mol of solid NaCNO, sodium cyanate, to 250 mL of 0.025 M cyanic acid, HCNO. Assume that no volume change occurs. Calculate the pH of the solution. The ionization constant of cyanic acid is 1.2×10^{-4}.

13. Strychnine is a weak base, and its aqueous ionization is represented by

$$S(aq) + H_2O \rightleftharpoons SH^+(aq) + OH^-(aq)$$

A 1.0 M solution of strychnine has a pH of 11.00. What is the ionization constant of strychnine?

14. What is the pH of a buffer that is 0.15 M in NaZ and 3.0 M in HZ if the ionization constant of the weak acid HZ is 5.0×10^{-6}?

15. How many moles of solid NH_4Cl must be dissolved in 500.0 mL of a solution that is originally 0.250 M in NH_3 in order to have a final pH of 10.0? Assume that the volume remains 500.0 mL. The ionization constant of NH_3 is 1.8×10^{-5}.

16. Calculate the pH of a solution composed of 50.0 mL of 1.0 M $NaC_2H_3O_2$ and 50.0 mL of 1.0 M $HC_2H_3O_2$. Assume the volume after mixing to be 100 mL. The ionization constant of acetic acid, $HC_2H_3O_2$, is 1.8×10^{-5}.

17. What is the $[H^+]$ in a 0.036 M H_2S solution? For H_2S the ionization constant of the primary ionization is 1.1×10^{-7}. Assume that effects due to secondary ionization are negligible.

18. Vitamin C, ascorbic acid, is a diprotic acid with the formula $C_6H_8O_6$. Calculate the $[H^+]$, $[C_6H_7O_6^-]$, and $[C_6H_6O_6^{2-}]$ of a 0.10 M solution of ascorbic acid. The ionization constants of ascorbic acid are $K_1 = 7.9 \times 10^{-5}$ and $K_2 = 1.6 \times 10^{-12}$.

19. A buffer solution is prepared by mixing 500.0 mL of 1.00 M $HC_2H_3O_2$ and 500.0 mL of 1.00 M $NaC_2H_3O_2$. The ionization constant of acetic acid, $HC_2H_3O_2$, is 1.8 × 10^{-5}. Calculate the following:

   (a) the pH of the buffer
   (b) the pH of the buffer solution after the addition of 0.005 mol of HCl

20. What is the pH of a 0.10 M $NaNO_2$ solution? The ionization constant of HNO_2, a weak acid, is 4.5 × 10^{-4}.

21. A 0.10 M solution of KCN has a pH of 11.00. What is the ionization constant, K, of HCN?

22. What is the pH of a 0.10 M Na_2S solution? The ionization constant for the primary ionization of H_2S is 1.1 × 10^{-7} and that for the secondary ionization is 1.0 × 10^{-14}.

23. Glutaramic acid has an ionization constant of 4.0 × 10^{-5}. Draw the titration curve for the titration of 50.0 mL of 0.200 N glutaramic acid with 0.200 N NaOH. The 0.200 N glutaramic acid solution is diluted to 100.0 mL before any base is added.

ANSWER TO EXERCISES

I. Concepts of ionic equilibrium

1. (c) 7.0

Since

$$K_w = 1.0 \times 10^{-14} = [H^+][OH^-]$$

and in pure water

$$[H^+] = [OH^-]$$

the pH is equal to 7:

$$[H^+][OH^-] = 1.0 \times 10^{-14}$$
$$[H^+]^2 = 1.0 \times 10^{-14}$$
$$[H^+] = 1.0 \times 10^{-7}$$
$$pH = -\log [H^+] = 7$$

2. (a) 6.0   Since HCl completely dissociates,

$$[H^+] \cong 10^{-6} \; M$$

Therefore,

$$pH = 6$$

3. (c) 7.0   The $[H^+]$ from HCl is $10^{-8} \; M$, but that from the dissociation of water is much greater, $\sim 10^{-7} \; M$. Thus, $pH = 7.0$.
The pH is slightly less than 7 because of the presence
of $[H^+]$ from HCl.

4. (d) 11.0   Since NaOH completely dissociates

$$NaOH \rightarrow Na^+ + OH^-$$

The pH is determined as follows:

$$pOH = -\log [OH^-] = 3$$
$$pH = 14 - 3$$
$$pH = 11$$

5. (c) 7.0   A $10^{-3} \; M$ NaCl solution is neutral. Since the cation is
from a strong base and the anion is from a strong acid
there is no hydrolysis.

6. (b) 4.0   If acetic acid were a strong acid, i.e., if it were completely dissociated in water, the pH would be 3. If it
were not dissociated at all, the pH would be 7. Acetic
acid actually lies somewhere between the limits; thus,
4.0 is the best choice.

7. (e) $HC_2H_3O_2$ In solutions of weak acids the most concentrated species
is usually the undissociated acid.

8. (a) H^+ In water HCl completely dissociates, giving $10^{-5} \; M \; H^+$
and $10^{-5} \; M \; Cl^-$. There is a small contribution of H^+ from
the ionization of water; thus, $[H^+]$ is slightly greater
than $[Cl^-]$.

9. (a) $HClO_2$ A larger dissociation constant corresponds to a greater
degree of dissociation.

10. (b) 4.7   The pH equals the pK_a since $[HC_2H_3O_2]$ equals $[C_2H_3O_2^-]$.

$$pH = pK_a + \log \frac{[C_2H_3O_2^-]}{[HC_2H_3O_2]}$$

$$= 4.7 + 0$$

$$= 4.7$$

11. (d) thymol blue   At the equivalence point the number of moles of acid is equal to the number of moles of base. Since HNO_2 is a weak acid, the conjugate base NO_2^- will hydrolyze, thus producing a basic solution. The indicator must therefore change color in the basic region.

12. (e) 9.50   Determine pOH as follows:

$$pH + pOH = 14.00$$

$$pOH = 14.00 - pH$$

$$= 14.00 - 4.50$$

$$= 9.50$$

13. (c) 100 ml   The number of moles of acid equals the number of moles of base:

$$\text{moles of acid} = \text{moles of base}$$

$$(VM)_{acid} = (VM)_{base}$$

$$(0.050 \text{ L}) \left(\frac{0.1 \text{ mol}}{1 \text{ L}}\right) = V_{base}\left(\frac{0.05 \text{ mol}}{1 \text{ L}}\right)$$

$$V_{base} = 100 \text{ mL}$$

14. (c) 7   There are equal concentrations of a weak acid, NH_4^+, and a weak base, $C_2H_3O_2^-$, and both ions have the same K values.

15. (b) less than 7   The pH is less than 7 because HNO_2 is a stronger acid ($K_a = 4.5 \times 10^{-4}$) than NH_3 ($K_b = 1.8 \times 10^{-5}$) is a base. NH_4^+ hydrolyzes more (producing H_3O^+) than NO_2^- does (producing OH^-).

II. Ionic equilibrium calculations

1. $pH = 0.810$ The definition of pH is

$$pH = -\log [H^+]$$

Substituting the value of $[H^+]$ into the preceding equation,

$$
\begin{aligned}
pH &= -\log [H^+] \\
&= -\log (1.55 \times 10^{-1}) \\
&= (-\log 1.55) + (-\log 10^{-1}) \\
&= 0.810
\end{aligned}
$$

Alternatively, use your calculator, but review the proper manipulation of logs in the instruction manual.

2. $pH = 10.68$ The water constant at 25°C is

$$[H^+][OH^-] = 1.0 \times 10^{-14}$$

If we solve the preceding equation for $[H^+]$ and substitute values into the equation, we find

$$[H^+] = \frac{1.0 \times 10^{-14}}{4.8 \times 10^{-4}}$$

$$= 2.08 \times 10^{-11} \; M$$

We determine pH:

$$
\begin{aligned}
pH &= -\log [H^+] \\
&= -\log [2.08 \times 10^{-11}) \\
&= (-\log 2.08) + (-\log 10^{-11}) \\
&= 0.32 + 11 \\
&= 10.68
\end{aligned}
$$

3. $[H^+] = 2.0 \times 10^{-6} \; M$ $[H^+]$ can be determined as follows:

$$
\begin{aligned}
-\log [H^+] &= pH \\
\log [H^+] &= -pH \\
&= -5.70 \\
&= 0.30 - 6.0 \\
[H^+] &= (\text{antilog } 0.30)(\text{antilog } -6.0) \\
&= 2.0 \times 10^{-6} \; M
\end{aligned}
$$

Another method of determining [H$^+$] can also be used:

$$pH = -\log [H^+]$$
$$[H^+] = 10^{-pH}$$
$$= 10^{-5.7}$$
$$= (10^{+0.3})(10^{-6})$$
$$= 2.0 \times 10^{-6} \ M$$

4. $pH = 12.30$
$pOH = 1.70$

Potassium hydroxide is a strong electrolyte. Thus,

$$[OH^-] = 2.0 \times 10^{-2} \ M$$

and

$$pOH = -\log [OH^-]$$
$$= -\log (2.0 \times 10^{-2})$$
$$= -0.30 + 2$$
$$= 1.70$$

The pH is determined as follows: The water constant is

$$[H^+][OH^-] = 1.0 \times 10^{-14}$$

Arrange the preceding equation and substitute values into it:

$$[H^+] = \frac{1.0 \times 10^{-14}}{[OH^-]}$$

$$= \frac{1.0 \times 10^{-14}}{2.0 \times 10^{-2}}$$

$$= 5.0 \times 10^{-13} \ M$$

Calculate pH in either of two ways:

$$pH = -\log [H^+]$$
$$= -\log (5.0 \times 10^{-13})$$
$$= -0.70 + 13$$
$$= 12.30$$

or

$$pH + pOH = 14.00$$
$$pH = 14.00 - pOH$$
$$= 14.00 - 1.7$$
$$= 12.30$$

5. $pH = 2.38$ The expression for the ionization constant is

$$K_a = 1.77 \times 10^{-4} = \frac{[H^+][HCOO^-]}{[HCOOH]}$$

Make the following assumptions:

Species	Equilibrium Concentration
H^+	$(x)M$
$HCOO^-$	$(x)M$
$HCOOH$	$((1.0 \times 10^{-1}) - x \cong 1.0 \times 10^{-1})M$

Substitute concentrations into the expression for the ionization constant:

$$K_a = 1.77 \times 10^{-4} = \frac{[H^+][HCOO^-]}{[HCOOH]}$$

$$1.77 \times 10^{-4} = \frac{x^2}{1.0 \times 10^{-1}}$$

$$x^2 = 1.77 \times 10^{-5}$$

$$x = 4.21 \times 10^{-3}$$

Thus, $[H^+] = 4.21 \times 10^{-3}$ M and the pH is

$$pH = -\log [H^+]$$
$$= -\log (4.21 \times 10^{-3})$$
$$= 2.38$$

6.
$$[H^+] = 5.7 \times 10^{-5} \ M$$
$$[ClO^-] = 5.7 \times 10^{-5} \ M$$
$$[HClO] = 1.0 \times 10^{-1} \ M$$

The expression for the ionization constant is

$$K_a = 3.2 \times 10^{-8} = \frac{[H^+][ClO^-]}{[HClO]}$$

Make the following assumptions:

Species	Original Concentration	Equilibrium Concentration
H^+	10^{-7} M	$(10^{-7} + x \cong x)M$
ClO^-	0	$(x)M$
$HClO$	1.0×10^{-1} M	$((1.0 \times 10^{-1}) - x \cong 1.0 \times 10^{-1})M$

Substitute concentrations into the expression for the ionization constant:

$$3.2 \times 10^{-8} = \frac{[H^+][ClO^-]}{[HClO]}$$

$$= \frac{x^2}{1.0 \times 10^{-1}}$$

$$x^2 = 3.2 \times 10^{-9}$$

$$x = 5.7 \times 10^{-5}$$

Thus

$$[H^+] = 5.7 \times 10^{-5} \; M$$

7. 3.2×10^{-3} mol HNO_2

The equation for the ionization of nitrous acid is

$$HNO_2 \rightleftharpoons H^+ + NO_2^-$$

A pH of 3.0 corresponds to a $[H^+]$ of 1.0×10^{-3} M:

$$[H^+] = 10^{-pH}$$

$$= 10^{-3.0}$$

$$= 1.0 \times 10^{-3} \; M$$

Make the following assumptions:

Species	Original Concentration	Equilibrium Concentration
H^+	10^{-7} M	$(10^{-7} + (1.0 \times 10^{-3}) = 1.0 \times 10^{-3})M$
NO_2^-	0	1.0×10^{-3} M
HNO_2	$(x)M$	$(x - (1.0 \times 10^{-3}))M$

Substitute concentrations into the expression of the ionization constant:

$$K_a = 4.5 \times 10^{-4} = \frac{[H^+][NO_2^-]}{[HNO_2]}$$

$$4.5 \times 10^{-4} = \frac{(1.0 \times 10^{-3})^2}{x - (1.0 \times 10^{-3})}$$

$$x = 3.2 \times 10^{-3}$$

Thus, 3.2×10^{-3} mol of HNO_2 must be added.

8. [HX] = 5.0   The ionization of HX is

$$HX \rightleftharpoons H^+ + X^-$$

Make the following assumptions:

Species	Original Concentration	Equilibrium Concentration
H^+	10^{-7} M	$(10^{-7} + 0.01x \cong 0.01x)M$
X^-	0	$(0.01x)M$
HX	$(x)M$	$(x - 0.01x \cong x)M$

Substitute concentrations into the expression for the ionization constant:

$$K_a = 5.0 \times 10^{-4} = \frac{[H^+][X^-]}{[HX]}$$

$$= \frac{(0.01x)^2}{x}$$

$$(5.0 \times 10^{-4})x = (1.0 \times 10^{-4})x^2$$

$$x = \frac{5.0 \times 10^{-4}}{1.0 \times 10^{-4}}$$

$$= 5.0$$

Thus,

[HX] = 5.0

9. $pK_a = 6.1$ The ionization of carbonic acid is

$$CO_2 + H_2O \rightleftharpoons H^+ + HCO_3^-$$

The expression for the ionization constant is

$$K_a = \frac{[H^+][HCO_3^-]}{[CO_2]}$$

a pH of 7.3 corresponds to a $[H^+]$ of $5. \times 10^{-8}$ M:

$$-\log [H^+] = pH$$
$$= 7.3$$
$$[H^+] = 10^{-7.3}$$
$$= (10^{0.7})(10^{-8.0})$$
$$= 5. \times 10^{-8} M$$

Substituting concentrations into the expression for the ionization constant,

$$K_a = \frac{[H^+][HCO_3^-]}{[CO_2]}$$

$$= (5. \times 10^{-8})(16)$$

$$= 8. \times 10^{-7}$$

and

$$pK_a = -\log K_a$$
$$= -\log (8. \times 10^{-7})$$
$$= -0.9 + 7$$
$$= 6.1$$

The pK_a can also be determined as follows:

$$pH = pK_a + \log \frac{[HCO_3^-]}{[CO_2]}$$

$$pK_a = pH - \log \frac{[HCO_3^-]}{[CO_2]}$$

$$= 7.3 - \log \left(\frac{16}{1}\right)$$

$$= 7.3 - \log (1.6 \times 10^1)$$

$$= 7.3 - 1.20$$

$$= 6.1$$

10. $pH = 7.7$ Substituting concentrations into the expression for the ionization constant,

$$K_a = \frac{[H^+][HCO_3^-]}{[CO_2]}$$

$$8. \times 10^{-7} = [H^+]40$$

$$[H^+] = \left(\frac{1}{40}\right)(8.0 \times 10^{-7})$$

$$= 2. \times 10^{-8} \ M$$

and

$$pH = -\log [H^+]$$

$$= -\log (2. \times 10^{-8})$$

$$= -0.3 + 8$$

$$= 7.7$$

The Henderson-Hasselbalch equation can also be used to solve the problem

$$pH = pK_a + \log \frac{[HCO_3^-]}{[CO_2]}$$

$$= 6.1 + \log\left(\frac{40}{1}\right)$$

$$= 6.1 + \log (4.0 \times 10^1)$$

$$= 6.1 + 1.60$$

$$= 7.7$$

11. pH range
   = 6.0 to 7.6

The ionization of HIn is

$$HIn \rightleftharpoons H^+ + In^-$$

and the expression for the ionization constant is

$$K_a = 5.0 \times 10^{-8} = \frac{[H^+] \, [In^-]}{[HIn]}$$

The yellow form is HIn and the blue form is In⁻. Thus, the ratio of yellow to blue is the ratio of HIn to In⁻ which is equal to 20 when the yellow color is visible. Calculate the pH corresponding to the ratio of HIn to In⁻ equal to 20:

$$K_a = 5.0 \times 10^{-8} = \frac{[H^+] [In^-]}{[HIn]}$$

$$[H^+] = (5.0 \times 10^{-8})(2.0 \times 10^1)$$

$$= 1.0 \times 10^{-6}$$

and

$$pH = -\log [H^+]$$

$$= -\log (1.0 \times 10^{-6}) M$$

$$= 6.0$$

The blue color is visible when the ratio of blue to yellow is 2, or when the ratio of In^- to HIn is 2. Calculate the pH corresponding to the ratio of HIn to In^- equal to 1/2:

$$K_a = 5.0 \times 10^{-8} = \frac{[H^+][In^-]}{[HIn]}$$

$$[H^+] = (5.0 \times 10^{-8}) \left(\frac{1}{2}\right)$$

$$= 2.5 \times 10^{-8} \, M$$

and

$$pH = -\log (2.5 \times 10^{-8})$$

$$= -0.40 + 8$$

$$= 7.6$$

12. $pH = 4.73$ The Henderson-Hasselbalch equation can also be used. The ionization of HCNO is

$$HCNO \rightleftharpoons H^+ + CNO^-$$

Since NaCNO is a strong electrolyte,

$$NaCNO \rightarrow Na^+ + CNO^-$$

Make the following assumptions:

Species	Original Concentration	Equilibrium Concentration
H^+	$10^{-7} \, M$	$(10^{-7} + x \cong x) M$
CNO^-	$\frac{0.040 \text{ mol}}{0.250 \text{ L}} = 0.16 M$	$(0.16 + x \cong 0.16) M$
HCNO	$0.025 \, M$	$(0.025 - x \cong 0.025) M$

Substituting concentrations into the expression for the ionization constant of HCNO, we find

$$K_a = 1.2 \times 10^{-4} = \frac{[H^+][CNO^-]}{[HCNO]}$$

$$1.2 \times 10^{-4} = \frac{x(0.16)}{0.025}$$

$$x = 1.88 \times 10^{-5}$$

and

$$pH = -\log [H^+]$$

$$= -\log (1.88 \times 10^{-5})$$

$$= 4.73$$

13. $K =$
    1.0×10^{-6}

The Henderson-Hasselbalch equation can also be used.   The expression for the constant ionization of strychnine is

$$K_b = \frac{[SH^+][OH^-]}{[S]}$$

Calculate $[OH^-]$ as follows:

$$[OH^-] = \frac{1.0 \times 10^{-14}}{[H^+]}$$

$$= \frac{1.0 \times 10^{-14}}{1.0 \times 10^{-11}}$$

$$= 1.0 \times 10^{-3}$$

Make the following assumptions:

Species	Equilibrium Concentration
S	$(1.0 - (1.0 \times 10^{-3}) \cong 1.0)M$
SH^+	1.0×10^{-3} M (since the only source is S in the ionization)
OH^-	1.0×10^{-3} M

Substituting concentrations into the expression for the ionization constant, we find

$$K = \frac{(1.0 \times 10^{-3})^2}{1.0}$$

$$K = 1.0 \times 10^{-6}$$

14. $pH = 5.0$

The ionization of HZ is

$$HZ \rightleftharpoons H^+ + Z^-$$

Since NaZ is a strong electrolyte,

$$NaZ \rightarrow Na^+ + Z^-$$

Make the following assumptions:

Species	Original Concentration	Equilibrium Concentration
H^+	10^{-7} M	$(10^{-7} + x \cong x)M$
HZ	3.0×10^{-1} M	$((3.0 \times 10^{-1}) - x \cong 3.0 \times 10^{-1})M$
Z^-	1.5×10^{-1} M	$((1.5 \times 10^{-1}) + x \cong 1.5 \times 10^{-1})M$

Substitute concentrations into the expression for the ionization constant:

$$K = 5.0 \times 10^{-6} = \frac{(1.5 \times 10^{-1})}{3.0 \times 10^{-1}}$$

$$x = \frac{(5.0 \times 10^{-6})(3.0 \times 10^{-1})}{1.5 \times 10^{-1}}$$

$$= 1.0 \times 10^{-5} \quad \text{(Note the assumption that}$$
$$x \ll 10^{-1} \text{ is valid.)}$$

Thus,

$$[H^+] = 1.0 \times 10^{-5} \ M$$

and

$$pH = -\log [H^+]$$
$$= -\log (1.0 \times 10^{-5})$$
$$= 5.0$$

The pH can also be determined as follows:

$$pH = pK_a + \log \frac{[Z^-]}{[HZ]}$$

$$= -\log (5.0 \times 10^{-6}) + \log \frac{0.15}{0.30}$$

$$= 0.70 + 6 + (-0.30)$$

$$= 5.0$$

15. 0.022 mol
NH$_4$Cl

The ionization of NH_3 is

$$NH_3 + H_2O \rightleftharpoons NH_4^+ + OH^-$$

The expression for the ionization constant is

$$K_a = 1.8 \times 10^{-5} = \frac{[NH_4^+][OH^-]}{[NH_3]}$$

The $[OH^-]$ is determined as follows:

$$pH = 10.0$$
$$[H^+] = 10^{-pH}$$
$$= 1.0 \times 10^{-10} \ M$$
$$[OH^-] = \frac{1.0 \times 10^{-14}}{[H^+]}$$
$$= \frac{1.0 \times 10^{-14}}{1.0 \times 10^{-10}}$$
$$= 1.0 \times 10^{-4} \ M$$

Make the following assumptions:

Species	Equilibrium Concentration
NH_3	$(0.250 - (1.0 \times 10^{-4}) \cong 0.250) \ M$
NH_4^+	$(x) M$
OH^-	$1.0 \times 10^{-4} \ M$

Substitute concentrations into the expression for the ionization constant:

$$K_a = 1.8 \times 10^{-5} = \frac{[NH_4^+][OH^-]}{[NH_3]}$$

$$1.8 \times 10^{-5} = \frac{x(1.0 \times 10^{-4})}{2.50 \times 10^{-1}}$$

$$x = \frac{(1.8 \times 10^{-5})(2.50 \times 10^{-1})}{1.0 \times 10^{-4}}$$

$$= 4.5 \times 10^{-2}$$

Thus,

$$[NH_4^+] = 0.045 \ M$$

and

$$? \text{ mol NH}_4^+ = (0.045 \ M \ \text{NH}_4^+)(0.500 \text{ liter})$$
$$= 0.022 \text{ mol NH}_4^+$$

16. $p\text{H} = 4.74$ The expression for the ionization constant of $\text{HC}_2\text{H}_3\text{O}_2$ is

$$K_a = 1.8 \times 10^{-5} = \frac{[\text{H}^+][\text{C}_2\text{H}_3\text{O}_2^-]}{[\text{HC}_2\text{H}_3\text{O}_2]}$$

We make the following assumptions:

Species	Original Concentration	Equilibrium Concentration
H^+	$10^{-7} \ M$	$(10^{-7} + x = x)M$
$\text{C}_2\text{H}_3\text{O}_2^-$	$1.0 \ M$	$\left(\dfrac{(50 \text{ mL})(1.0)}{(100 \text{ mL})} + x = 0.50 + x\right)M \cong 0.50$
$\text{HC}_2\text{H}_3\text{O}_2$	$1.0 \ M$	$\left(\dfrac{(50 \text{ mL})(1.0)}{(100 \text{ mL})} - x = 0.50 - x\right)M \cong 0.50$

Substitute concentrations into the expression for the ionization constant:

$$K_a = 1.8 \times 10^{-5} = \frac{[\text{H}^+][\text{C}_2\text{H}_3\text{O}_2^-]}{[\text{HC}_2\text{H}_3\text{O}_2]}$$

$$1.8 \times 10^{-5} = \frac{x(0.50)}{0.50}$$

$$x = 1.8 \times 10^{-5}$$

Thus,

$$[\text{H}^+] = 1.8 \times 10^{-5} \ M$$

and

$$p\text{H} = -\log [\text{H}^+]$$
$$= -\log (1.8 \times 10^{-5})$$
$$= -0.26 + 5$$
$$= 4.74$$

The Henderson-Hasselbalch equation can also be used to determine the $p\text{H}$.

17. $[H^+] =$
    6.3×10^{-5} M

The expression for the primary ionization constant of H_2S is

$$K_a = 1.1 \times 10^{-7} = \frac{[H^+][HS^-]}{[H_2S]}$$

We make the following assumptions:

Species	Equilibrium Concentration
H^+	$(x)M$
HS^-	$(x)M$
H_2S	$((3.6 \times 10^{-2}) - x \cong 3.6 \times 10^{-2})M$

We substitute concentrations into the expression for the ionization constant:

$$K_a = 1.1 \times 10^{-7} = \frac{[H^+][HS^-]}{[H_2S]}$$

$$1.1 \times 10^{-7} = \frac{x^2}{3.6 \times 10^{-2}}$$

$$x^2 = 40 \times 10^{-10}$$

$$x = 6.3 \times 10^{-5}$$

Thus,

$$[H^+] = 6.3 \times 10^{-5}\ M$$

18.

$[H^+] = 2.8 \times 10^{-3}\ M$
$[C_6H_7O_6^-] = 2.8 \times 10^{-3}\ M$
$[C_6H_6O_6^{2-}] = 1.6 \times 10^{-12}\ M$

The ionizations of ascorbic acid are

$$C_6H_8O_6 \rightleftharpoons H^+ + C_6H_7O_6^-$$

$$C_6H_7O_6^- \rightleftharpoons H^+ + C_6H_6O_6^{2-}$$

For the first ionization make the following assumptions:

Species	Equilibrium Concentration
$C_6H_8O_6$	$((1.0 \times 10^{-1}) - x = 1.0 \times 10^{-1})M$
$C_6H_7O_6^-$	$(x)M$
H^+	$(x)M$

Substitute concentrations into the expression for the ionization constant of ascorbic acid K_{a_1}:

$$K_{a_1} = 7.9 \times 10^{-5} = \frac{[H^+][C_6H_7O_6^-]}{[C_6H_8O_6]}$$

$$7.9 \times 10^{-5} = \frac{x^2}{1.0 \times 10^{-1}}$$

$$x^2 = 7.9 \times 10^{-6}$$

$$x = 2.8 \times 10^{-3}$$

Thus,

$$[H^+] = [C_6H_7O_6^-] = 2.8 \times 10^{-3} \ M$$

For the second ionization of ascorbic acid make the following assumptions:

Species	Equilibrium Concentration
H^+	$2.8 \times 10^{-3} \ M$
$C_6H_7O_6^-$	$2.8 \times 10^{-3} \ M$
$C_6H_6O_6^{2-}$	$(y) M$

Substitute concentrations into the expression for the ionization constant of ascorbic acid K_{a_2}:

$$K_{a_2} = 1.6 \times 10^{-12} = \frac{[H^+][C_6H_6O_6^{2-}]}{[C_6H_7O_6^-]}$$

$$1.6 \times 10^{-12} = \frac{(2.8 \times 10^{-3})(y)}{2.8 \times 10^{-3}}$$

$$y = 1.6 \times 10^{-12}$$

Thus,

$$[C_6H_6O_6^{2-}] = 1.6 \times 10^{-12} \ M$$

19. (a) $pH = 4.74$ The ionization of $HC_2H_3O_2$ is

$$HC_2H_3O_2 \rightleftarrows H^+ + C_2H_3O_2^-$$

The expression for the ionization constant is

$$K_a = 1.8 \times 10^{-5} = \frac{[H^+][C_2H_3O_2^-]}{[HC_2H_3O_2]}$$

Make the following assumptions:

Species	Equilibrium Concentration
H^+	$(x)\,M$
$C_2H_3O_2^-$	$\left(\dfrac{(500\ mL)\ (1.0)}{1000\ mL} + x \cong 0.50\right) M$
$HC_2H_3O_2$	$\left(\dfrac{(500\ mL)\ (1.0)}{1000\ mL} - x \cong 0.50\right) M$

Substitute concentrations into the expression for the ionization constant:

$$K_a = 1.8 \times 10^{-5} = \frac{[H^+][C_2H_3O_2^-]}{[HC_2H_3O_2]}$$

$$1.8 \times 10^{-5} = \frac{x\,(0.50)}{0.50}$$

$$x = 1.8 \times 10^{-5} \quad \text{(Note the assumption that}$$
$$\phantom{x = 1.8 \times 10^{-5} \quad } x \ll 0.50 \text{ is valid.)}$$

Thus,

$$[H^+] = 1.8 \times 10^{-5}\ M$$

and

$$
\begin{aligned}
pH &= -\log\ [H^+]\\
&= -\log\ (1.8 \times 10^{-5})\\
&= -0.26 + 5\\
&= 4.74
\end{aligned}
$$

This can also be solved as follows:

$$pH = pK_a + \log \frac{[C_2H_3O_2^-]}{[HC_2H_3O_2]}$$

$$= -\log (1.8 \times 10^{-5}) + \log \left(\frac{0.50}{0.50}\right)$$

$$= -0.26 - (-5) + 0$$

$$= 4.74$$

(b) $pH = 4.73$ The H^+ from the completely dissociated HCl will react with $C_2H_3O_2^-$ to form $HC_2H_3O_2$:

$$H^+ + C_2H_3O_2^- \longrightarrow HC_2H_3O_2$$

The protonation of $C_2H_3O_2^-$ has a K equal to $1/(1.8 \times 10^{-5})$, or 5.6×10^4. Since this constant is large, we assume that all the H^+ reacts to form $HC_2H_3O_2$. Thus, we make the following assumptions:

Species	Equilibrium Concentration
H^+	$(y)M$
$C_2H_3O_2^-$	$(0.50 - 0.005 = 0.495)M$
$HC_2H_3O_2$	$(0.50 + 0.005 = 0.505)M$

Substituting into the Henderson-Hasselbalch equation:

$$pH = pK_a + \log \frac{[C_2H_3O_2^-]}{[HC_2H_3O_2]}$$

$$= 4.74 + \log \left(\frac{0.495}{0.505}\right)$$

$$= 4.74 + \log (9.80 \times 10^{-1})$$

$$= 4.74 + (-0.009)$$

$$= 4.73$$

The buffer solution closely maintains its pH in spite of the addition of hydrochloric acid.

20. $pH = 8.17$ Sodium nitrite is a strong electrolyte.   Thus,

$$NaNO_2(s) \rightarrow Na^+(aq) + NO_2^-(aq)$$

The hydrolysis of NO_2^- is

$$NO_2^-(aq) + H_2O \rightleftharpoons HNO_2(aq) + OH^-(aq)$$

The equilibrium constant for the hydrolysis is obtained as follows:

$$K_b = \frac{[HNO_2][OH^-]}{[NO_2^-]} = \frac{K_w}{K_a} = \frac{1.0 \times 10^{-14}}{4.5 \times 10^{-4}}$$

$$= 2.2 \times 10^{-11}$$

We make the following assumptions:

Species	Equilibrium Concentration
HNO_2	$(x)M$
OH^-	$(x)M$
NO_2^-	$(0.10 - x \cong 0.10)M$

Substitute concentrations into the equilibrium expression for the hydrolysis of NO_2^-:

$$K_b = 2.2 \times 10^{-11} = \frac{[HNO_2][OH^-]}{[NO_2^-]}$$

$$= \frac{x^2}{0.10}$$

$$x^2 = 2.2 \times 10^{-12}$$

$$x = 1.5 \times 10^{-6}$$

Thus,

$$[OH^-] = 1.5 \times 10^{-6}$$

Calculate pH as follows:

$$[H^+][OH^-] = 1.0 \times 10^{-14}$$

$$[H^+] = \frac{1.0 \times 10^{-14}}{[OH^-]}$$

$$= \frac{1.0 \times 10^{-14}}{1.5 \times 10^{-6}}$$

$$= 6.7 \times 10^{-9}$$

$$pH = -\log (6.7 \times 10^{-9})$$

$$= 8.17$$

21. $K = 1.0 \times 10^{-9}$

Potassium cyanide is a strong electrolyte:

$$KCN \rightarrow K^+ \ CN^-$$

The cyanide ion is hydrolyzed:

$$CN^- + H_2O \rightleftharpoons HCN + OH^-$$

The equilibrium expression for the hydrolysis is

$$K_b = \frac{K_w}{K_a} = \frac{[HCN][OH^-]}{[CN^-]}$$

Calculate the $[OH^-]$:

$$pOH = 14 - pH$$

$$= 14 - 11.00$$

$$= 3.00$$

$$[OH^-] = 10^{-pOH}$$

$$= 1.00 \times 10^{-3}$$

Make the following assumptions:

Species	Equilibrium Concentration
OH^-	$1.00 \times 10^{-3} \ M$
HCN	$1.00 \times 10^{-3} \ M$
CN^-	$(1.0 \times 10^{-1} \ M) - (1.0 \times 10^{-3} \ M) \cong 1.0 \times 10^{-1} \ M$

Substitute concentrations into the expression for the K_b:

$$K_b = \frac{K_w}{K_a} = \frac{[HCN][OH^-]}{[CN^-]}$$

$$\frac{1.0 \times 10^{-14}}{K_a} = \frac{(1.00 \times 10^{-3})(1.00 \times 10^{-3})}{1.0 \times 10^{-1}}$$

$$K_a = \frac{(1.0 \times 10^{-14})(1.0 \times 10^{-1})}{(1.0 \times 10^{-3})^2}$$

$$= 1.0 \times 10^{-9}$$

22. pH = 12.96

We consider the following equilibria:

$$S^{2-} + H_2O \rightleftharpoons HS^- + OH^- \quad K_b = \frac{K_w}{K_{a2}} = \frac{1.0 \times 10^{-14}}{1.0 \times 10^{-14}} = 1.00$$

$$HS^- + H_2O \rightleftharpoons H_2S + OH^- \quad K_b = \frac{K_w}{K_{a1}} = \frac{1.0 \times 10^{-14}}{1.1 \times 10^{-7}} = 9.09 \times 10^{-8}$$

$$H_2O \rightleftharpoons H^+ + OH^- \quad K_w = 1.0 \times 10^{-14}$$

Since the K of the first equilibrium is much larger than either of the other two, the first equilibrium is the most important and the only one that must be considered.

The expression for the hydrolysis of S^{2-} is

$$K_b = 1.00 = \frac{[HS^-][OH^-]}{[S^{2-}]}$$

Make the following assumptions:

Species	Equilibrium Concentration
OH^-	$(x)M$
HS^-	$(x)M$
S^{2-}	$(0.10 - x)M$

Substitute concentrations into the expression for the hydrolysis constant of S^{2-}:

$$K_b = 1.00 = \frac{[HS^-][OH^-]}{[S^{2-}]}$$

$$1.00 = \frac{x^2}{(0.10 - x)}$$

$$x^2 = 0.100 - 1.00x$$

$$x^2 + 1.00x - 0.100 = 0$$

Use the quadratic formula to solve for

$$x = \frac{-1.00 \pm \sqrt{(1.00)^2 - 4(1.00)(-0.100)}}{2(1.00)}$$

$$x = 9.16 \times 10^{-2} = OH^-$$

The pH is then calculated:

$$[H^+] = \frac{1.0 \times 10^{-14}}{9.16 \times 10^{-2}}$$

$$= 1.09 \times 10^{-13}$$

$$p\text{H} = 12.96$$

23. The titration curve is

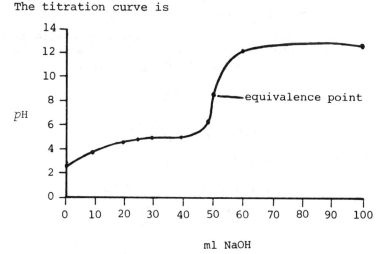

ml NaOH

The titration curve for the titration of glutaramic acid, a weak acid, with sodium hydroxide, a strong base, is determined as follows:

For simplicity we will use HA to designate glutaramic acid. At the initial point of the titration curve (i.e., the point at which no base has been added) the solution contains only HA, H^+, and A^-. Make the following assumptions:

Species	Equilibrium Concentration
H^+	$(x)M$
A^-	$(x)M$
HA	$\left(\dfrac{(50.0 \text{ mL}) (0.200 \text{ } M)}{100 \text{ mL}} - x\right)M \cong 0.100 \text{ } M$

Substitute concentrations into the expression for the ionization constant of HX:

$$K_a = 4.0 \times 10^{-5} = \frac{[H^+][X^-]}{[HX]}$$

$$= \frac{x^2}{0.100}$$

Thus,

$$[H^+] = 2.0 \times 10^{-3} \text{ } M$$

and

$$pH = -\log [H^+] = 2.7$$

Addition of 10.0 mL of NaOH

With this addition, 10/50 of the acid is neutralized and 40/50 remains unneutralized. A buffer exists. Therefore, we can determine the pH as follows:

$$pH = pK_a + \log \frac{[A^-]}{[HA]}$$

$$= 4.40 + \log \frac{(10/50)}{(40/50)}$$

$$= 3.80$$

Addition of 20.0 mL of NaOH

$$pH = pK_a + \log \frac{(20/50)}{(30/50)}$$

$$= 4.22$$

Addition of 25.0 mL of NaOH

$$pH = pK_a + \log \frac{(25/50)}{(25/50)}$$

$$= 4.40$$

Addition of 30.0 mL of NaOH

$$pH = 4.58$$

Addition of 40.0 mL of NaOH

$$pH = 5.00$$

Addition of 49.0 mL of NaOH

$$pH = pK_a + \log \frac{(49/50)}{(\ 1/50)}$$

$$= 4.40 + (1.69)$$

$$= 6.09$$

Addition of 50.0 mL of NaOH

The acid is completely neutralized. The solution contains only the salt of a weak base so that the pH is determined from the expression for hydrolysis:

$$A^- + H_2O \rightleftharpoons HA + OH^-$$

$$K_b = \frac{K_w}{K_a} = \frac{[HA][OH^-]}{[A^-]}$$

Substituting and remembering that the concentration of A^- is now one-third the original concentration of HA because of dilution, we obtain:

$$K_b = \frac{K_w}{K_a} = \frac{[HA][OH^-]}{[A^-]}$$

$$[OH^-]^2 = \frac{K_w[A^-]}{K_a}$$

$$= \frac{(1.0 \times 10^{-14})(6.67 \times 10^{-3})}{(4.0 \times 10^{-5})}$$

$$= 1.67 \times 10^{-11}$$

$$[OH^-] = 4.08 \times 10^{-6}$$

278 Chapter 17 Ionic Equilibrium, Part I

Finally,

$pH = 8.61$

Addition of 60.0 mL of NaOH

$$[OH^-] = \frac{\text{moles of excess base}}{\text{total volume of solution}}$$

$$= \frac{(0.010 \text{ L})(0.200 \text{ } M)}{(0.160 \text{ L})}$$

$$= 0.0125 \text{ } M$$

or

$pH = 12.10$

Addition of 100.0 mL of NaOH

$$[OH^-] = \frac{(0.050 \text{ mL})(0.200 \text{ } M)}{(0.200 \text{ mL})}$$

$$= 0.0500 \text{ } M$$

$$pH = 12.70$$

SELF-TEST   Complete the test in 40 minutes:

1. What is the pH of a solution that has an OH^- concentration of 0.075 M?

2. A weak acid, HX, is 0.20 percent ionized in 0.50 M solution. What is the ionization constant of the acid?

3. The ionization constant of a weak acid HX is 6.0×10^{-6}. What concentration of NaX must be present in a 0.30 M solution of HX to prepare a buffer with a pH of 5.00?

4. What is the pH at the equivalence point of a titration of 50.0 mL of 0.10 M trimethylacetic acid with 0.10 M NaOH? The pK of the ionization of trimethylacetic acid is 5.0.

5. What is the pH of the solution described in problem 4 after 30.0 mL of 0.10 M NaOH is added?

IONIC EQUILIBRIUM, PART II

CHAPTER

18

OBJECTIVES

I. You should be able to demonstrate your knowledge of the following terms by defining them, describing them, or giving specific examples of them:

amphoterism [18.5]
common-ion effect [18.2]
complex ion [18.4]
formation constant [18.4]
instability constant [18.4]
ion product [18.2]
ligand [18.4]
molar solubility [18.1]
salt effect [18.1]
solubility product, K_{SP} [18.1]

II. You should be able to determine K_{SP} from experimental data and calculate the solubility of a compound from the value of K_{SP}.

III. You should be able to use K_{SP} to predict whether or not precipitation will occur in a given solution.

IV. You should be able to determine the concentrations of species in a salt solution or a solution that contains complex ions.

V. You should be able to determine the pH of a solution containing a known concentration of a compound that will hydrolyze.

UNITS,   I. The only new symbol in this chapter is K_{SP}, which is
SYMBOLS,
MATHEMATICS   the symbol for the solubility product.

II. The methods used for solving the problems in this chapter
are very similar to those used for solving problems in
Chapter 17.

EXERCISES   I. Answer each of the following questions without performing
any calculations. This exercise will increase your
qualitative understanding of equilibria.

_____

1. The most soluble of the following carbonates is
   (a) $BaCO_3$, $K_{SP} = 1.6 \times 10^{-9}$
   (b) $CdCO_3$, $K_{SP} = 5.2 \times 10^{-12}$
   (c) $CaCO_3$, $K_{SP} = 4.7 \times 10^{-9}$
   (d) $PbCO_3$, $K_{SP} = 1.5 \times 10^{-15}$
   (e) $MgCO_3$, $K_{SP} = 1 \times 10^{-15}$

_____

2. The K_{SP} of $CdCO_3$ is 5.2×10^{-12}, and the solubility of
   $CdCO_3$ calculated with that value is 2.6×10^{-6} mol/liter.
   The experimentally determined solubility is
   (a) much lower because of temperature effects
   (b) much higher because of carbonate hydrolysis
   (c) much higher because of temperature effects
   (d) much higher because of supersaturation
   (e) the same

_____

3. The value of the solubility constant of silver(I)
   sulfate is approximately the same as that of calcium
   sulfate. Which of the following statements is true?
   (a) Calcium sulfate is more soluble than silver
       sulfate.
   (b) Silver sulfate is more soluble than calcium
       sulfate.
   (c) Silver sulfate is as soluble as calcium sulfate.
   (d) Nothing can be said about the relative solubili-
       ties of silver sulfate and calcium sulfate.
   (e) None of the preceding is true.

II. Answer the following:

1. At 25°C a saturated solution of $Ca(OH)_2$ has a pH of
   12.45.
   (a) What is the K_{SP} of $Ca(OH)_2$?
   (b) What is the molar solubility of $Ca(OH)_2$?

2. Will a precipitate form in a solution that is
   1.00×10^{-3} M in Ba^{2+} and 1.00×10^{-3} M in SO_4^{2-}?
   The K_{SP} of $BaSO_4$ is 1.5×10^{-9}.

3. How many moles of $Ag_2C_2O_4$ will dissolve in 100 ml of 0.050 M $Na_2C_2O_4$? Neglect the hydrolysis of ions. The K_{SP} of $Ag_2C_2O_4$ is 1.1×10^{-11}.

4. Which is the least soluble: $AgCl$ or Ag_2CrO_4? The K_{SP} of $AgCl$ is 1.7×10^{-10} and that of Ag_2CrO_4 is 1.9×10^{-12}.

5. A solution is 0.20 M in Ca^{2+} and 0.10 M in Sr^{2+}.
   (a) If solid Na_2SO_4 is added very slowly to the solution, which will precipitate first: $CaSO_4$ or $SrSO_4$? Neglect volume changes. The K_{SP} of $CaSO_4$ is 2.4×10^{-5} and that of $SrSO_4$ is 7.6×10^{-7}.
   (b) The addition of Na_2SO_4 is continued until the second cation starts to precipitate as the sulfate. What is the concentration of the first cation at that point?

6. What are the final concentrations of Fe^{2+} and Cd^{2+} in a solution originally 0.10 M in Fe^{2+} and 0.10 M in Cd^{2+}, which is buffered with 1.0 M SO_4^{2-} and 1.0 M HSO_4^- and saturated with H_2S? For a saturated solution of H_2S

$$[H^+]^2[S^{2-}] = 1.1 \times 10^{-22}$$

The ionization constant for the secondary ionization of H_2SO_4 is 1.3×10^{-2}. The K_{SP} of FeS is 4.0×10^{-19} and that of CdS is 1.0×10^{-28}.

7. A solution is prepared by mixing 50.0 mL of 0.20 M Ag^+ with 50.0 mL of 0.20 M NH_3. The principal equilibrium is

$$Ag^+ + 2NH_3 \rightleftharpoons Ag(NH_3)_2^+$$

and the equilibrium expression is

$$K = \frac{[Ag(NH_3)_2^+]}{[Ag^+][NH_3]^2} = 1.67 \times 10^7$$

What are the final concentrations of Ag^+, NH_3, $Ag(NH_3)_2^+$, H^+, OH^-, NH_3, and NH_4^+ in the solution? The equilibrium constant for the hydrolysis of NH_3 is 1.8×10^{-5}.

ANSWERS TO   I. Concepts of ionic equilibria
EXERCISES

1.
(c) $CaCO_3$

The most soluble of the carbonates is the one with the largest value of K_{SP}:

$$K_{SP} = [M^{2+}][CO_3^{2-}]$$

The largest value of K_{SP} corresponds to the compound that gives the most ions in solution.

2.
(b) much
higher
because of
carbonate
hydrolysis

3.
(a) Calcium
sulfate
is more
soluble
than
silver
sulfate

If we indicate molar solubility by S, we find that for Ag_2SO_4:

$$K_{SP} = S(2S)^2$$
$$= 4S^3$$

and for $CaSO_4$

$$K_{SP} = S^2$$

If the value of K_{SP} is the same for both compounds, we find that for Ag_2SO_4

$$S = \sqrt[3]{(1/4)K_{SP}}$$

and for $CaSO_4$

$$S = \sqrt{K_{SP}}$$

Thus, the solubility of $CaSO_4$ is larger.

II. Ionic equilibria calculations

1.
(a) $K_{SP} =$
1.1×10^{-5}

The equilibrium between solid $Ca(OH)_2$ and a saturated solution of $Ca(OH)_2$ is

$$Ca(OH)_2(s) \rightleftharpoons Ca^{2+}(aq) + 2OH^-(aq)$$

and the K_{SP} is

$$K_{SP} = [Ca^{2+}][OH^-]^2$$

Calculate the $[OH^-]$ as follows:

$$pH + pOH = 14$$
$$pOH = 14 - pH$$
$$= 14 - 12.45$$
$$= 1.55$$
$$[OH^-] = 2.8 \times 10^{-2} M$$

According to the balanced equation, 2 moles of OH^- are produced for every mole of Ca^{2+}. Thus,

$$[Ca^{2+}] = \frac{[OH^-]}{2}$$
$$= \frac{(2.8 \times 10^{-2})}{2}$$
$$= 1.4 \times 10^{-2}$$

Calculate the K_{SP}:

$$K_{SP} = [Ca^{2+}][OH^-]^2$$
$$= (1.4 \times 10^{-2})(2.8 \times 10^{-2})^2$$
$$= 1.1 \times 10^{-5}$$

(b) 1.4×10^{-2} The molar solubility of $Ca(OH)_2$ at 25°C is the number of moles of $Ca(OH)_2$ dissolved per liter of solution at that temperature. Since 1 mol of Ca^{2+} is present per mole of $Ca(OH)_2$ dissolved,

$$\text{molar solubility of } Ca(OH)_2 = [Ca^{2+}]$$
$$= 1.4 \times 10^{-2}$$

2. yes   The K_{SP} of $BaSO_4$ is

$$K_{SP} = 1.5 \times 10^{-9} = [Ba^{2+}][SO_4^{2-}]$$

In a solution that is 0.001 M in Ba^{2+} and 0.001 M in SO_4^{2-}, the ion product is

$$\text{ion product} = [Ba^{2+}][SO_4^{2-}]$$

$$= (1.00 \times 10^{-3})(1.00 \times 10^{-3})$$

$$= 1.00 \times 10^{-6}$$

If the ion product is greater than K_{SP}, precipitation will occur until the ion product equals the K_{SP}. Since the ion product of $BaSO_4$ is greater than the K_{SP}, precipitation will occur.

3. 7.4×10^{-7} mol

Sodium oxalate is a strong electrolyte:

$$Na_2C_2O_4 \rightarrow 2Na^+ + C_2O_4^{2-}$$

In a saturated solution the ion product is equal to the K_{SP} and no more solid will dissolve. The equilibrium between solid $Ag_2C_2O_4$ and a saturated solution of $Ag_2C_2O_4$ is

$$Ag_2C_2O_4 \rightleftharpoons 2Ag^+ + C_2O_4^{2-}$$

and the K_{SP} for $Ag_2C_2O_4$ is

$$K_{SP} = 1.1 \times 10^{-11} = [Ag^+]^2[C_2O_4^{2-}]$$

Make the following assumptions:

Species	Equilibrium Concentration
Ag^+	$(2x)M$ (where x = molar solubility of $Ag_2C_2O_4$)
$C_2O_4^{2-}$	$((5.0 \times 10^{-2}) + x \cong 5.0 \times 10^{-2})M$

Substitute concentrations into the expression for the K_{SP}:

$$K_{SP} = 1.1 \times 10^{-11} = [Ag]^2[C_2O_4^{2-}]$$

$$= (2x)^2(5.0 \times 10^{-2})$$

$$= (2.0 \times 10^{-1})(x^2)$$

$$x^2 = \frac{1.1 \times 10^{-11}}{2.0 \times 10^{-1}}$$

$$x^2 = 5.5 \times 10^{-11}$$

$$x = 7.4 \times 10^{-6}$$

Thus, 7.4×10^{-6} mol of $Ag_2C_2O_4$ will dissolve per liter of solution. Calculate the number of moles that will dissolve in 100 mL of solution:

$$? \ mol = \left(\frac{7.4 \times 10^{-6} \ mol}{1000 \ ml} \right) 100 \ mol$$

$$= 7.4 \times 10^{-7} \ mol$$

4. AgCl

The K_{SP} of AgCl is 1.7×10^{-10} and that of Ag_2CrO_4 is 1.9×10^{-12}. The compound with the smaller molar solubility is the less soluble. Calculate the molar solubility of each compound.

For AgCl, the equilibrium between solid silver chloride and a saturated solution of silver chloride is

$$AgCl \rightleftharpoons Ag^+ + Cl^-$$

Since 1 mol of Ag^+ and 1 mol of Cl^- are produced per mole of AgCl dissolved, the molar solubility of AgCl, x, is

$$x = [Ag^+] = [Cl^-]$$

Calculate x from the K_{SP} of AgCl:

$$K_{SP} = 1.7 \times 10^{-10} = [Ag^+][Cl^-]$$

$$1.7 \times 10^{-10} = x^2$$

$$x = 1.3 \times 10^{-5}$$

For Ag_2CrO_4, the equilibrium between solid silver chromate and a saturated solution of silver chromate is

$$Ag_2CrO_4 \rightleftharpoons 2Ag^+ + CrO_4^{2-}$$

Since 2 mol of Ag^+ and 1 mol of CrO_4^{2-} are produced per 1 mol of Ag_2CrO_4 dissolved, the molar solubility of Ag_2CrO_4, y, is

$$y = \frac{[Ag^+]}{2} = [CrO_4^{2-}]$$

Calculate y from the K_{SP} of Ag_2CrO_4:

$$K_{SP} = 1.9 \times 10^{-12} = [Ag^+]^2[CrO_4^{2-}]$$
$$= [2y]^2[y]$$
$$= 4y^3$$
$$y = 7.8 \times 10^{-5}$$

Thus, AgCl is the less soluble.

5. (a) $SrSO_4$ The K_{SP} of $CaSO_4$ is 2.4×10^{-5} and that of $SrSO_4$ is 7.6×10^{-7}:

$$K_{SP} = 2.4 \times 10^{-5} = [Ca^{2+}][SO_4^{2-}]$$
$$K_{SP} = 7.6 \times 10^{-7} = [Sr^{2+}][SO_4^{2-}]$$

Precipitation begins when the ion product just exceeds the K_{SP}. A greater concentration of SO_4^{2-} is necessary to exceed the K_{SP} of $CaSO_4$ than to exceed that of $SrSO_4$.

(b) $[Sr^{2+}] = 6.3 \times 10^{-3}$ M It is known that $[SO_4^{2-}] = 1.2 \times 10^{-4}$ M when Ca^{2+} begins to precipitate as $CaSO_4$:

$$[Ca^{2+}][SO_4^{2-}] = 2.4 \times 10^{-5}$$
$$[SO_4^{2-}] = \frac{2.4 \times 10^{-5}}{[Ca^{2+}]}$$
$$= \frac{2.4 \times 10^{-5}}{2.0 \times 10^{-1}}$$
$$= 1.2 \times 10^{-4}$$

$[Sr^{2+}]$ when Ca^{2+} begins to precipitate can be determined as follows:

$$K_{SP} = 7.6 \times 10^{-7} = [Sr^{2+}][SO_4^{2-}]$$
$$[Sr^{2+}] = \frac{7.6 \times 10^{-7}}{1.2 \times 10^{-4}}$$
$$= 6.3 \times 10^{-3}$$

6. $[Fe^{2+}] =$ 0.10 M

$[Cd^{2+}] =$ 1.5 $\times$ 10^{-10} M

To determine whether a metal sulfide precipitates, calculate the ion product for each compound. First determine $[S^{2-}]$ in a saturated H_2S solution:

$$[H^+]^2 [S^{2-}] = 1.1 \times 10^{-22}$$

The $[H^+]$ is controlled by the HSO_4^-/SO_4^{2-} buffer:

$$HSO_4^- \rightleftharpoons H^+ + SO_4^{2-}$$

Thus, calculate $[H^+]$ from the ionization expression for HSO_4^-:

$$K_a = 1.3 \times 10^{-2} = \frac{[H^+][SO_4^{2-}]}{[HSO_4^-]}$$

Make the following assumptions:

Species	Equilibrium Concentration
H^+	$(x)M$
SO_4^{2-}	$(1.0 + x \cong 1.0)M$
HSO_4^-	$(1.0 - x \cong 1.0)M$

Substitute concentrations into the expression for the ionization constant of HSO_4^-:

$$K_a = 1.3 \times 10^{-2} = \frac{[H^+][SO_4^{2-}]}{[HSO_4^-]}$$

$$= \frac{(x)(1.0)}{1.0}$$

$$x = 1.3 \times 10^{-2}$$

Thus, $[H^+] = 1.3 \times 10^{-2}$ M

Determine the $[S^{2-}]$:

$$[H^+]^2 [S^{2-}] = 1.1 \times 10^{-22}$$

$$[S^{2-}] = \frac{1.1 \times 10^{-22}}{[H^+]^2}$$

$$= \frac{1.1 \times 10^{-22}}{(1.3 \times 10^{-2})^2}$$

$$= 6.5 \times 10^{-19}$$

The ion product just exceeds K_{SP} when the metal sulfide begins to precipitate. Thus, for FeS the ion product is

$$[Fe^{2+}][S^{2-}] = (1.0 \times 10^{-1})(6.5 \times 10^{-19})$$
$$= 6.5 \times 10^{-20}$$

and the K_{SP} is 4.0×10^{-19}. Since the K_{SP} of FeS is greater than the ion product, FeS will not precipitate and the final $[Fe^{2+}]$ is 0.10 M.

For CdS the ion product is

$$[Cd^{2+}][S^{2-}] = (1.0 \times 10^{-1})(6.5 \times 10^{-19})$$
$$= 6.5 \times 10^{-20}$$

and the K_{SP} is 1.0×10^{-28}.

Since the K_{SP} of CdS is less than the ion product, CdS will precipitate:

$$Cd^{2+} + H_2S \rightarrow 2H^+ + CdS$$

Calculate the concentration of Cd^{2+} after precipitation of CdS:

$$K_{SP} = 1.0 \times 10^{-28} = [Cd^{2+}][S^{2-}]$$
$$[Cd^{2+}] = \frac{1.0 \times 10^{-28}}{[S^{2-}]}$$
$$= \frac{1.0 \times 10^{-28}}{6.5 \times 10^{-19}}$$
$$= 1.5 \times 10^{-10}$$

7. $[Ag^+] =$ 9.4×10^{-8} M

$[NH_3] =$ 0.08 M

$[Ag(NH_3)_2^+] =$ 1.0×10^{-2} M

$[NH_4^+] =$ 1.2×10^{-3} M

$[OH^-] =$ 1.2×10^{-3} M

$[H^+] =$ 8.3×10^{-12} M

Make the following assumptions:

Species	Equilibrium Concentration
Ag^+	$(x)M$
NH_3	$(0.10 - .02 + 2x \cong 0.08)M$
$Ag(NH_3)_2^+$	$(0.010 - x \cong 0.010)M$

Substitute concentrations into the expression for the formation constant of $Ag(NH_3)_2^+$:

$$K = 1.67 \times 10^7 = \frac{[Ag(NH_3)_2^+]}{[Ag^+][NH_3]^2}$$

$$= \frac{0.010}{(x)(0.08)^2}$$

$$x = \frac{0.010}{(0.08)^2(1.67 \times 10^7)}$$

$$= 9.4 \times 10^{-8}$$

Thus,

$$[NH_3] = 0.08\ M + 2(9.4 \times 10^{-8}\ M)$$

$$= 0.08$$

$$[Ag(NH_3)_2^+] = (0.010\ M) - (9.4 \times 10^{-8}\ M)$$

$$= 0.010\ M$$

$$[Ag^+] = 9.4 \times 10^{-8}\ M$$

Use the equilibrium expression for the hydrolysis of NH_3 to determine $[OH^-]$ and $[NH_4^+]$:

$$NH_3(aq) + H_2O \rightleftharpoons NH_4^+(aq) + OH^-(aq)$$

$$K_b = 1.8 \times 10^{-5} = \frac{[NH_4^+][OH^-]}{[NH_3]}$$

Make the following assumptions:

Species	Equilibrium Concentration
OH^-	$(x)M$
NH_4^+	$(x)M$
NH_3	$(0.08 - x \cong 0.08)M$

Substitute concentrations into the expression for the hydrolysis constant of NH_3:

$$K_H = 1.8 \times 10^{-5} = \frac{[NH_4^+][OH^-]}{[NH_3]}$$

$$= \frac{x^2}{0.08}$$

$$x^2 = 1.4 \times 10^{-6}$$

$$x = 1.2 \times 10^{-3}$$

Thus,

$[NH_3] = 0.08 \ M$

$[NH_4^+] = 1.2 \times 10^{-3} \ M$

$[OH^-] = 1.2 \times 10^{-3} \ M$

Determine $[H^+]$:

$[H^+][OH^-] = 1.0 \times 10^{-14}$

$$[H^+] = \frac{1.0 \times 10^{-14}}{1.2 \times 10^{-3}}$$

$$= 8.3 \times 10^{-12}$$

Thus,

$[H^+] = 8.3 \times 10^{-12} \ M$

SELF-TEST   Complete the test in 30 minutes.

1. How many grams of Ag^+ are in 250 ml of a 0.100 M K_2CrO_4 solution that is saturated with Ag_2CrO_4? The K_{SP} of Ag_2CrO_4 is 1.9×10^{-12}.

2. (a) A solution is 0.15 M in Pb^{2+} and 0.20 M in Ag^+. If solid Na_2SO_4 is very slowly added to this solution, which will precipitate first, $PbSO_4$ or Ag_2SO_4? Neglect volume changes.
   (b) The addition of Na_2SO_4 is continued until the second cation just starts to precipitate as the sulfate. What is the concentration of the first cation at this point?

3. A solution that is 0.10 M in H^+ and 0.10 M in Cu^{2+} is saturated with H_2S. What is the concentration of Cu^{2+} after the CuS has precipitated? Note that it is necessary to take into account the increase in acidity caused by the precipitation. Neglect the hydrolysis of ions.

ELEMENTS OF CHEMICAL THERMODYNAMICS

CHAPTER

19

OBJECTIVES
    I. You should be able to demonstrate your knowledge of the following terms by defining them, describing them, or giving specific examples of them:

enthalpy, H [19.2]
entropy, S [19.3]
equilibrium constant, K [19.7]
first law of thermodynamics [19.1]
Gibbs free energy [19.4]
internal energy, E [19.1]
joule, J [19.1]
second law of thermodynamics [19.3]
spontaneous [19.4]
standard absolute entropy, $S°$ [19.6]
standard free energy of formation, $\Delta G_f°$ [19.5]
state function [19.1]
surroundings [19.1]
system [19.1]
thermodynamics [introduction]
third law of thermodynamics [19.6]

  II. You should be able to calculate ΔH and ΔE for a reaction from enthalpies of formation or bomb calorimeter data.

 III. You should be able to calculate a boiling point from the enthalpy and entropy of vaporization.

 IV. Using tabulated values of the Gibbs free energy of formation $\Delta G_f°$, you should be able to calculate $\Delta G°$ for a reaction.

> V. You should be able to determine equilibrium constants from Gibbs free energies.
>
> VI. You should be able to calculate equilibrium constants for a given reaction at various temperatures given an equilibrium constant for the reaction at one temperature and the value of $\Delta H°$ in that temperature range.
>
> VII. Given equilibrium constants at two different temperatures for a reaction, you should be able to calculate the value of $\Delta H°$ in that temperature range.

UNITS, SYMBOLS, MATHEMATICS

One of the most difficult problems for the novice in thermodynamics is being able to understand the many new symbols. Take the time to study all of the following:

- ° is the symbol for the degree sign. It is used to represent standard states, i.e., $S°$ is the absolute entropy of a system at 25°C and 1 atm and $G°$ represents the Gibbs free energy of a material in its standard state at 25°C and 1 atm.
- Δ is the symbol used to represent a change in a particular value, i.e., ΔE represents a change in internal energy, E, and ΔH represents a change in enthalpy.
- E is the symbol used to represent the internal energy of a system.
- G is the symbol used to represent the Gibbs free energy of a system, $G = H - TS$.
- H is the symbol used to represent the enthalpy of a system, $H = E - PV$.
- J is the abbreviation for the joule, the SI unit for work.
- K is the symbol used to represent the equilibrium constant.
- K_p is the symbol used to represent the equilibrium constant when pressures are used instead of concentrations for gas phase reactions.
- N is the abbreviation for the newton, a unit of force.
- n is the symbol used to represent the number of moles of material in mathematical equations such as $PV = nRT$, and $\Delta H = \Delta E + (\Delta n)RT$.
- P is the symbol used to represent pressure in atmospheres in equations such as $PV = nRT$, and $\Delta H = \Delta E + P\Delta V$.
- q is the symbol used to represent the heat which is either absorbed or evolved from a system. The quantity is usually measured in joules.
- q_p is the symbol used to represent the heat which is either absorbed or evolved from a system when a process occurs at constant pressure.

- q_V is the symbol used to represent the heat which is either absorbed or evolved from a system when a process occurs at constant volume.
- R is the symbol used to represent the gas constant, $R = 0.082056$ liter·atm/(K·mol).
- S is the symbol used to represent the entropy in equations such as $G = H - TS$, and $\Delta G = \Delta H - T\Delta S$.
- T is the symbol used to represent temperature in Kelvin, K, in equations such as $PV = nRT$, and $\Delta G = \Delta H - T\Delta S$.
- V is the symbol used to represent volume in equations such as $PV = nRT$, and $H = E + PV$.
- w is the symbol used to represent work in equations such as $\Delta E = q - w$.

EXERCISES

I. Answer each of the following with *true* or *false*. If the statement is false, correct it.

_____

1. Consider a 1-mL sample of pure water at 25°C and 1 atm (state A). The sample is cooled to 1°C and then the pressure is reduced to 0.1 atm (state B). It takes 5 hours to carry out the change from state A to state B. The sample is then heated and the pressure is raised to 1 atm. In 20 sec the water is at 25°C (a return to state A). The internal energy change in going from state A to B is equal to, but opposite in sign from, the change from B to A.

_____

2. The work done in changing from A to B and that done in changing from B to A in statement 1 of this section are numerically the same but opposite in sign.

_____

3. At constant pressure the amount of heat absorbed or evolved by a system is called the enthalpy change, ΔH.

_____

4. If volume does not change, the amount of heat released during a change of state of a system is equal to the decrease in internal energy of that system.

_____

5. If ΔS for a change in a system is positive, the change is said to be thermodynamically spontaneous.

_____

6. The more positive the value of ΔS_{total} for a change, the more rapid the change.

_____

7. For a spontaneous change the Gibbs free energy change is less than zero.

_____

8. The standard free energy of formation of any element in its standard state is zero.

_____   9. The value of ΔG of a reaction is independent of
pressure.

_____   10. The value of Gibbs free energy of formation is positive
for every compound.

_____   11. The entropy of a mole of solid sodium is less than that
of a mole of gaseous sodium.

_____   12. More heat is released to the surroundings when a mole
of H_2 is burned at constant pressure (an open flame)
than when it is burned at constant volume (a bomb
calorimeter).

II. Complete each of the following statements with an entry
from the list on the right. An entry may be used more
than once.

1. The energy of the universe is constant,   (a) enthalpy
   but the _____ tends toward a
   maximum.

2. The _____ of a reaction can be   (b) entropy
   measured indirectly with a bomb
   calorimeter.

3. The ΔS of dissolution of NaCl in H_2O (c) zero
   is a _____ value.

4. The amount of work done by a process   (d) negative
   carried out at constant volume is

   _____ .

5. By convention, in the equation describ-   (e) positive
   ing the change in internal energy of a
   system, $\Delta E = q - w$, work done by the
   system is always _____ .   (f) gram

6. Entropy, enthalpy, and internal energy   (g) atmosphere
   are _____ functions.

7. For any thermodynamically spontaneous   (h) newton
   change the total change in entropy must
   be a _____ value.

8. For any thermodynamically spontaneous   (i) activity
   change the Gibbs free energy must be a
   _____ value.

9. For a system at equilibrium $\Delta G°$ equals   (j) state
   _____ .

10. The thermodynamic function that does not necessitate a consideration of the changes in surroundings is _____ .

    (k) Gibbs free energy

III. Answer each of the following:

1. Use the information in Table 5.1 of your text and the heat of combustion of CH_4 at 25°C and 1 atm, $\Delta H = 803.3$ kJ/mol, to calculate ΔH for the following reaction at 25°C and 1 atm:

   $$CO_2(g) + 2H_2(g) \rightarrow 2H_2O(g) + C(s)$$

2. What is ΔE for the reaction described in problem 1 above?

3. An unknown reaction is carried out in a bomb calorimeter containing only carbon and water. The heat of reaction at 25°C is found to be 131.4 kJ/mol. Which of the following reactions occurred?

   $$C(s) + 2H_2O(g) \rightarrow CO_2(g) + 2H_2(g)$$
   $$C(s) + H_2O(g) \rightarrow CO(g) + H_2(g)$$

4. Use values of ΔG_f° to determine ΔG for the oxidation of aluminum at 25°C and 1 atm:

   $$4Al(s) + 3O_2(g) \rightarrow 2Al_2O_3(s)$$

5. Calculate K_p for the reaction

   $$C(s) + H_2O(g) \rightarrow CO(g) + H_2(g)$$

6. Use the data in Table 5.2 of your text to determine the standard enthalpy of formation of $CF_4(g)$. The heat of atomization of $C(s)$ is 720 kJ/mol.

7. Use the values of ΔH_f° and ΔG_f° given in Tables 5.1 and 19.4 of your text to determine the change in entropy for the following reaction at 25°C and 1 atm:

8. Compare the value of ΔS calculated in problem 7 above with the value computed from the data in Table 19.6 of your text.

9. When 1.46 g of adipic acid, $C_6H_{10}O_4$, is burned at 25°C and constant pressure, the products are $CO_2(g)$ and $H_2O(l)$. If 27.991 kJ are released in the reaction, what is ΔH for the reaction of 1 mole of adipic acid?

10. For the reaction

$$2\ HI \rightleftharpoons H_2 + I_2$$

$\Delta H°$ is 12.3 kJ. The equilibrium constant K_p at 669 K is 1.63.   What is K_p at 769 K?

ANSWERS TO
EXERCISES

I. Concepts of chemical thermodynamics

1. True

Internal energy is a state function.   The difference in the internal energies of the two states is independent of the path taken between states.

2. False

Work depends on the path, i.e., the method of going from one state to another.

3. True

The relationship between ΔH and ΔE is

$$\Delta H = \Delta E + P\Delta V$$

At constant volume the amount of heat absorbed or evolved by a system is the internal energy change, ΔE.

4. True

At constant volume

$$\Delta H = \Delta E$$

and

$$q_V = q_P$$

5. False

For a spontaneous change ΔS_{total} is positive.   Since $\Delta S_{total} = \Delta S_{system} + \Delta S_{surroundings}$, it is possible to have a negative value of ΔS_{system} and still have a thermodynamically spontaneous reaction.

6. False

A positive value of ΔS_{total} indicates that the change is thermodynamically favorable.   The change, however, may not occur at an observable rate.

7. True

8. True

9. False

The values of ΔG and ΔS and the amount of work done by the system are not independent of pressure.

10. False

Values of $\Delta G_f°$ can be positive or negative.

11. True   The solid is more ordered than the gas.

12. True   At constant volume the heat released is

$$\Delta E = q_V$$

No pressure-volume work is done. At constant pressure

$$q_p = \Delta E + P\Delta V$$

Since the ΔV for the reaction is negative, the heat released to the surroundings by the burning of H_2 is more at constant pressure than at constant volume:

$$H_2(g) + 1/2\ O_2(g) \rightarrow H_2O(g)$$

II. Conventions and terms of thermodynamics

1. (b) entropy
2. (a) enthalpy
3. (e) positive
4. (c) zero
5. (e) positive
6. (j) state
7. (e) positive
8. (d) negative
9. (c) zero
10. (k) Gibbs free energy

III. Chemical thermodynamic calculations

1. $\Delta H = -89.1$ kJ   Add the appropriate equations and the corresponding enthalpy values from Table 5.1 of your text:

$$
\begin{array}{lr}
CH_4(g) \rightarrow C(s) + 2H_2(g) & \Delta H = 74.85\ \text{kJ} \\
CO_2(g) + 2H_2O(g) \rightarrow 2O_2(g) + CH_4(g) & \Delta H = 803.3\ \text{kJ} \\
4H_2(g) + 2O_2(g) \rightarrow 4H_2O(g) & \Delta H = 4(-241.8)\ \text{kJ} \\
\hline
CO_2(g) + 2H_2(g) \rightarrow 2H_2O(g) + C(s) & \Delta H = -89.05\ \text{kJ}
\end{array}
$$

2. $\Delta E = -86.6$ kJ   The relationship between ΔE and ΔH is

$$\Delta E = \Delta H - \Delta nRT$$

Substituting values into the equation, we find

$$\Delta E = \Delta H - \Delta nRT$$
$$= -89.05\ \text{kJ} - (-1)(8.314 \times 10^{-3}\ \text{kJ/(K·mol)})(298\ \text{K})$$
$$= -86.57\ \text{kJ}$$

3. $C(s)$ +
   $2H_2O(g) \rightarrow$
   $CO(g)$ +
   $2H_2(g)$

Use the information in Table 5.1 of your text to calculate the theoretical value of ΔH for each reaction and the information in the problem to calculate the experimental value of ΔH for each reaction. For the reaction

$$C(s) + 2H_2O(g) \rightarrow CO_2(g) + 2H_2(g)$$

first determine the value of ΔH from values of ΔH_f°:

$$\Delta H = \Delta H_f^\circ(CO_2) - 2\Delta H_f^\circ(H_2O)$$

$$= 393.5 \text{ kJ} - 2(-241.8 \text{ kJ})$$

$$= 90.1 \text{ kJ}$$

For the reaction

$$C(s) + H_2O(g) \rightarrow CO(g) + H_2(g)$$

determine the value of ΔH from values of ΔH_f°:

$$\Delta H = \Delta H_f^\circ(CO) - \Delta H_f^\circ(H_2O)$$

$$= -110.5 \text{ kJ} - (-241.8 \text{ kJ})$$

$$= 131.3 \text{ kJ}$$

Then determine the value of ΔH from experimental data:

$$\Delta H = \Delta E + \Delta nRT$$

$$= 131.3 \text{ kJ} + (1)(8.314 \times 10^{-3} \text{ kJ}/(K \cdot mol))(298 \text{ K})$$

$$= 133.8 \text{ kJ}$$

The values of ΔH determined from values of ΔH_f° and from experimental data for the reaction

$$C(s) + 2H_2O(g) \rightarrow CO(g) + 2H_2(g)$$

are in closer agreement than those values of ΔH determined for the reaction

$$C(s) + H_2O(g) \rightarrow CO_2(g) + H_2(g)$$

Therefore, the water and carbon react in the calorimeter to form carbon monoxide and hydrogen.

4. $\Delta G =$
   -3152.8 kJ

The equation

$$4Al(s) + 3O_2(g) \rightarrow 2Al_2O_3(s)$$

represents the formation of 2 mol of Al_2O_3 at 25°C and 1 atm. From Table 19.4 of your text we note

$$2Al(s) + 3/2\ O_2(g) \rightarrow Al_2O_3(s) \qquad \Delta G_f^\circ = -1576.4\ \text{kJ}$$

Thus,

$$4Al(s) + 3O_2(g) \rightarrow 2Al_2O_3(s) \qquad \Delta G_f^\circ = -3152.8\ \text{kJ}$$

5. $K_p =$
   1.01×10^{-16}

Since

$$\Delta G^\circ = -2.303RT \log K_p$$

we must determine ΔG° for the reaction:

$$\Delta G^\circ = \Delta G_f^\circ(CO) - \Delta G_f^\circ(H_2O)$$

$$= -137.28\ \text{kJ} - (-228.61\ \text{kJ})$$

$$= 91.33\ \text{kJ}$$

Rearranging

$$\Delta G^\circ = -2.303RT \log K_p$$

and substituting values into it, we find

$$\log K_p = -\frac{\Delta G^\circ}{2.303RT}$$

$$= -\frac{91.33\ \text{kJ/mol}}{(2.303)(8.314 \times 10^{-3}\ \text{kJ/(K·mol)})(298.2\ \text{K})}$$

$$= -15.996$$

and

$$K_p = 1.01 \times 10^{-16}$$

6. -910 kJ

We add the appropriate equations and corresponding values of ΔH:

$C(s) \rightarrow C(g)$	$\Delta H = 720\ \text{kJ}$
$2F_2(g) \rightarrow 4F(g)$	$\Delta H = 2(155\ \text{kJ})$
$C(g) + 4F(g) \rightarrow CF_4(g)$	$\Delta H = 4(-485\ \text{kJ})$
$C(s) + 2F_2(g) \rightarrow CF_4(g)$	$\Delta H = -910\ \text{kJ}$

The value of ΔH_f° for $CF_4(g)$ from Table 5.1 of your text is -913.4 kJ

7. -4.83 J/K   First determine ΔH for the reaction:

$$\Delta H = \Delta H_f^\circ(CO_2) + 2\Delta H_f^\circ(H_2O) - \Delta H_f^\circ(CH_4)$$

$$= -393.5 \text{ kJ} + 2(-241.8 \text{ kJ}) - (-74.85 \text{ kJ})$$

$$= -802.25 \text{ kJ}$$

Then determine ΔG for the reaction:

$$\Delta G = \Delta G_f^\circ(CO_2) + 2\Delta G_f^\circ(H_2O) - \Delta G_f^\circ(CH_4)$$

$$= -394.38 \text{ kJ} + 2(-228.61 \text{ kJ}) - (-50.79 \text{ kJ})$$

$$= -800.81 \text{ kJ}$$

Finally calculate ΔS for the reaction:

$$\Delta S = \frac{\Delta H - \Delta G}{T}$$

$$= \frac{(-802.25 \text{ kJ}) - (-800.81 \text{ kJ})}{298.2\text{K}}$$

$$= -4.83 \text{ J/K}$$

8. -5.3 J/K   First calculate $\Delta S°$:

$$\Delta S = [S°(CO_2) + 2S°(H_2O)] - [S°(CH_4) + 2S°(O_2)]$$

$$= 213.6 \text{ J/K} + 2(188.7 \text{ J/K}) - 186.2 \text{ J/K} - 2(205.03 \text{ J/K})$$

$$= -5.3 \text{ J/K}$$

9. -2800 kJ   First write a balanced equation for the reaction:

$$2C_6H_{10}O_4(s) + 13O_2(g) \rightarrow 10H_2O(l) + 12CO_2(g)$$

Then compute the number of moles of $C_6H_{10}O_4$ in 1.46 g of $C_6H_{10}O_4$:

$$? \text{ mol } C_6H_{10}O_4 = 1.46 \text{ g } C_6H_{10}O_4\left(\frac{1 \text{ mol } C_6H_{10}O_4}{146.1 \text{ g } C_6H_{10}O_4}\right)$$

$$= 1.00 \times 10^{-2} \text{ mol } C_6H_{10}O_4$$

We know that 27.991 kJ is released when 1.00×10^{-2} mol of $C_6H_{10}O_4$ is burned. Therefore, we can calculate the amount of heat released by 1 mol of $C_6H_{10}O_4$:

$$\Delta H = \left(\frac{27.991 \text{ kJ}}{0.01 \text{ mol } C_6H_{10}O_4}\right)(1 \text{ mol } C_6H_{10}O_4)$$

$$= 2800 \text{ kJ}$$

10. $K_p = 2.07$ Solve for K_2 in the equation which relates the change in equilibrium constants to temperature:

$$\log \frac{K_2}{K_1} = \left(\frac{\Delta H^\circ}{2.303R}\right)\left(\frac{T_2 - T_1}{T_2 T_1}\right)$$

After substituting the values from this problem, the equation looks like this:

$$\log \frac{K_2}{1.63} = \left(\frac{12,300 \text{ J/mol}}{(2.303)[8.314\text{J}/(\text{K} \cdot \text{mol})]}\right)\left(\frac{769 \text{ K} - 669 \text{ K}}{(769 \text{ K})(669 \text{ K})}\right)$$

Performing the operations within the large parentheses yields:

$$\log \frac{K_2}{1.63} = (642 \text{ K})(1.94 \times 10^{-4})$$

$$= 0.125$$

Then rearrange the logarithm:

$$\log(K_2) - \log(1.63) = 0.125$$

or

$$\log(K_2) = 0.125 + \log(1.63)$$

$$= 0.125 + 0.212$$

$$= 0.337$$

Finally, take the antilogarithm:

$$K_2 = 2.07$$

SELF-TEST   Complete the test in 15 minutes:

1. The heat of combustion of oxalic acid, $H_2C_2O_4$, is −253 kJ/mol at 25°C and constant volume. What is ΔH?

2. Using standard entropies and enthalpies of formation at 25°C determine if the reaction

$$CO(g) + H_2O(g) \rightarrow CO_2(g) + H_2(g)$$

is spontaneous as written and calculate K_p.

3. Use the values for absolute entropy from Table 19.5 of your text and calculate the standard entropy for formation of HgO(s). Since the standard enthalpy of formation of HgO(s) is -90.7 kJ/mol, calculate the standard free energy of formation of HgO(s).

4. For the reaction

$$CO(g) + H_2O(g) \longrightarrow H_2(g) + CO_2(g)$$

K_p = 5.10 at 800 K; what is K_p at 900 K?   The value of $\Delta H°$ for the reaction is -33 kJ/mol.

ELECTROCHEMISTRY

CHAPTER

20

OBJECTIVES   I. You should be able to demonstrate your knowledge of the
following terms by defining them, describing them, or
giving specific examples of them:

ampere [20.1]
anode [20.2]
cathode [20.2]
concentration cell [20.10]
conduction [20.1, 20.2]
corrosion [20.12]
coulomb [20.1]
coulometer [20.4]
current [20.1]
disproportionation [20.7]
electrolysis [20.3, 20.4, 20.11]
electrolytic cell [introduction, 20.2]
electromotive force, emf [20.5]
equilibrium constant [20.8]
faraday [20.4]
Faraday's laws [20.4]
fuel cell [20.12]
galvanic cell [introduction]
Gibbs free energy [20.8]
Nernst equation [20.9, 20.10]
ohm [20.1]
Ohm's law [20.1]
overvoltage [20.11]
resistance [20.1]
salt bridge [20.7]
standard electrode potential [20.7]

standard emf [20.6]
standard hydrogen electrode [20.7]
volt [20.1]
voltaic cell [introduction, 20.5, 20.13]

II. You should be able to use Faraday's laws.

III. You should be able to determine standard electrode potentials for reactions from tabulated values.

IV. Using a table of standard electrode potentials, you should be able to determine the potential of an electrochemical cell for any set of concentrations of reactants and products.

V. You should be able to diagram electrochemical cells and determine the electrode signs, the direction of the electron movement, and the direction of ion migration.

VI. You should be able to determine $\mathscr{E}°$ and K from Gibbs free energy data.

VII. You should be able to predict $\mathscr{E}$ for a cell from $\mathscr{E}°$ data and concentrations of all reactants and products by using the Nernst equation.

UNITS, SYMBOLS, MATHEMATICS

I. You should understand and be able to manipulate logarithms in order to work problems using the Nernst equation. Please read the appendix in your text and the material in Chapter 14 of the Study Guide.

Do not depend on your calculator entirely. There are limits to its ability, but it will cause you no difficulty if you understand what you are doing. Common logarithms are powers, i.e., exponents, to which the number 10 must be raised to obtain the number. For example:

$$100 = 10^2$$

so

$$\log 100 = 2$$

or 10 must be raised to the second power to obtain the number 100. Also remember that

$1 = 10^0$

$\log 1 = 0$

and

$10 = 10^1$

$\log 10 = 1$

To avoid difficulty in determining the logarithm of a number other than 1, 10, 100, and so on, you should always write the number in proper scientific notation. (See Chapter 1 of the study guide.) When this is done, you will have a number between 1 and 10, which is multiplied by 10 raised to a whole number exponent. For example, the number 106 is written as

1.06×10^2

Then take the log of the number written in scientific notation:

$\log (1.06 \times 10^2) = \log 1.06 + \log 10^2$

Remember that the log of a number between 1 and 10 is a number between 0 and 1, so

$\log 1.06 + \log 10^2 = 0.025 + 2 = 2.025$

There are only three significant figures in 2.025. The number before the decimal, called the characteristic, indicates the power to which 10 is raised when the original number is written in proper scientific notation, and the decimal portion, called the mantissa, gives the rest of the number. Each number, including the zero in a mantissa, is significant. Therefore, the following logarithms are reported with three significant figures:

$\log (1.06 \times 10^2) = 2.025$

$\log (1.06 \times 10^{24}) = 24.025$

$\log (1.06 \times 10^{124}) = 124.025$

The same rules apply for small numbers. Three significant figures are reported here:

$$\log (1.06 \times 10^{-24}) = 0.025 - 24 = -23.975$$
$$\log (1.06 \times 10^{-124}) = -123.975$$

Do the following on your calculator, but make sure that you can report the result properly:

$$\log 76.23 = 1.8821$$
$$\log \left(\frac{7.623}{22.4}\right) = -0.468$$
$$\log (6.023 \times 10^{24}) = 24.7798$$
$$\log \left(\frac{6.023 \times 10^{24}}{1.7 \times 10^{-86}}\right) = 110.55$$
$$\log 7674 = \log (7.674 \times 10^{3}) = 3.8850$$
$$\log (0.0243) = -1.614$$
$$\log (1.0243) = 0.01043$$

II. The following symbols were used in this chapter; you should be familiar with them.

- A is the abbreviation for ampere, the SI unit for current.
- C is the abbreviation for coulomb, the SI unit for charge: 1 A = 1 C/s, exactly.
- ΔG is the symbol used to represent the change in Gibbs free energy of a system in equations such as $\Delta G = \Delta G° + 2.30RT \log Q$.
- $\Delta G°$ is the symbol used to represent the change in the Gibbs free energy of a system in its standard state, 298 K and 1 atm. $\Delta G° = -nF\mathscr{E}°$.
- $\Delta S°$ is the symbol used to represent the change in entropy of a system in its standard state.
- $\Delta H°$ is the symbol used to represent the change in enthalpy of a system in its standard state.
- $\mathscr{E}$ is the symbol used to represent a potential difference in expressions such as Ohms law, $\mathscr{E} = IR$.
- $\mathscr{E}°$ is the symbol used to represent standard electrode potentials.
- emf is the abbreviation for electromotive force.
- F is the abbreviation for the faraday, 1 F = 96,000 C.

- I is the symbol used to represent current in mathematical expressions such as the one stating Ohms law, $\mathscr{E} = IR$.
- M is the abbreviation for moles per liter, mol/L a commonly used unit of concentration.
- n is the symbol used to represent the number of moles of material in mathematical expressions such as $PV = nRT$ and $\Delta G = -nF\mathscr{E}$.
- Ω is the abbreviation for ohm, the unit of resistance.
- R is the symbol used to represent resistance in mathematical expressions such as $\mathscr{E} = IR$.
- STP is the abbreviation for standard temperature and pressure, 1 atm and 273 K.

EXERCISES

I. Answer each of the following with *true* and *false*. If a statement is false, correct it.

_____ 1. In an electrolytic cell, oxidation occurs at the positive electrode.

_____ 2. In a galvanic cell, oxidation occurs at the positive electrode.

_____ 3. The anode of a cell is indicated on the left in the cell notation.

_____ 4. In both galvanic and electrolytic cells, reduction occurs at the cathode.

_____ 5. A positive cell potential indicates that the reaction should occur spontaneously as written.

_____ 6. A battery is an electrolytic cell.

_____ 7. One faraday is required to reduce 1 mole of Cu^{2+} ions to $Cu(s)$.

_____ 8. An Avogadro's number of Ni^{2+} ions can be reduced to $Ni(s)$ with 2 faradays.

_____ 9. Four faradays produce 22.4 L of O_2 at STP in the electrolysis of water:

$$2H_2O(l) \rightarrow O_2(g) + 4H^+(aq) + 4e^-$$

_____ 10. Electrode potentials are temperature dependent.

_____ 11. The best oxidizing agent in Table 20.1 of the study guide is F_2.

_____ 12. The best reducing agent in Table 20.1 is Li^+.

_____ 13. According to Ohm's law, potential difference is directly proportional to current, I.

_____ 14. In the electrolysis of a unit activity solution of Na_2SO_4 using Pt electrodes, the reaction occurring at the anode is

$$2H_2O(l) \rightarrow O_2(g) + 4H^+(aq) + 4e^-$$

_____ 15. According to Faraday's law, 1 F would liberate 2 g of H_2 in the reduction of hydrogen ion.

_____ 16. The standard hydrogen electrode consists of hydrogen gas at 1 atm pressure bubbling over a Pt electrode that is immersed in an acid solution containing $H^+(aq)$ at unit activity.

_____ 17. The addition of the standard electrode potentials of the half reactions gives the value of the emf of a cell.

_____ 18. In the cell

$$Cd \mid Cd^{2+} \| Ni^{2+} \mid Ni$$

The Cd electrode gains weight.

_____ 19. For the cell

$$Cr \mid Cr^{3+}(0.001 \ M) \| Fe^{2+}(0.01 \ M) \mid Fe$$

the Nernst equation is

$$\mathscr{E} = +0.304 \ V - (0.0592/6) \ \log \left(\frac{(10^{-3})^2}{(10^{-2})^3} \right)$$

_____ 20. If $G°$ is positive then $\mathscr{E}°$ is also positive for the system.

_____ 21. An ampere is a coulomb/sec.

_____ 22. A faraday equals 0.0592 coulomb.

II. Answer each of the following:

1. Using data from Table 20.1 in the study guide, calculate the standard electrode potential for the half reaction $Cr_2O_7^{2-} \to Cr$.

2. Corrosion is an oxidation process. Often iron is coated with other metals to protect it from rusting. Determine what would happen to iron if it were coated with tin, as in tin cans, and the coating cracked, exposing the iron to air and water. (Hint: Assume that the metal forms an electrochemical cell in solution, the cathode reaction being the reduction of O_2 to OH^-.)

3. If the iron of problem 2 above were galvanized, i.e., coated with zinc, what would happen if the coating cracked and the iron became exposed to air and water. (Hint: Assume that the metal forms an electrochemical cell in solution, the cathode reaction being the reduction of O_2 to OH^-.)

4. During cellular respiration, oxidation-reduction reactions take place. If the free energy of the reaction NAD to FAD (nicotinamide adenine dinucleotide to flavin adenine dinucleotide) is -12.00 kcal/mol and the standard emf is +0.26 V, how many electrons are involved in this oxidation step in the respiratory chain?

TABLE 20.1 Standard Electrode Potentials at 25°C

Half Reaction	$\mathscr{E}°$ (volts)
$Li^+ + e^- \rightleftharpoons Li$	-3.045
$K^+ + e^- \rightleftharpoons K$	-2.925
$Ba^{2+} + 2e^- \rightleftharpoons Ba$	-2.906
$Ca^{2+} + 2e^- \rightleftharpoons Ca$	-2.866
$Na^+ + e^- \rightleftharpoons Na$	-2.714
$Mg^{2+} + 2e^- \rightleftharpoons Mg$	-2.363
$Al^{3+} + 3e^- \rightleftharpoons Al$	-1.662
$2H_2O + 2e^- \rightleftharpoons H_2 + 2OH^-$	-0.82806
$Zn^{2+} + 2e^- \rightleftharpoons Zn$	-0.7628
$Cr^{3+} + 3e^- \rightleftharpoons Cr$	-0.744
$Fe^{2+} + 2e^- \rightleftharpoons Fe$	-0.4402
$Cd^{2+} + 2e^- \rightleftharpoons Cd$	-0.4029
$Ni^{2+} + 2e^- \rightleftharpoons Ni$	-0.250
$Sn^{2+} + 2e^- \rightleftharpoons Sn$	-0.136
$Pb^{2+} + 2e^- \rightleftharpoons Pb$	-0.126
$2H^+ + 2e^- \rightleftharpoons H_2$	0
$Cu^{2+} + 2e^- \rightleftharpoons Cu$	+0.337
$Cu^+ + e^- \rightleftharpoons Cu$	+0.521
$I_2 + 2e^- \rightleftharpoons 2I^-$	+0.5355
$Fe^{3+} + e^- \rightleftharpoons Fe^{2+}$	+0.771
$Ag^+ + e^- \rightleftharpoons Ag$	+0.7991
$Br_2 + 2e^- \rightleftharpoons 2Br^-$	+1.0652
$O_2 + 4H^+ + 4e^- \rightleftharpoons 2H_2O$	+1.229
$Cr_2O_7^{2-} + 14H^+ + 6e^- \rightleftharpoons 2Cr^{3+} + 7H_2O$	+1.33
$Cl_2 + 2e^- \rightleftharpoons 2Cl^-$	+1.3595
$MnO_4^- + 8H^+ + 5e^- \rightleftharpoons Mn^{2+} + 4H_2O$	+1.51
$F_2 + 2e^- \rightleftharpoons 2F^-$	+2.87

5. In the Dow process for obtaining magnesium, the magnesium
   is precipitated from seawater as the hydroxide, which is
   dissolved in HCl after purification. The resulting
   magnesium chloride solution is evaporated, and the magnesium chloride is melted and electrolyzed.
   Answer the following:
   (a) How many hours would it take to obtain 5.0 kg of
       magnesium metal using 1.0×10^4 amperes?
   (b) What volume of chlorine gas at STP would be evolved
       in the time it takes to obtain the 5.0 kg of Mg
       metal using 1.0×10^4 amperes?

6. Predict whether or not each of the following reactions will occur spontaneously and write a balanced chemical equation for each reaction that is predicted to occur. Assume that each reactant and product is present at unit activity in aqueous solution at 25°C.
   (a) Cd^{2+} ion reduced to Cd by Ni
   (b) Sn oxidized to Sn^{2+} by Cu^{2+} ion
   (c) I_2 reduced to I^- ion by Ag
   (d) MnO_4^- ion reduced to Mn^{2+} ion by Br^- ion

7. Write the shorthand cell notation for the spontaneous reactions in problem 6 above.

8. What is the potential of the cell?

$$Pt\,|\,Fe^{3+}\,(0.010\ M),\ Fe^{2+}\,(0.12\ M)\,\|\,Fe^{3+}\,(0.13\ M),\ Fe^{2+}\,(0.71\ M)\,|\,Pt$$

9. Use Table 20.1 of the study guide to predict what will happen when a bar of cadmium is added to a solution of Fe^{3+}.

10. The salt $Co_2(SO_4)_3 \cdot 18H_2O$ decomposes when added to water. Why?

11. If 17.6 liters of O_2 at STP are evolved at the anode in an electrochemical cell and the only reaction at the cathode is the reduction of Cu^{2+} to Cu(s), how many grams of Cu are formed?

12. Use the standard electrode potentials of bromine-containing compounds in basic solution

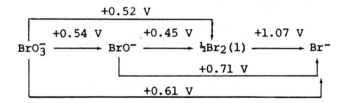

to answer the following:
   (a) Does BrO^- disproportionate in basic solution?
   (b) Does Br_2 disproportionate in basic solution?

13. The potential difference between two hydrogen cells $(H_2 \rightarrow 2H^+ + 2e^-)$ is 0.076 V. One of the cells contains 1.0 M H^+ and H_2(g) at 1 atm. The other contains 1 atm H_2(g). What is the concentration of H^+ in the second cell?

$$Pt\,|\,H_2\,|\,H^+\,(?\ M)\,\|\,H^+\,(1.0\ M)\,|\,H_2\,|\,Pt$$

14. Calculate the concentration of Fe^{2+} and Fe^{3+} in a half cell that contains a total iron concentration of 1.00×10^{-3} M if the potential difference of the cell with respect to a standard hydrogen electrode is -0.694 V.

$$Pt\,|\,Fe^{2+}(?\ M)\,||\,H^+(1\ M)\,|\,H_2(1\ atm)\,|\,Pt$$

15. For 3.00 hours a current of 10.0 amp plated out 13.6 g of a substance. How many grams are plated by a mole of electrons?

16. In a fuel cell hydrogen and oxygen can react to produce electricity. The process is represented by the cell in Figure 20.1 of the study guide.

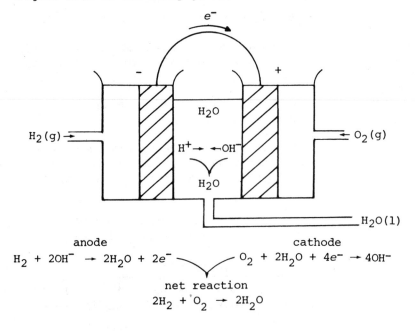

anode
$$H_2 + 2OH^- \rightarrow 2H_2O + 2e^-$$

cathode
$$O_2 + 2H_2O + 4e^- \rightarrow 4OH^-$$

net reaction
$$2H_2 + O_2 \rightarrow 2H_2O$$

Figure 20.1 Fuel Cell

In the process hydrogen gas is oxidized to H^+ ion at the anode, and the oxygen gas is reduced to OH^- at the cathode. The excess water is removed. Answer the following:
(a) If 47.8 L of hydrogen at STP react in 10.0 minutes, how many liters of oxygen at STP react in the same time?
(b) How many moles of water are produced in 10.0 minutes?
(c) What is the average current produced during this time?

17. Wind generators are attractive sources of energy except for the fact that the wind occasionally ceases to blow. It has been proposed that the generator be used to electrolyze water during off-peak power times during these hours and that the hydrogen and oxygen generated during these hours could be used in a fuel cell to generate electricity when the wind stops blowing. If the efficiency is 100 percent—i.e., no energy losses occur—what size of tank would be needed to store sufficient hydrogen and oxygen at STP to supply a single residence with 20 amps for 5.0 hours?

18. Calculate the equilibrium constant for the reaction

$$H_2O \rightarrow H^+(aq) + OH^-(aq)$$

from the following electrode potentials:

$$2H_2O + 2e^- \rightarrow H_2(g) + 2OH^-(aq) \qquad \mathscr{E}° = -0.828 \text{ V}$$
$$2H^+(aq) + 2e^- \rightarrow H_2(g) \qquad \mathscr{E}° = 0.000 \text{ V}$$

19. Use values of $\Delta G_f°$ to determine $\mathscr{E}°$ for the reaction

$$2Al(s) + 3/2\ O_2(g) \rightarrow Al_2O_3(s) \qquad \Delta G_f° = -1576.4 \text{ kJ}$$

III. Do the following for each of the electrochemical cells shown on pages 289, 290, and 291.

(a) Label the anode and the cathode.
(b) Write the reaction occurring at the anode and that occurring at the cathode.
(c) Show the direction of the electron movement in the external connection.
(d) Indicate the sign of the cathode and that of the anode.
(e) Show the direction of ion movement in the cell.
(f) Diagram the cell in the shorthand notation described in Section 20.7 of your text. Assume that all concentrations are 1 M.
(g) Determine the half-cell reactions from the standard electrode potentials in Table 20.1 of the study guide.
(h) Calculate the cell potential.

1.

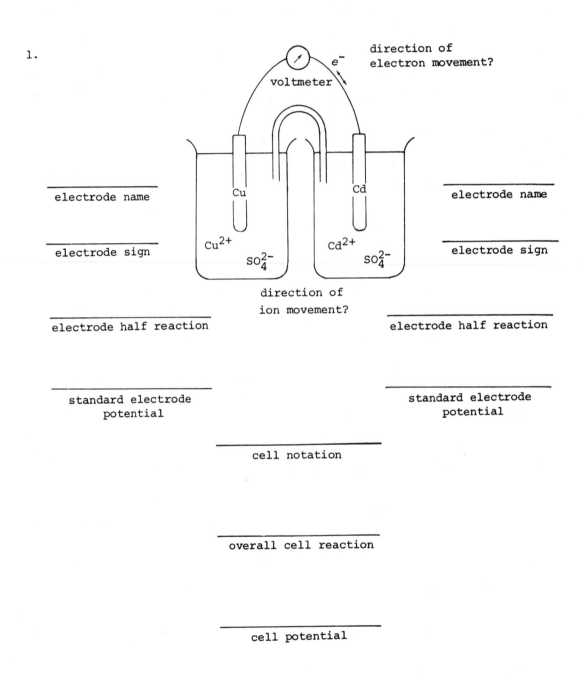

2.

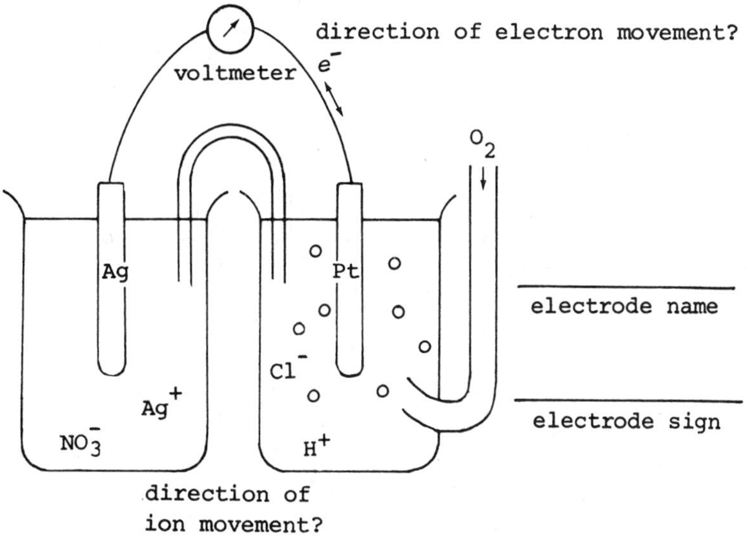

direction of electron movement?

voltmeter e^-

O_2

Ag

Pt

O O O O O

electrode name

Cl^-

Ag^+

O O

electrode name

electrode sign

NO_3^-

H^+

electrode sign

direction of
ion movement?

_____

electrode half reaction

_____

electrode half reaction

_____

standard electrode
potential

_____

standard electrode
potential

_____

cell notation

_____

overall cell reaction

_____

cell potential

3.

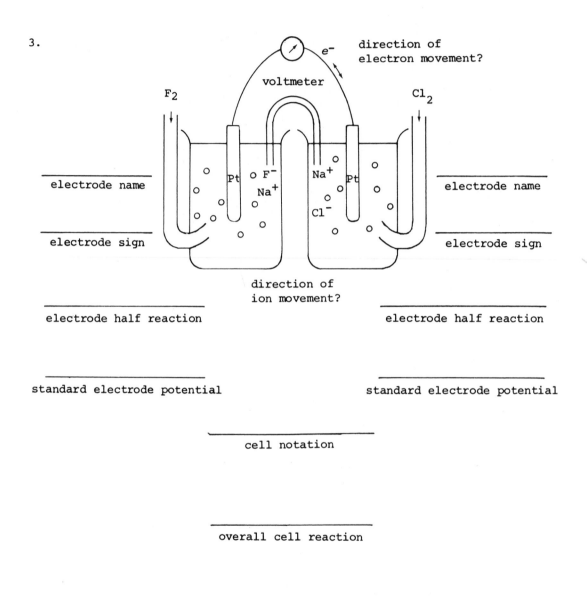

electrode name

electrode sign

electrode half reaction

standard electrode potential

electrode name

electrode sign

electrode half reaction

standard electrode potential

cell notation

overall cell reaction

cell potential

ANSWERS TO
EXERCISES

1. True

2. False

I. Basic electrochemical principles

In a galvanic cell the positive electrode is the cathode, and reduction occurs at this electrode. The conventions relating to the terms anode and cathode are outlined in Table 20.2 of the study guide.

TABLE 20.2 Electrode conventions

	Cathode	Anode
ions attracted	cations	anions
direction of electron movement	into cell	out of cell
half reaction	reduction	oxidation
sign		
electrolysis cell	negative	positive
galvanic cell	positive	negative

You may find it easier to remember the conventions relating to a galvanic, or voltaic, cell by using the mnemonic device outlined in Table 20.3 of the study guide. Note that when the opposite electrochemical terms are alphabetized, the first term describes the anode and the second describes the cathode.

For an electrolysis cell, i.e., a cell in which an external power supply causes the reaction to proceed in a non-spontaneous direction, the mnemonic device is valid except that the electrode signs are reversed. In an electrolysis cell, oxidation occurs at the positive electrode, and electrons move away from this electrode in the external circuit. Also, cell notations are not used for electrolysis cells.

TABLE 20.3 Electrode Conventions for Galvanic Cells

	Anode	Cathode
ions attracted	anions	cations
electrode sign	negative	positive
electrode process	oxidation	reduction
direction of electron movement	away	toward
location in cell notation	left	right

3. True

4. True

5. True

6. False

Batteries are sources of energy and are therefore voltaic cells. Electrolytic cells require an energy source to operate.

7. False   The balanced equation

$$Cu^{2+} + 2e^- \rightarrow Cu(s)$$

indicates that 2 mol of electrons, or 2 F, are required to produce 1 mol of $Cu(s)$.

8. True

9. True   The oxygen-producing half reaction

$$2H_2O \rightarrow O_2 + 4H^+ + 4e^-$$

indicates that 4 faradays are required to produce 1 mol of O_2, or 22.4 liters of O_2, at STP.

10. True   Notice that the Nernst equation

$$\mathscr{E} = \mathscr{E}° - \frac{RT}{nF} \ln Q$$

includes temperature, T. The standard electrode potentials in Table 20.1 of the study guide are given for a specific temperature, 25°C.

11. True

12. False   A positive cell emf indicates that the reaction occurs spontaneously as written. In Table 20.1 of the study guide, oxidizing agents are on the left. The stronger the oxidizing agent, the more positive the value of $\mathscr{E}°$. Reducing agents are on the right. The stronger the reducing agent, the more negative the value of $\mathscr{E}°$. Thus, we diagram Table 20.1 of the study guide as follows:

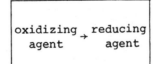

| increasing strength of oxidizing agent | oxidizing agent $\rightarrow$ reducing agent | increasing strength of reducing agent |

The best reducing agent in Table 20.1 of the study guide is Li. The Li^+ ion is reduced; thus, it is an oxidizing agent with an $\mathscr{E}°$ of -3.045 V. The lithium atom is a reducing agent, being oxidized to Li^+ with an $\mathscr{E}°$ of +3.045 V.

13. True   Ohm's law states

$$\mathscr{E} = IR$$

The proportionality constant R is called the resistance.

14. True

15. False   According to the reaction

$$2H^+ + 2e^- \rightarrow H_2$$

1 F, i.e., 1 mol of electrons, would liberate $\frac{1}{2}$ mol of H_2, or $\frac{1}{2}$ mol H_2 (2 g H_2/1 mol H_2) = 1 g H_2.

16. True

17. False   The cell emf is the sum of the half-cell potential for the oxidation reaction and the half-cell potential for the reduction reaction. In the table of standard electrode potentials (Table 20.1 of the study guide), all reactions are written as reduction reactions. Thus, both the sign of one electrode potential and the corresponding reaction must be reversed to obtain the necessary half-cell potential and oxidation reaction.

18. False   By convention, the anode is indicated on the left. The cell reaction as diagrammed is

$$Cd(s) + Ni^{2+} \rightarrow Cd^{2+} + Ni(s)$$

Thus, Ni is deposited on the electrode, the Ni electrode gains weight, and the Cd electrode loses weight.

19. True   The balanced equation

$$2Cr + 3Fe^{2+} \rightarrow 2Cr^{3+} + 3Fe$$

indicates that six electrons are exchanged, $n = 6$, and that

$$Q = \frac{[Cr^{3+}]^2}{[Fe^{2+}]^3} = \frac{[10^{-3}]^2}{[10^{-2}]^3}$$

The metals Cr and Fe have unit activity and are not included in Q. The value of $\mathscr{E}°$ is +0.304 V:

$$3[Fe^{2+} + 2e^- \rightleftharpoons Fe] \qquad\qquad \mathscr{E}° = -0.4402 \text{ V}$$
$$\underline{ 2[Cr \rightleftharpoons Cr^{3+} + 3e^-] \qquad\qquad \mathscr{E}° = +0.744 \text{ V}}$$
$$2\ Cr + 3Fe^{2+} \rightarrow 2Cr^{3+} + 3Fe \qquad \mathscr{E}° = +0.304 \text{ V}$$

20. False   Since $\Delta G° = -nF\mathscr{E}°$, $\Delta G°$ and $\mathscr{E}°$ always have the opposite sign.

21. True

22. False   One faraday equals 96,487 coulombs.

      II. Electrochemical calculations

1. +0.293 V   First write the necessary half reactions, balance them,
                 and add them. Multiply any equation and the corresponding
                 value of ΔG by integers so that unwanted intermediate ions
                 or molecules cancel. The desired result is thus obtained
                 in the same manner as you used in the law of Hess.

$$Cr_2O_7^{2-} + 14H^+ + 6e^- \rightleftharpoons 2Cr^{3+} + 7H_2O$$

$$\Delta G = -6F(+1.33 \text{ V})$$

$$2Cr^{3+} + 6e^- \rightleftharpoons 2Cr$$

$$= -7.98F$$

$$\Delta G = 2(-3F(-0.744 \text{ V})$$
$$= +4.464F$$

$$Cr_2O_7^{2-} + 14H^+ + 12e^- \rightleftharpoons 2Cr + 7H_2O$$

$$\Delta G = -3.516F$$

Rearranging so that 12 electrons appear in the equation
for ΔG:

$$\Delta G = -nF\mathscr{E}°$$
$$-3.516F = -12F\mathscr{E}°$$
$$\mathscr{E}° = +0.293 \text{ V}$$

2. rust   For the tin-coated iron, possible oxidation reactions (see
                 Table 20.1 of the study guide) are

$$Fe \rightarrow Fe^{2+} + 2e^- \qquad \mathscr{E}° = +0.4402 \text{ V}$$
$$Sn \rightarrow Sn^{2+} + 2e^- \qquad \mathscr{E}° = +0.136 \text{ V}$$
$$2H_2O \rightarrow O_2 + 4H^+ + 4e^- \qquad \mathscr{E}° = -1.229 \text{ V}$$

The reduction reaction is

$$O_2 + 2H_2O + 4e^- \rightarrow 4OH^- \qquad \mathscr{E}° = +0.401 \text{ V}$$

Since iron is the best of the three reducing agents, it
will be oxidized—i.e., it will rust—when not protected.

3. no rust

For galvanized iron, possible oxidation reactions (see Table 20.1 of the study guide) are

$$Fe \rightarrow Fe^{2+} + 2e^- \qquad \mathscr{E}° = +0.4402 \text{ V}$$
$$Zn \rightarrow Zn^{2+} + 2e^- \qquad \mathscr{E}° = +0.7628 \text{ V}$$
$$2H_2O \rightarrow O_2 + 4H^+ + 4e^- \qquad \mathscr{E}° = -1.229 \text{ V}$$

The reduction reaction is

$$O_2 + 2H_2O + 4e^- \rightarrow 4OH^- \qquad \mathscr{E}° = +0.401 \text{ V}$$

The iron is protected since zinc is a better reducing agent than iron. The zinc will be oxidized by the O_2 and the H_2O.

4. 2

Use the relationship between standard Gibbs free energy and standard emf:

$$\Delta G° = -nF\mathscr{E}°$$

$$n = \frac{-\Delta G°}{F\mathscr{E}°}$$

$$n = \frac{-(-50.21 \text{ kJ/mol})}{(96.487 \text{ kJ/V})(+0.26 \text{ V})}$$

$$n = 2.0$$

5. (a) 1.1 hr

The reduction reaction is

$$Mg^{2+} + 2e^- \rightarrow Mg$$

Thus, 2 mol of electrons, or 2 F, are required per mole of Mg. We can use conversion factors to solve:

? hr

$$= 5.0 \text{ kg Mg}\left(\frac{10^3 \text{ g}}{1 \text{ kg}}\right)\left(\frac{1 \text{ mol Mg}}{24.3 \text{ g Mg}}\right)\left(\frac{2F}{1 \text{ mol Mg}}\right)\left(\frac{96{,}500 \text{ coulombs}}{1F}\right)$$

$$\times \left(\frac{1}{1.0 \times 10^4 \text{ coulomb/sec}}\right)\left(\frac{1 \text{ min}}{60 \text{ sec}}\right)\left(\frac{1 \text{ hr}}{60 \text{ min}}\right)$$

$$= 1.1 \text{ hr}$$

(b) 4.6×10^3 L Cl_2 The oxidation reaction is

$$2Cl^- \rightarrow Cl_2 + 2e^-$$

Thus, for every mole of Mg produced, 1 mol of Cl_2, or 22.4 L of Cl_2, is obtained. Therefore,

? L Cl_2

$$= 5.0 \text{ kg Mg}\left(\frac{10^3 \text{ g}}{1 \text{ kg}}\right)\left(\frac{1 \text{ mol Mg}}{24.3 \text{ g Mg}}\right)\left(\frac{1 \text{ mol } Cl_2}{1 \text{ mol Mg}}\right)\left(\frac{22.4 \text{ L } Cl_2}{1 \text{ mol } Cl_2}\right)$$

$$= 4.6 \times 10^3 \text{ L } Cl_2$$

The problem can also be solved as follows:

? L Cl_2

$$= 1.1 \text{ hr}\left(\frac{60 \text{ min}}{1 \text{ hr}}\right)\left(\frac{60 \text{ sec}}{1 \text{ min}}\right)\left(\frac{1.0 \times 10^4 \text{ coulombs}}{1 \text{ sec}}\right)$$

$$\times \left(\frac{1F}{96,500 \text{ coulombs}}\right)\left(\frac{1 \text{ mol } Cl_2}{2F}\right)\left(\frac{22.4 \text{ L } Cl_2}{1 \text{ mol } Cl_2}\right)$$

$$= 4.6 \times 10^3 \text{ L } Cl_2$$

6.   Spontaneous reactions are reactions with a positive cell potential.

(a) no   Reaction (a) is not spontaneous since $\mathscr{E}^\circ_{cell} = -0.153$ V:

$$
\begin{array}{ll}
Cd^{2+} + 2e^- \rightarrow Cd & \mathscr{E}^\circ = -0.4029 \text{ V} \\
\underline{\text{Ni} \rightarrow Ni^{2+} + 2e^-} & \underline{\mathscr{E}^\circ = +0.250 \text{ V}} \\
Cd^{2+} + \text{Ni} \rightarrow Cd + Ni^{2+} & \mathscr{E}^\circ_{cell} = -0.153 \text{ V}
\end{array}
$$

(b) yes   Reaction (b) is spontaneous since $\mathscr{E}^\circ_{cell} = +0.473$ V:

$$
\begin{array}{ll}
\text{Sn} \rightarrow Sn^{2+} + 2e^- & \mathscr{E}^\circ = +0.136 \text{ V} \\
\underline{Cu^{2+} + 2e^- \rightarrow Cu} & \underline{\mathscr{E}^\circ = +0.337 \text{ V}} \\
\text{Sn} + Cu^{2+} \rightarrow Sn^{2+} + Cu & \mathscr{E}^\circ_{cell} = +0.473 \text{ V}
\end{array}
$$

(c) no   Reaction (c) is not spontaneous since $\mathscr{E}^\circ_{cell} = -0.2636$ V:

$$
\begin{array}{ll}
I_2 + 2e^- \rightarrow 2I^- & \mathscr{E}^\circ = +0.5355 \text{ V} \\
\underline{2Ag \rightarrow 2Ag^+ + 2e^-} & \underline{\mathscr{E}^\circ = -0.7991 \text{ V}} \\
I_2 + 2Ag \rightarrow 2I^- + 2Ag^+ & \mathscr{E}^\circ_{cell} = -0.2636 \text{ V}
\end{array}
$$

Note that the half reaction for the oxidation of Ag must be multiplied by 2 before addition so that the electrons lost and gained in the half reactions will cancel. The $\mathscr{E}°$ for the Ag/Ag$^+$ electrode, however, is *not* multiplied by 2. The $\mathscr{E}°$ value of a half reaction is independent of the number of electrons lost or gained.

(d) yes

Reaction (d) is spontaneous since

$$2[MnO_4^- + 8H^+ + 5e^- \rightarrow Mn^{2+} + 4H_2O] \qquad \mathscr{E}° = +1.51 \text{ V}$$
$$5[2Br^- \rightarrow Br_2 + 2e^-] \qquad \mathscr{E}° = -1.0652 \text{ V}$$
$$\overline{2MnO_4^- + 16H^+ + 10Br^- \rightarrow 2Mn^{2+} + 8H_2O + 5Br_2 \qquad \mathscr{E}°_{cell} = +0.44 \text{ V}}$$

7.

In standard cell notation the anode is listed first. Oxidation always occurs at the anode. Thus, for reaction (b) the cell is noted

(b) $Sn|Sn^{2+}||Cu^{2+}|Cu$

and for reaction (d) the cell is noted

(d) $Pt|Br^-|Br_2||MnO_4^-|Mn^{2+}|Pt$

The inert platinum electrodes are noted in the cell notation for reaction (d) since they are necessary to make an external electrical connection.

8. +0.079 V

From the Nernst equation

$$\mathscr{E} = \mathscr{E}° - \frac{0.0592}{n} \log Q$$

$$\mathscr{E} = 0.00 - 0.0592 \log \frac{(0.010 \text{ M})(0.071 \text{ M})}{(0.13 \text{ M})(0.12 \text{ M})}$$

$$= -0.0592 \log (4.55 \times 10^{-2})$$

$$= 0.0592(-1.342)$$

$$= +0.079 \text{ V}$$

9.

From Table 20.1 of the study guide choose reactions that contain any of the possible reactants, Cd, Fe^{3+}, and H$_2$O:

Oxidation Reactions

$$Cd \rightarrow Cd^{2+} + 2e^- \qquad \mathscr{E}° = +0.4029 \text{ V}$$
$$2H_2O \rightarrow O_2 + 4H^+ + 4e^- \qquad \mathscr{E}° = -1.229 \text{ V}$$

Reduction Reactions

$$Fe^{3+} + e^- \rightarrow Fe^{2+} \qquad \mathscr{E}° = +0.771 \text{ V}$$

$$Fe^{2+} + 2e^- \rightarrow Fe \qquad \mathscr{E}° = -0.4402 \text{ V}$$

$$2H_2O + 2e^- \rightarrow H_2 + 2OH^- \qquad \mathscr{E}° = 0.82806 \text{ V}$$

The only pair of reactions that has a positive cell potential is

$$Cd \rightarrow Cd^{2+} + 2e^- \qquad \mathscr{E}° = +0.4029 \text{ V}$$

$$\underline{2[Fe^{3+} + e^- \rightarrow Fe^{2+}] \qquad \mathscr{E}° = +0.771 \text{ V}}$$

$$Cd + 2Fe^{3+} \rightarrow Cd^{2+} + 2Fe^{2+} \qquad \mathscr{E}°_{cell} = +1.174 \text{ V}$$

10.

In Appendix E of your text the reduction potential of Co^{3+} is given:

$$Co^{3+} + e^- \rightarrow Co^{2+} \qquad \mathscr{E}° = +1.808 \text{ V}$$

The oxidation potential of water can be obtained from Table 20.1 of the study guide.

$$2H_2O \rightarrow O_2 + 4H^+ + 4e^- \qquad \mathscr{E}° = -1.229 \text{ V}$$

Thus, when the reduction of Co^{3+} and the oxidation of water are coupled, we find that $\mathscr{E}°$ for the oxidation-reduction reaction is positive:

$$4[Co^{3+} + e^- \rightarrow Co^{2+}] \qquad \mathscr{E}° = +1.808 \text{ V}$$

$$\underline{2H_2O \rightarrow O_2 + 4H^+ + 4e^- \qquad \mathscr{E}° = -1.229 \text{ V}}$$

$$4Co^{3+} + 2H_2O \rightarrow 4Co^{2+} + O_2 + 4H^+ \qquad \mathscr{E}° = +0.579 \text{ V}$$

Thus, when the Co^{3+} salt is added to water, the reduction of Co^{3+} and the oxidation of water occur spontaneously.

11. 99.8 g Cu

The balanced equation is

$$2Cu^{2+} + 2H_2O \rightarrow 2Cu + O_2 + 4H^+$$

Using conversion factors:

$$? \text{ g Cu} = 17.6 \text{ L O}_2 \left(\frac{1 \text{ mol O}_2}{22.41 \text{ L O}_2}\right)\left(\frac{2 \text{ mol Cu}}{1 \text{ mol O}_2}\right)\left(\frac{63.55 \text{ g Cu}}{1 \text{ mol Cu}}\right)$$

$$= 99.8 \text{ g Cu}$$

12. (a) yes

$$2[BrO^- + 2e^- + 2H^+ \rightarrow Br^- + H_2O] \qquad \mathscr{E}° = +0.71 \text{ V}$$
$$BrO^- + 2H_2O \rightarrow BrO_3^- + 4H^+ + 4e^- \qquad \mathscr{E}° = -0.54 \text{ V}$$
$$\overline{3BrO^- \rightarrow 2Br^- + BrO_3^-} \qquad \mathscr{E}° = +0.17 \text{ V}$$

(b) yes

$$5[\tfrac{1}{2}Br_2 + e^- \rightarrow Br^-] \qquad \mathscr{E}° = +1.07 \text{ V}$$
$$\tfrac{1}{2}Br_2 + 6OH^- \rightarrow BrO_3^- + 3H_2O + 5e^- \qquad \mathscr{E}° = -0.52 \text{ V}$$
$$\overline{3Br_2 + 6OH^- \rightarrow 5Br^- + BrO_3^- + 3H_2O} \qquad \mathscr{E}° = +0.55 \text{ V}$$

13. 5.2×10^{-2} M

The reaction taking place

$$H_2(1 \text{ atm}) + 2H^+(1.0 \text{ } M) \rightarrow H_2(1 \text{ atm}) + 2H^+(? \text{ } M) \quad \mathscr{E} = +0.076 \text{ V}$$

Two faradays are involved in the cell reaction, and therefore $n = 2$. Since the same electrode is in each half cell, $\mathscr{E}°$ for the cell is zero.

We can obtain the unknown $[H^+]$ from the Nernst equation:

$$\mathscr{E} = \mathscr{E}° - \frac{0.0592}{n} \log Q$$

$$0.076 = 0 - \frac{0.0592}{2} \log \left(\frac{(1)[H^+]^2}{(1)(1)^2} \right)$$

$$\log [H^+]^2 = 0.076 \left(-\frac{2}{0.0592} \right)$$

$$= -2.57$$

$$[H^+]^2 = \text{antilog } -2.57$$

$$= 2.7 \times 10^{-3}$$

$$[H^+] = 5.2 \times 10^{-2} \text{ } M$$

14. $[Fe^{2+}] = 9.52 \times 10^{-4}$ M
    $[Fe^{3+}] = 4.8 \times 10^{-5}$ M

Since the total iron concentration is 1.00×10^{-3},

$$[Fe^{2+}] + [Fe^{3+}] = 1.00 \times 10^{-3}$$

If $x = [Fe^{2+}]$, then $((1.00 \times 10^{-3}) - x)M = [Fe^{3+}]$.

Use the Nernst equation to solve the problem:

$$\mathscr{E} = \mathscr{E}° - \frac{0.0592}{n} \log Q$$

$$\log Q = (\mathscr{E} - \mathscr{E}°) \frac{-n}{0.0592}$$

$$\log \frac{((1.00 \times 10^{-3}) - x)M}{x} = (-0.694 - (-0.771)) \frac{-1}{0.0592}$$

Solving for x yields:

$$x = 9.52 \times 10^{-4}\ M = [Fe^{2+}]$$

$$((1.00 \times 10^{-3}) - x)M = 4.8 \times 10^{-5}\ M = [Fe^{3+}]$$

15. 12.1 g

Faraday's laws and the conversion factor method can be used to solve this problem:

$$\text{? moles of electrons} = 3.00\ \text{hr} \left(\frac{60\ \text{min}}{1\ \text{hr}}\right)\left(\frac{60\ \text{sec}}{1\ \text{min}}\right)$$

$$\times\ (10.0\ \text{coulomb/sec}) \left(\frac{1\ F}{96{,}500\ \text{coulombs}}\right)\left(\frac{1\ \text{mole of electrons}}{1\ F}\right)$$

$$= 1.12\ \text{moles of electrons}$$

$$\text{? g} = 1\ \text{mole of electrons} \left(\frac{13.6\ \text{g}}{1.12\ \text{moles of electrons}}\right)$$

$$= 12.1\ \text{g}$$

16. (a) 23.9 L O$_2$

The net reaction shows that 1 mol of O_2 reacts with 2 mol of H_2; therefore, half as much O_2 as H_2, or **23.9 L of O_2**, is required at STP:

$$\text{? L } O_2$$

$$= 47.8\ \text{L } H_2 \left(\frac{1\ \text{mol } O_2}{2\ \text{mol } H_2}\right)\left(\frac{22.4\ \text{L } O_2/1\ \text{mol } O_2}{22.4\ \text{L } H_2/1\ \text{mol } H_2}\right)$$

$$= 23.9\ \text{L } O_2$$

(b) 2.13 mol H$_2$O

From the net reaction we know that 2 mol of H_2 reacts with 1 mol of O_2 to produce 2 mol of H_2O. Thus

$$\text{? L } H_2O$$

$$= 47.8\ \text{L } H_2 \left(\frac{2\ \text{mol } H_2O}{2\ \text{mol } H_2}\right)\left(\frac{1\ \text{mol } H_2}{22.4\ \text{L } H_2}\right)$$

$$= 2.13\ \text{mol } H_2O$$

(c) 686 coulombs/sec

To determine the current, expressed in amperes or coulombs/sec, we use Faraday's laws.

$$\text{? coulomb/sec}$$

$$= \left(\frac{47.8\ \text{L } H_2}{10\ \text{min}}\right)\left(\frac{1\ \text{mol } H_2}{22.4\ \text{L } H_2}\right)\left(\frac{2\ F}{1\ \text{mol } H_2}\right)$$

$$\text{mol } H_2$$
$$\text{reacted per min}$$

$$\times \left(\frac{96,500 \text{ coulombs}}{1 \text{ F}} \right) \left(\frac{1 \text{ min}}{60 \text{ sec}} \right)$$

current in coulombs/sec
calculated at this point

= 686 coulombs/sec

17. 42 L H_2
    21 L O_2

To determine the size of the tanks, we use Faraday's laws:

? liters H_2

$$= 20 \text{ coulombs/sec} \left(\frac{60 \text{ sec}}{1 \text{ min}} \right) \left(\frac{60 \text{ min}}{1 \text{ hr}} \right) \left(\frac{22.4 \text{ L } H_2}{1 \text{ mol } H_2} \right)$$

$$\times \left(\frac{1 \text{ mol } H_2}{2 \text{ F}} \right) \left(\frac{1 \text{ F}}{96,500 \text{ coulombs}} \right) (5 \text{ hr})$$

$$= 42 \text{ L } H_2$$

Since the cell reaction

$$2H_2(g) + O_2(g) \rightarrow 2H_2O(l)$$

indicates that 1 mol of O_2 reacts with 2 mol of H_2, only 21 L of O_2 are required.

18. $K_p =$
    1.0×10^{-14}

Add the following half reactions and corresponding values of $\mathscr{E}°$ to determine the value of $\mathscr{E}°$ for the ionization of water

$$2H_2O + 2e^- \rightarrow H_2 + 2OH^- \qquad\qquad \mathscr{E}° = -0.828 \text{ V}$$

$$\underline{ H_2 \rightarrow 2H^+ + 2e^- \qquad\qquad\quad \mathscr{E}° = 0.000 \text{ V}}$$

$$2H_2O \rightarrow 2H^+ + 2OH^- \qquad\qquad \mathscr{E}° = -0.828 \text{ V}$$

Since

$$\Delta G° = -nF\mathscr{E}°$$

and

$$\Delta G° = -2.303RT \log K_p$$

then

$$-nF\mathscr{E}° = -2.303RT \log K_p$$

and

$$\log K_p = \frac{nF\mathscr{E}°}{2.303RT}$$

Substituting values into the preceding equation, we find

$$\log K_p = \frac{nF\mathscr{E}°}{2.303RT}$$

$$= -14.00$$

$$K_p = 1.0 \times 10^{-14}$$

19. 2.7230 V

The relationship between $\Delta G°$ and $\mathscr{E}°$ is $\Delta G° = -nF\mathscr{E}°$. Rearranging the preceding equation and substituting values into it, we find

$$\mathscr{E}° = -\frac{\Delta G°}{nF}$$

$$= -\frac{(-1.5764 \times 10^6 \text{ J})}{(6)(96487 \text{ C}}$$

$$= 2.7230 \text{ V}$$

III. Electrochemical cells

In solving problems such as this, check the table of standard electrode potential (Table 20.1 of the study guide) first to determine which species will be oxidized and which will be reduced. The couple with the more positive electrode potential will occur as written, a reduction; the other will occur in reverse, an oxidation.

1.

In this cell Cu^{2+} is reduced and Cd is oxidized. The electrons travel toward the positive electrode. The ions that are reduced, the cations, migrate toward the cathode, and the anions migrate toward the anode. In the cell notation the anode is represented to the left of the salt bridge, which is represented by the double bar, and the cathode is represented to the right of the salt bridge. The cell potential is

$$\mathscr{E}°_{\text{right}} - \mathscr{E}°_{\text{left}} = +0.740 \text{ V}$$

1.

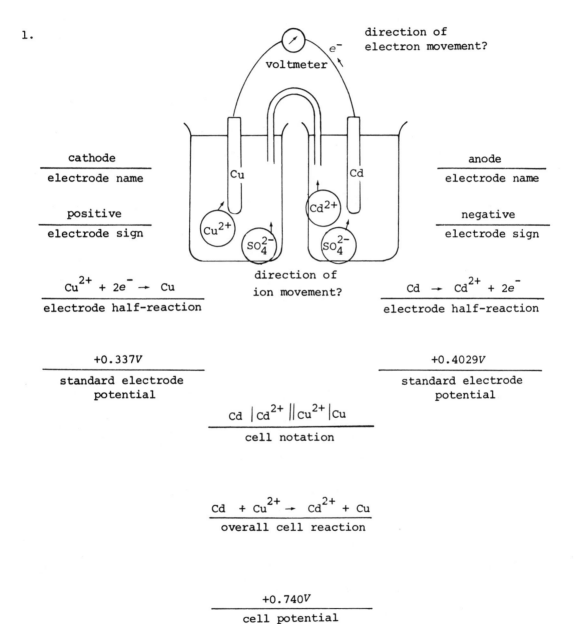

direction of
electron movement?

voltmeter

cathode
_____
electrode name

anode
_____
electrode name

positive
_____
electrode sign

negative
_____
electrode sign

$$Cu^{2+} + 2e^- \rightarrow Cu$$
electrode half-reaction

$$Cd \rightarrow Cd^{2+} + 2e^-$$
electrode half-reaction

+0.337V
_____
standard electrode
potential

+0.4029V
_____
standard electrode
potential

direction of
ion movement?

$$Cd \mid Cd^{2+} \parallel Cu^{2+} \mid Cu$$
cell notation

$$Cd + Cu^{2+} \rightarrow Cd^{2+} + Cu$$
overall cell reaction

+0.740V
_____
cell potential

2.

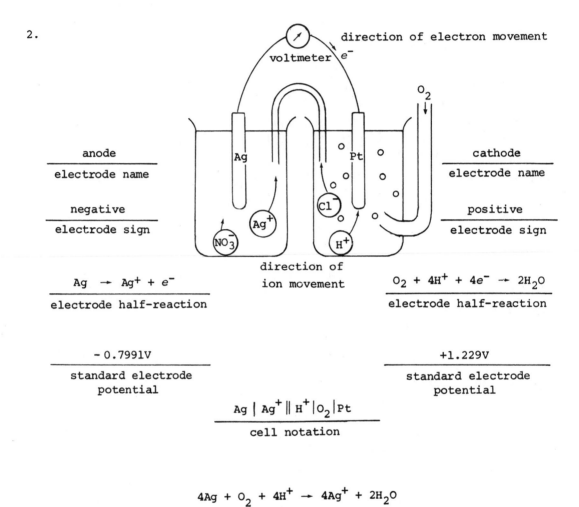

direction of electron movement

voltmeter e^-

O_2

anode	cathode
electrode name	electrode name
negative	positive
electrode sign	electrode sign

Ag   Pt   Cl⁻   Ag⁺   NO₃⁻   H⁺

direction of
ion movement

$Ag \rightarrow Ag^+ + e^-$	$O_2 + 4H^+ + 4e^- \rightarrow 2H_2O$
electrode half-reaction	electrode half-reaction

$-0.7991V$	$+1.229V$
standard electrode potential	standard electrode potential

$$Ag \mid Ag^+ \parallel H^+ \mid O_2 \mid Pt$$

cell notation

$$4Ag + O_2 + 4H^+ \rightarrow 4Ag^+ + 2H_2O$$

overall cell reaction

$$+0.430V$$

cell potential

3.

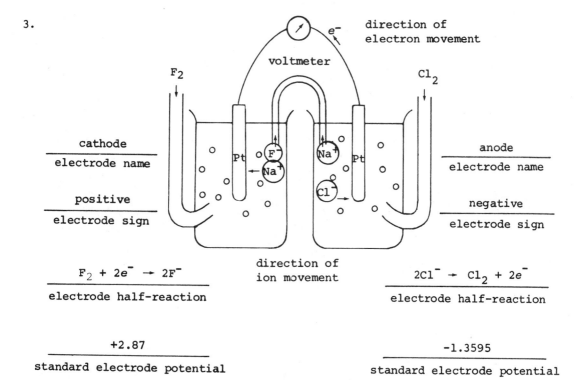

cathode	anode
electrode name	electrode name
positive	negative
electrode sign	electrode sign

$$F_2 + 2e^- \rightarrow 2F^-$$

electrode half-reaction

$$2Cl^- \rightarrow Cl_2 + 2e^-$$

electrode half-reaction

+2.87

standard electrode potential

−1.3595

standard electrode potential

$$Pt \left| Cl_2 \right| Cl^- \left\| F^- \right| F_2 \left| Pt \right.$$

cell notation

$$F_2 + 2Cl^- \rightarrow Cl_2 + 2F^-$$

overall cell reaction

+1.51V

cell potential

SELF-TEST   Complete the test in 20 minutes:

I. Do the following:

1. Calculate the emf of the following cell at 25°C:

   $$Cd(s) | Cd^{2+}(0.050 \ M) \| Ag^+(0.50 \ M) | Ag(s)$$

   The pertinent electrode potentials are

   $$2e^- + Cd^{2+} \rightleftharpoons Cd(s) \qquad \mathcal{E}° = -0.40 \ V$$
   $$e^- + Ag^+ \rightleftharpoons Ag(s) \qquad \mathcal{E}° = +0.80 \ V$$

   The Nernst equation is

   $$\mathcal{E} = \mathcal{E}° - \frac{0.0592}{n} \log Q$$

2. What weight of Mg is obtained in 10.0 minutes from the electrolysis of dry, molten $MgCl_2$ using a current of 10.0 amp? The atomic weight of Mg is 24.3 and that of Cl is 35.5.

3. Given the electrode potentials

   $$Mg^{2+} + 2e^- \rightleftharpoons Mg \qquad \mathcal{E}° = -2.36 \ V$$
   $$Cu^{2+} + 2e^- \rightleftharpoons Cu \qquad \mathcal{E}° = +0.34 \ V$$

   answer each of the following:
   (a) What is the best oxidizing agent of the chemical species in the half reactions?
   (b) In a voltaic cell consisting of standard Mg^{2+}/Mg and Cu^{2+}/Cu half cells, which half cell is the cathode?
   (c) Is the Cu electrode of the voltaic cell in part (b) above positive or negative?
   (d) Will Mg metal spontaneously reduce Cu^{2+} ions in 1.0 M concentration at 25°C?
   (e) In an electrolytic cell is the cathode positive or negative?

II. Answer each of the following. The questions refer to a galvanic cell that involves the reaction

$$2VO^{2+} + 4H^+ + Ni(s) \rightarrow 2V^{3+} + Ni^{2+} + 2H_2O$$

Pertinent electrode potentials are

$$2e^- + Ni^{2+} \rightleftharpoons Ni \qquad\qquad \mathscr{E}° = -0.25 \text{ V}$$
$$e^- + 2H^+ + VO^{2+} \rightleftharpoons V^{3+} + H_2O \qquad \mathscr{E}° = +0.36 \text{ V}$$

1. The standard emf of the cell is
   - (a) 0.11 V
   - (b) 0.61 V
   - (c) 0.97 V
   - (d) 0.47 V

2. The emf of the cell could be increased by
   - (a) increasing the Ni^{2+} concentration
   - (b) lowering the pH
   - (c) reducing VO^{2+} concentration
   - (d) increasing V^{3+} concentration

3. The anode could be made of
   - (a) Ni
   - (b) V
   - (c) Pt
   - (d) H_2

4. The cathode could be made of
   - (a) Ni
   - (b) V
   - (c) Pt
   - (d) H_2

5. The cell must be constructed so that the mixing of the following is avoided:
   - (a) V^{3+} and Ni^{2+}
   - (b) V^{3+} and Ni(s)
   - (c) VO^{2+} and Ni^{2+}
   - (d) VO^{2+} and Ni(s)

6. The best oxidizing agent(s) is (are)
   - (a) Ni^{2+}
   - (b) Ni(s)
   - (c) VO^{2+}
   - (d) V^{3+}

THE NONMETALS, PART I: HYDROGEN AND THE HALOGENS

CHAPTER 21

OBJECTIVES

I. You should be able to demonstrate your knowledge of the following terms by defining them, describing them, or giving specific examples of them:

cracking [21.2]
displacement reaction [21.3]
halogens [21.6]
hydrides [21.4]
hydrocarbons [21.1]
interhalogen compounds [21.9]
water gas [21.2]

II. You should be familiar with the physical properties of hydrogen and the halogens.

III. You should be familiar with the abundance of these elements in nature, the ways in which they can be obtained from natural sources, and their uses. These elements are shown in the shaded area of the periodic table, Figure 21.1 of the study guide.

IV. You should know the fundamental chemistry of these elements, i.e., the types of compounds they form, how such compounds are prepared, and how such compounds react.

V. You should be able to apply the basic information of all previous chapters of your text toward an understanding of the chemistry of these elements.

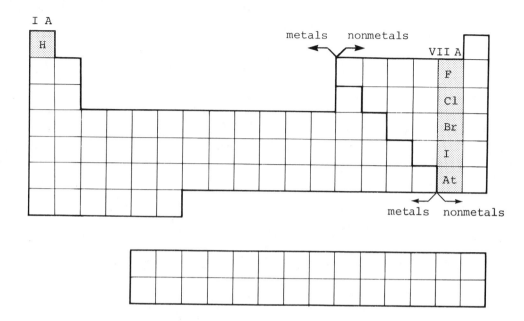

Figure 21.1   Periodic Table Showing Those Nonmetals Which Are
Discussed in This Chapter

EXERCISES   I. Hydrogen

Hydrogen composes 63 percent of the total number of atoms
in living organisms on the earth and 91 percent of the
number of atoms in our universe. The ability of certain
molecules to form hydrogen bonds is important. Without
hydrogen bonding life would be different; for example,
water would boil at a much lower temperature, and DNA
would not be a double helix.

1. Is the following statement *true* or *false*? The density
   of hydrogen gas is lower than that of any other gas.

2. The mass of the earth's atmosphere is approximately
   5.0×10^{15} tons (1 ton = 2000 lb). If 0.50 percent
   of the mass is H_2, how many tons of H_2 are in the
   atmosphere?

3. The earth contains approximately 1.5×10^{24} g of free
   water. How many tons of hydrogen combined with oxygen
   to form the earth's free water?

4. Which of the following metals should be capable of displacing hydrogen from an aqueous solution of an acid? Refer to Appendix B in your text.
   (a) Zn
   (b) Ag
   (c) Fe
   (d) Ca
   (e) Sn
   (f) Au

5. Complete each of the following by writing a balanced equation. If no reaction occurs, write NR in the space provided for products:

   (a) H_2O $\xrightarrow{\text{electrolysis}}$

   (b) $Ca(s) + H_2O \rightarrow$

   (c) $Al(s) + H^+(aq) \rightarrow$

   (d) $H^- + H_2O \rightarrow$

   (e) $H_2(g) + S(g) \rightarrow$

   (f) $FeO(s) + H_2(g) \rightarrow$

II. Halogens

1. Chlorine can be prepared by the reaction of concentrated HCl with solid MnO_2. Write the balanced equation for the reaction:

   $MnO_2(s) + HCl(aq) \rightarrow$

2. Use the ion-electron method to balance the equation for the reaction of iodate ion and sulfite ion to form iodine and sulfate ion in acid solution.

3. What is the physical state at STP of each of the following: fluorine, chlorine, bromine, and iodine?

4. Write the notation for the electronic configuration of each of the following:
   (a) Cl   (b) Cl^- (c) Br   (d) Br^-

5. From the following standard electrode potential diagram of chlorine, chloride, and the chlorine oxyanions in basic solution, predict which compounds are stable:

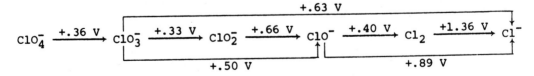

6. Complete each of the following by writing a balanced equation. If no reaction occurs, write *NR* in the space provided for products:

   (a) $SiO_2(s) + 6HF(aq) \rightarrow$

   (b) $Br_2(l) + 2I^-(aq) \rightarrow$

   (c) $Cl_2(g) + Bi(s) \rightarrow$

   (d) $Cl_2(g) + Hg(l) \rightarrow$

   (e) $CaF_2(s) + H_2SO_4(g) \xrightarrow{heat}$

   (f) $Br_2(g) + H_2(g) \rightarrow$

   (g) $Br_2(l) + 2Cl^-(aq) \rightarrow$

7. Predict the shape of each of the following interhalogens:
   (a) ClF_2^+ (b) ICl_2^- (c) ClF_3 (d) ICl_4^-

8. Draw the molecular orbital energy-level diagram for diatomic fluorine and explain why F_2 is more stable than two free fluorine atoms.

9. Hydrofluoric acid, HF, is a weak electrolyte, but the others of the series HF, HCl, HBr, and HI are strong. Why?

10. Determine the oxidation number of iodine in each of the following compounds. Name each compound in the space provided.
    (a) HOI
    (b) IF_7
    (c) I_2
    (d) NaI
    (e) I_2O_5
    (f) H_5IO_6
    (g) AgI
    (h) ICl_3
    (i) I_2O_4
    (j) As_2I_4
    (k) I_2Cl_6

11. How many grams of MnO_2 are needed to prepare 1.00 liter of dry Cl_2 gas measured at 25°C and 1.00 atm by the following reaction?

$$MnO_2(s) + 4HCl(aq) \rightarrow Cl_2(g) + MnCl_2(aq) + 2H_2O(l)$$

12. What would you expect the shape of the ion IO_6^{5-} to be? Draw the Lewis structure.

13. Chlorine is contained in DDT, $C_{14}H_9Cl_5$, the common insecticide. What is the percentage of chlorine in DDT?

ANSWERS TO EXERCISES

I. Hydrogen

1. True

If we consider hydrogen to be an ideal gas, 1 mol of that gas would occupy 22.4 L. Since 1 mol of H_2 is 2 g of H_2, there would be 2 g of H_2 in 22.4 L. Similarly, the density of helium would be 4 g/22.4 L. All other gases are heavier than hydrogen or helium and would have a larger density.

2. 2.5×10^{13} tons H_2

Calculate the number of tons of H_2 in the atmosphere of the earth:

$$? \text{ tons } H_2$$

$$= 5.0 \times 10^{15} \text{ tons atmosphere} \left(\frac{5.0 \text{ tons } H2}{1000 \text{ tons atmosphere}} \right)$$

$$= 2.5 \times 10^{13} \text{ tons } H_2$$

3. 1.8×10^{17} tons H_2

Calculate the number of tons of H_2 as follows:

$$? \text{ tons } H_2$$

$$= 1.5 \times 10^{24} \text{ g } H_2O \left(\frac{1 \text{ mol } H2O}{18.0 \text{ g } H_2O} \right) \left(\frac{2.02 \text{ g } H2}{1 \text{ mol } H_2O} \right) \left(\frac{1 \text{ lb } H2}{454. \text{ g } H_2} \right) \left(\frac{1 \text{ ton } H2}{2000 \text{ lb } H_2} \right)$$

$$= 1.8 \times 10^{17} \text{ tons } H_2$$

4. (a) Zn
   (b) Fe
   (c) Ca
   (d) Sn

Since the $\mathscr{E}°$ of the H^+/H_2 couple is zero by convention, any metal with a positive value for $\mathscr{E}°_{ox}$ (i.e., a negative standard electrode potential) should be able to displace hydrogen.

5. (a) $2H_2O \xrightarrow{\text{electrolysis}} 2H_2(g) + O_2(g)$

   (b) $Ca(s) + 2H_2O \rightarrow Ca^{2+}(aq) + 2OH^-(aq) + H_2(g)$

   (c) $2Al(s) + 6H^+(aq) \rightarrow 2Al^{3+}(aq) + 3H_2(g)$

   (d) $H^- + H_2O \rightarrow H_2(g) + OH^-(aq)$

(e) $H_2(g) + S(g) \rightarrow H_2S(g)$

(f) $FeO(s) + H_2(g) \rightarrow Fe(s) + H_2O(g)$

II. Halogens

To determine the amount of descriptive chemistry you should know, rely on your class notes and the comments of your instructor.

1. $MnO_2(s) + 4HCl(aq) \rightarrow Cl_2(g) + MnCl_2(aq) + 2H_2O(l)$

2. $2IO_3^- + 2H^+ + 5SO_3^{2-} \rightarrow 5SO_4^{2-} + I_2 + H_2O$

3.

Element	Color	Physical State at STP
fluorine	pale yellow	gas
chlorine	green-yellow	gas
bromine	red	liquid
iodine	violet-black	solid

4.

(a) $Cl: 1s^2 2s^2 2p^6 3s^2 3p^5$

(b) $Cl^-: 1s^2 2s^2 2p^6 3s^2 3p^6$

(c) $Br: 1s^2 2s^2 2p^6 3s^2 3p^6 3d^{10} 4s^2 4p^5$

(d) $Br^-: 1s^2 2s^2 2p^6 3s^2 3p^6 3d^{10} 4s^2 4p^6$

5. The chloride ion, Cl^-, and the perchlorate ion, ClO_4^-, are stable to disproportionation in basic solution. The perchlorate ion is stable because it cannot be further oxidized, and the chloride ion is stable because it cannot be further reduced. Neither of these ions can disproportionate because neither can undergo oxidation and reduction spontaneously and simultaneously. The chlorate ion, ClO_3^-, however, can disproportionate:

$$6e^- + 3H_2O + ClO_3^- \rightarrow Cl^- + 6OH^- \qquad \mathscr{E}° = +0.63 \text{ V}$$

$$3[2OH^- + ClO_3^- \rightarrow ClO_4^- + H_2O + 2e] \qquad \mathscr{E}° = -0.36 \text{ V}$$

$$\overline{4ClO_3^- \rightarrow Cl^- + 3ClO_4^-} \qquad \mathscr{E}° = +0.27 \text{ V}$$

Also, the chlorite ion, ClO_2^-, can disproportionate:

$$2e^- + ClO_2^- + H_2O \rightarrow ClO^- + 2OH^- \qquad \mathscr{E}° = +0.66 \text{ V}$$

$$ClO_2^- + 2OH^- \rightarrow ClO_3^- + H_2O + 2e^- \qquad \mathscr{E}° = -0.33 \text{ V}$$

$$\overline{2ClO_2^- \rightarrow ClO^- + ClO_3^-} \qquad \mathscr{E}° = +0.33 \text{ V}$$

Both of the products of the disproportionation of ClO_2^- can disproportionate. The ClO^- can disproportionate as follows:

$$2[H_2O + ClO^- + 2e^- \rightarrow Cl^- + 2OH^-] \qquad \mathscr{E}° = +0.89 \text{ V}$$
$$ClO^- + 4OH^- \rightarrow ClO_3^- + 2H_2O + 4e^- \qquad \mathscr{E}° = -0.50 \text{ V}$$
$$\overline{3ClO^- \rightarrow 2Cl^- + ClO_3^- \qquad\qquad\qquad \mathscr{E}° = +0.39 \text{ V}}$$

Chlorine gas, Cl_2, disproportionates as follows:

$$2OH^- + Cl_2 \rightarrow ClO^- + Cl^- + H_2O \qquad \mathscr{E}° = +0.96 \text{ V}$$

6. (a) $SiO_2(s) + 6HF(aq) \rightarrow 2H^+(aq) + SiF_6^{2-}(aq) + 2H_2O(1)$

   (b) $Br_2(1) + 2I^-(aq) \rightarrow 2Br^-(aq) + I_2(s)$

You should learn some of the important halogen reactions summarized in Table 21.4 of the text.

   (c) $3Cl_2(g) + 2Bi(s) \rightarrow 2BiCl_3(s)$

   (d) $Cl_2(g) + Hg(1) \rightarrow HgCl_2(s)$

   (e) $CaF_2(s) + H_2SO_4 \xrightarrow{\text{heat}} CaSO_4(s) + 2HF(g)$

Hydrogen fluoride is prepared commercially in this way.

   (f) $Br_2(g) + H_2(g) \rightarrow 2HBr(1)$

   (g) $Br_2(1) + 2Cl^-(aq) \rightarrow NR$

The standard electrode potential of Cl_2 is larger than that of Br_2: $F_2 > Cl_2 > Br_2 > I_2$.

7.

(a)   +

angular

(b)

linear

(c)

T-shaped

(d)

square planar

8.   See Figure 21.2 of the study guide for the molecular
       orbital diagram of F_2.

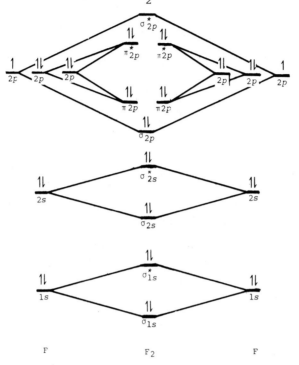

Figure 21.2   Molecular Orbital Diagram of F_2

9.   The bond of HF is stronger than that of any other hydrogen
       halide; therefore, HF is less easily dissociated. When
       dissociation does occur, HF_2^-(aq) species form. Hydrogen
       bonding stabilizes HF in solution.

10.   Write the formula of the compound in the space provided:

	Oxidation Number	Name	Formula
(a)	+1	hypoiodous acid	
(b)	+7	iodine heptafluoride	
(c)	0	iodine	
(d)	-1	sodium iodide	
(e)	+5	diiodine pentoxide	
(f)	+7	periodic acid	
(g)	-1	silver(I) iodide or silver iodide	
(h)	+3	iodine trichloride	
(i)	+4	diiodine tetraoxide	
(j)	-1	diarsenic tetraiodide	
(k)	+3	diiodine hexachloride	

11. 3.56 g   Calculate the number of liters of Cl_2 at STP:

$$? \text{ liter } Cl_2 = 1.00 \text{ liter}\left(\frac{273 \text{ K}}{298 \text{ K}}\right) = 0.9161 \text{ liter}$$

Calculate the number of grams of MnO_2:

$$? \text{ g } MnO_2 = 0.9161 \text{ liter } Cl_2\left(\frac{1 \text{ mol } Cl_2}{22.4 \text{ liter } Cl_2}\right)\left(\frac{1 \text{ mol } MnO_2}{1 \text{ mol } Cl_2}\right)$$

$$\times \frac{86.94 \text{ g } MnO_2}{1 \text{ mol } MnO_2}$$

$$= 3.56 \text{ g } MnO_2$$

12.   The Lewis structure is

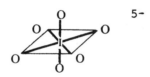

From a consideration of electron pair repulsions, we
predict the ion to be octahedral:

13. 50.006%   First determine the molecular weight of DDT:

14 carbon atoms:	14(12.011) =	168.15
9 hydrogen atoms:	9(1.0079) =	9.0711
5 chloride atoms:	5(35.453) =	177.26
molecular weight of DDT	=	354.48

Then we determine the percentage of chlorine in DDT:

$$\% \text{ Cl} = \left(\frac{177.26 \text{ g/mol DDT}}{354.48 \text{ g/mol DDT}}\right)100$$

$$= 50.006$$

SELF-TEST   I. We suggest that you go through your lecture notes and
                  compile a list of chemical reactions, formulas, and names
                  of compounds. Check your material to make sure that you
                  did not copy it incorrectly, and then prepare your own
                  self-test. Write a series of incomplete reactions with
                  either the reactants or products given and a list of
                  formulas without names or vice versa. The next day
                  complete the reactions and give either the name or the
                  formula of the compound without referring to any material.

If your instructor has emphasized the necessity of learning all chemical reactions in Chapter 21 of your text, complete the following test in 15 minutes.

1. Complete and balance each of the following equations. If no reaction occurs, write *NR*.

(a) $Mg(s) + H_2O(g) \rightarrow$

(b) $H_2(g) + Cl_2(g) \rightarrow$

(c) $Na(s) + H_2O(l) \rightarrow$

(d) $OH^-(aq) + Cl_2(aq) \rightarrow$

(e) $FeS(s) + H^+(aq) + Cl^-(aq) \rightarrow$

(f) $Br_2(g) + Ag_2O(s) + H_2O(l) \rightarrow$

(g) $PCl_3(l) + H_2O(l) \rightarrow$

(h) $Cl^-(aq) + H_2O(l) \xrightarrow[\text{heat}]{\text{electrolysis}}$

(i) $Cl^-(aq) + H^+(aq) + H_2O(l) \rightarrow$

(j) $Cl_2(g) + I^-(aq) \rightarrow$

THE NONMETALS, PART II: THE GROUP IV A ELEMENTS

OBJECTIVES

I. You should be able to demonstrate your knowledge of the following terms by defining them, describing them, or giving specific examples of them.

allotrope [22.6]
contact process [22.2]
Frasch process [22.9]
ozone [22.6]
ozonide [22.6]
peroxide [22.6]
peroxy acid [22.12]
photochemical pollutants [22.7]
superoxide [22.4]

II. You should be familiar with the physical properties of the group VI A elements. The elements discussed in Chapter 22 of your text are shown in the shaded area of Figure 22.1 of the study guide.

III. You should be familiar with the abundance of these elements in nature, the ways in which they can be obtained from natural sources, and their uses.

IV. You should know the fundamental chemistry of these elements, i.e., the types of compounds they form, how such compounds are prepared, and how such compounds react.

V. You should be able to apply the basic information of all previous chapters of your text toward an understanding of the chemistry of these elements.

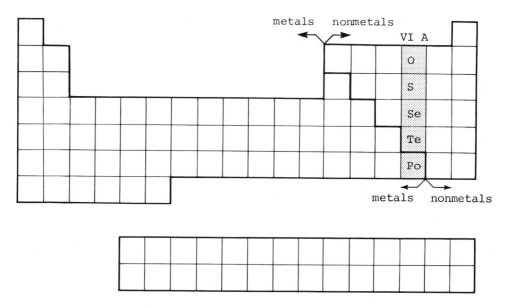

Figure 22.1   Periodic Table Showing Those Nonmetals Which Are Discussed in This Chapter

EXERCISES

Answer the following:

1. Oxygen in the atmosphere reacts in the presence of ultraviolet light to produce ozone, O_3. The ozone absorbs most of the ultraviolet light in the upper atmosphere and protects the earth from the damaging rays. Write a balanced equation for the production of ozone and give the oxidation number of oxygen in ozone.

2. Oxygen is consumed in plant and animal respiration, and carbon dioxide and water are produced. For example,

$$C_6H_{12}O_6 (aq) + O_2 (g) \rightarrow CO_2 (g) + H_2O (l)$$
glucose

$$C_{12}H_{22}O_{11} (aq) + 12O_2 (g) \rightarrow 12CO_2 (g) + 11H_2O (l)$$
maltose

The CO_2 that is produced enters the carbon dioxide cycle, and the H_2O enters the water cycle. Carbon dioxide dissolves in the oceans where it can react with Ca^{2+} to form carbonate deposits:

$$Ca^{2+} (aq) + H_2O (l) + CO_2 (g) \rightarrow CaCO_3 (s) + 2H^+ (aq)$$

Photosynthesis by the phytoplankton in the sea returns oxygen to the atmosphere:

$$6CO_2 (g) + 6H_2O (l) \rightarrow 6O_2 (g) + C_6H_{12}O_6 (aq)$$

Which of the preceding processes would irreversibly reduce the amount of O_2 in the atmosphere under ordinary conditions?

3. Oxygen oxidizes the volcanic gases CO, SO_2, and H_2. Write a balanced equation for the complete oxidation of each of the volcanic gases.

4. Which compound would you expect to have a higher normal boiling point: O_2 or O_3?

5. Draw a Lewis structure of the peroxide ion, O_2^{2-}, and the superoxide ion, O_2^-.

6. Write the molecular-orbital configuration of O_2^{2-}.

7. What is the bond order of the superoxide anion, O_2^-.

8. Which type of magnetic behavior would you expect for a solid containing the superoxide anion?

9. How many protons are in the nucleus of an oxygen atom?

10. Write the quantum numbers of each electron of an oxygen atom.

11. Which type of crystalline solid would oxygen form?

12. Which of the following molecules should contain the least polar bond: OF_2, H_2O_2, or SO_2?

13. What is the orbital hybridization of sulfur in each of the following? Draw structures of the molecules.
    (a) SF_4 (b) SF_6

14. Draw a Lewis structure of the thiosulfate ion, $S_2O_3^{2-}$.

15. What are the color and the physical state of each of the following at STP: oxygen, sulfur, selenium, and tellurium?

16. Write the electronic configuration of tellurium.

17. Draw a Lewis structure of S_8.

18. Write the formula for each of the following compounds in the space provided:
    (a) selenous acid
    (b) tellurium dioxide
    (c) sodium bisulfate
    (d) pyrosulfuric acid
    (e) peroxydisulfuric acid
    (f) sodium thiosulfate
    (g) ozone
    (h) hydrogen peroxide
    (i) telluric acid
    (j) potassium selenate

19. Complete and balance the following equations. If no reaction occurs, write *NR* in the space provided for the products.
    (a) $Ag_2O \xrightarrow{\text{heat}}$
    (b) $Rb(s) + O_2(g) \rightarrow$
    (c) $C_4H_4(g) + O_2(g) \rightarrow$
    (d) $Se(s) + H_2(g) \xrightarrow{\text{heat}}$
    (e) $FeS(s) + H^+(aq) \rightarrow$
    (f) $PbS(s) + O_2(g) \rightarrow$
    (g) $SeO_2(s) + H_2O(l) \rightarrow$
    (h) $BaSO_4(s) + H_2O(l) \rightarrow$

20. The chemistry of oxygen is different from that of the other group VI A elements. What reasons can be given for these differences?

ANSWERS TO
EXERCISES

1.   The equation for the conversion of oxygen into ozone is

$$3O_2 \xrightarrow{h\nu} 3O_3$$

The oxidation number of an oxygen atom in O_3 is 0.

2.   The process described by the equation

$$Ca^{2+}(aq) + H_2O\ (1) + CO_2(g) \rightarrow CaCO_3(s) + 2H^+(g)$$

ties up oxygen as a solid carbonate deposit. Respiration
consumes oxygen, but the oxygen can be released by
photosynthesis.

3.   The equations for the oxidations of the volcanic gases are

$$2CO(g) + O_2(g) \rightarrow 2CO_2(g)$$
$$2SO_2(g) + O_2(g) \rightarrow 2SO_3(g)$$
$$2H_2(g) + O_2(g) \rightarrow 2H_2O(g)$$

4. O_3 The molecular weight of O_3 is greater than that of O_2,
and the attractive forces of O_3, which has a dipole moment,
are stronger than those of O_2, which has no dipole moment.

5. $[:\overset{..}{O}:\overset{..}{O}:]^{2-}$
   $[:\overset{..}{O}:\overset{..}{O}.]^{-}$

6.   The molecular-orbital configuration of O_2^{2-} is

$$(\sigma 1s)^2 (\sigma*1s)^2 (\sigma 2s)^2 (\sigma*2s)^2 (\sigma 2p)^2 (\pi 2p)^4 (\pi*2p)^4.$$

7. $1\frac{1}{2}$

8. paramagnetic   There is one unpaired electron in O_2^-.

9. 8 protons

10.

Electron	n	l	m	s
1	1	0	0	$+\frac{1}{2}$
2	1	0	0	$-\frac{1}{2}$
3	2	0	0	$+\frac{1}{2}$
4	2	0	0	$-\frac{1}{2}$
5	2	1	+1	$+\frac{1}{2}$
6	2	1	0	$+\frac{1}{2}$
7	2	1	-1	$+\frac{1}{2}$
8	2	1	+1	$-\frac{1}{2}$

11.   Oxygen would form a nonpolar molecular solid with van der Waals forces holding the molecules together.

12. H_2O_2 In peroxide the O—O is nonpolar.

13.   (a)   (b)

                   dsp^3 d^2sp^3

14.   The thiosulfate anion has the same structure as the sulfate ion except that one oxygen atom is replaced by a sulfur atom:

15.

Element	Color	Physical State at STP
Oxygen	Colorless	Gas consisting of O_2 molecules
Sulfur	Yellow	Solid consisting of S_8 rings
Selenium	Red to black	Solid consisting of Se_8 rings and/or Se_n chains
Tellurium	Silver to white	Solid consisting of Te_n chains

16.   Te: $1s^2 2s^2 2p^6 3s^2 3p^6 3d^{10} 4s^2 4p^6 4d^{10} 5s^2 5p^4$

17.

or

18. (a) H_2SeO_3

(b) TeO_2

(c) $NaHSO_4$

(d) $H_2S_2O_7$

(e) $H_2S_2O_8$

(f) $Na_2S_2O_3$

(g) O_3

(h) H_2O_2

(i) H_6TeO_6

(j) K_2SeO_4

19. (a) $2Ag_2O(s) \xrightarrow{heat} 4Ag(s) + O_2(g)$

(b) $Rb(s) + O_2(g) \rightarrow RbO_2(s)$

(c) $C_4H_4(g) + 5O_2(g) \rightarrow 4CO_2(g) + 2H_2O(g)$

(d) $Se(s) + H_2(g) \xrightarrow{heat} H_2Se(g)$

(e) $FeS(s) + 2H^+(aq) \rightarrow Fe^{2+}(aq) + H_2S(g)$

(f) $2PbS(s) + 3O_2(g) \rightarrow 2PbO(s) + 2SO_2(g)$

(g) $SeO_2(s) + H_2O(l) \rightarrow H_2SeO_3(aq)$

(h) $BaSO_4(s) + H_2O(l) \rightarrow NR$

20.   Since it is a first row element, oxygen follows the octet rule. S, Se, Te and Po have d orbitals which can be used for bonding, thus these elements can form more than four bonds. For example, S can form SF_6 and tellurium can form TeF_6.

Oxygen has a higher electronegativity than the rest of the group VI A elements. Thus oxygen compounds have more ionic character than their group VI A analogs.

SELF-TEST   I. We suggest that you go through your lecture notes and compile a list of chemical reactions, formulas, and names of compounds. Check your material to make sure that you did not copy it incorrectly, and then prepare your own self-test. Write a series of incomplete reactions with either the reactants or products given and a list of formulas without names or vice versa. The next day complete the reactions and give either the name or the formula of the compound without referring to any material.

If your instructor has emphasized the necessity of learning all chemical reactions in Chapter 22 of your text, complete the following test in 15 minutes.

1. Complete and balance each of the following equations. If no reaction occurs, write *NR*.

   (a) $FeS(s) + H^+(aq) + Cl^-(aq) \rightarrow$

   (b) $Ba^{2+}(aq) + SO_4^{2-}(aq) \rightarrow$

   (c) $C_{12}H_{22}O_{11}(s) + H_2SO_4(conc.) \rightarrow$

   (d) $S_2O_3^{2-}(aq) + H^+(aq) \rightarrow$

   (e) $H_2O \xrightarrow{\text{electrolysis}}$

   (f) $Na_2O_2(s) + H_2O(l) \rightarrow$

   (g) $S(s) + O_2(g) \rightarrow$

   (h) $H_2S(g) + O_2(g) \rightarrow$

   (i) $Se(s) + O_2(g) \rightarrow$

THE NONMETALS, PART III: THE GROUP V A ELEMENTS

CHAPTER
23

OBJECTIVES

I. You should be able to demonstrate your knowledge of the following terms by defining them, describing them, or giving specific examples of them:

arc process [23.7]
azide [23.3]
cyanamid process [23.5]
Haber process [23.2]
isomers [23.6]
nitrogen cycle [23.2]
nitrogen fixation [23.2]
Ostwald process [23.7]

II. You should be familiar with the physical properties of the elements of group V A. These elements are shown in the shaded area of the periodic table of Figure 23.1 of the study guide.

III. You should be familiar with the abundance of these elements in nature, the ways in which they can be obtained in pure form, and their uses.

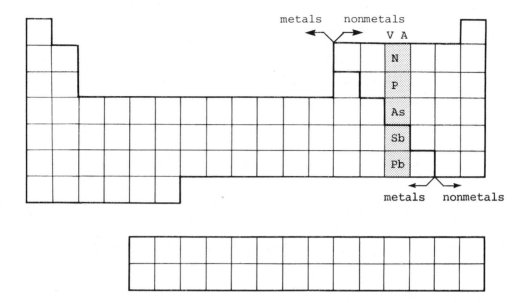

Figure 23.1 Periodic Table Showing the Group V A Elements Which Are
Discussed in This Chapter

IV. You should know the fundamental chemistry of these elements,
i.e., the types of compounds they form, how such compounds
are prepared, and how such compounds react.

V. You should be able to apply the basic information of all
previous chapters of your text toward an understanding of
the chemistry of these elements.

EXERCISES   I. Answer the following:

1. What are the color and physical state of each element
of group V A at STP?

2. Write the electronic configuration of each of the
following:
   (a) P
   (b) P^{3-}
   (c) As

   (d) As^{3+}
   (e) As^{5+}

3. Which element of group V A is most abundant on earth?

4. What is the oxidation number of nitrogen in each of the following compounds? Name each compound.
   (a) N_2O_4
   (b) NO
   (c) N_2O
   (d) N_2O_5
   (e) N_2O_3

5. Write the formula of each of the following in the space provided:
   (a) ammonia
   (b) barium azide
   (c) hydroxylamine
   (d) antimony(III) sulfide
   (e) calcium nitride
   (f) phosphorus(V) oxide
   (g) triphosphoric acid
   (h) calcium cyanamide
   (i) phosphorus(III) oxide
   (j) hydrazine
   (k) tetranitrogen tetrasulfide
   (l) phosphine
   (m) nitric oxide

6. Complete and balance the following equations. If no reaction occurs, write *NR* in the space provided for the products.
   (a) PCl_3 (l) + Cl_2 (g) →
   (b) NaN_3 (s) $\xrightarrow{\text{heat}}$
   (c) P_4 (s) + O_2 (g) →
   (d) NH_3 (g) + O_2 (g) $\xrightarrow[\text{Pt}]{\text{heat}}$
   (e) P_4O_{10} (s) + H_2O (l) →
   (f) NO_2 (g) + H_2O (l) →
   (g) Sb_4O_6 (s) + H_2O (l) →
   (h) Sb_2S_3 (s) + O_2 (g) →
   (i) NH_4NO_3 (s) $\xrightarrow{\text{heat}}$
   (j) $AsCl_3$ (l) + H_2O (l) →

(k) $Sb(s) + O_2(g) \xrightarrow{\text{heat}}$

(l) $PCl_5(s) + Cl_2(g) \rightarrow$

(m) $B(s) + N_2(g) \xrightarrow{\text{heat}}$

7. Predict the structure of NO_3^-. Draw the resonance forms of the ion and include the formal charges.

8. Given the following electrode potential diagram for common nitrogen compounds in acid solution, predict which compounds disproportionate:

$$NO_3^- \xrightarrow{+0.94\ V} HNO_2 \xrightarrow[+1.12\ V]{+1.00\ V} NO \xrightarrow{+1.59\ V} N_2O \xrightarrow{+1.77\ V} N_2 \xrightarrow{+0.27\ V} NH_4^+$$

with $+0.96\ V$ spanning NO_3^- to NO.

9. List two homopolar molecules that are isoelectronic with the cyanide ion.

10. How many grams of water must be added to 1.00 g P_4O_{10} to product tetrametaphosphoric acid?

$$P_4O_{10}(s) + 2H_2O(l) \xrightarrow{0°C} H_4P_4O_{12}(s)$$

11. Which of the following compounds is most basic?
    (a) PH_3 (c) SbH_3
    (b) AsH_3 (d) BiH_3

12. Which of the following compounds has the lowest boiling point?
    (a) NH_3 (c) AsH_3
    (b) PH_3 (d) SbH_3

ANSWERS TO EXERCISES

I. Properties of nonmetals

1.

Element	Color	Physical State at STP
Nitrogen	Colorless	Gas consisting of N_2 molecules
Phosphorus	White, red, black	Solid consisting of P_4 (white) or P_n (black)
Arsenic	Gray metallic, yellow	Solid consisting of As_4 (yellow) or As_n (gray metallic)

Element	Color	Physical State of STP
Antimony	Gray metallic, yellow	Solid consisting of Sb_4 (yellow) or Sb_n (gray metallic
Bismuth	Gray metallic	Solid consisting of Bi_n

2. (a)   P: $1s^2 2s^2 2p^6 3s^2 3p^3$

(b) P^{3-}: $1s^2 2s^2 2p^6 3s^2 3p^6$

(c)   As: $1s^2 2s^2 2p^6 3s^2 3p^6 3d^{10} 4s^2 4p^3$

(d) As^{3+}: $1s^2 2s^2 2p^6 3s^2 3p^6 3d^{10} 4s^2$

(e) As^{5+}: $1s^2 2s^2 2p^6 3s^2 3p^6 3d^{10}$

3. nitrogen

4.   After checking the answers, write the formula of the compound in the space provided:

	Oxidation Number	Name	Formula
(a)	+4	dinitrogen tetraoxide	
(b)	+2	mononitrogen monoxide, or nitric oxide	
(c)	+1	dinitrogen monoxide, or nitrous oxide	
(d)	+5	dinitrogen pentoxide	
(e)	+3	dinitrogen trioxide	

5.   After checking the answers, write the name of the compound in the space provided:

(a) NH_3, _____

(b) $Ba(N_3)_2$, _____

(c) NH_2OH, _____

(d) Sb_2S_3, _____

(e) Ca_3N_2, _____

(f) P_4O_{10}, _____

(g) $H_5P_3O_{10}$, _____

(h) $CaNCN$, _____

(i) P_4O_6, _____

(j) N_2H_4, _____

(k) H_4S_4, _____

(l) PH_3, _____

(m) NO, _____

6. (a) $PCl_3(l) + Cl_2(g) \rightarrow PCl_5(s)$

   (b) $2NaN_3(s) \xrightarrow{\text{heat}} 2Na(l) + 3N_2(g)$

   (c) $P_4(s) + 5O_2(g) \rightarrow P_4O_{10}(s)$

   (d) $4NH_3(g) + 5O_2(g) \xrightarrow[\text{Pt}]{\text{heat}} 4NO(g) + 6H_2O(g)$

   (e) $P_4O_{10}(s) + 6H_2O(l) \rightarrow 4H_3PO_4(aq)$

   (f) $3NO_2(g) + H_2O(l) \rightarrow 2H^+(aq) + 2NO_3^-(aq) + NO(g)$

   (g) $Sb_4O_6(s) + H_2O(l) \rightarrow NR$

   (h) $2Sb_2S_3(s) + 9O_2 \xrightarrow{\text{heat}} Sb_4O_6(g) + 6SO_2(g)$

   (i) $NH_4NO_3 \xrightarrow{\text{heat}} N_2O(g) + 2H_2O(g)$

   (j) $AsCl_3(l) + 3H_2O(l) \rightarrow H_3AsO_3(aq) + 3H^+(aq) + 3Cl^-(aq)$

   (k) $4Sb(s) + 3O_2(g) \xrightarrow{\text{heat}} Sb_4O_6(s)$

   (l) $PCl_5(s) + Cl_2(g) \rightarrow NR$

   (m) $2B(s) + N_2(g) \xrightarrow{\text{heat}} 2BN(s)$

7.   The molecule is predicted to be triangular planar. The resonance structures are

8. NO_3^-: stable
      HNO_2: disproportionates

      $3HNO_2 \rightarrow 2NO + H_2O + NO_3^- + H^+$ $\mathscr{E}° = +0.06$ V

      NO: disproportionates

      $4NO + H_2O \rightarrow N_2O + 2HNO_2$ $\mathscr{E}° = +0.59$ V

      Write another disproportionation reaction giving stable products.

      N_2O: disproportionates
      $5N_2O + H_2O \rightarrow 4N_2 + 2NO_3^- + 2H^+$ $\mathscr{E}° = +0.65$ V

N_2: stable

NH_4^+: stable

9. N_2, CO

10. 0.127 g

Calculate the number of grams of water:

$$? \text{ g H}_2\text{O} = 1.00 \text{ g P}_4\text{O}_{10} \left(\frac{1 \text{ mol P}_4\text{O}_{10}}{283.9 \text{ g P}_4\text{O}_{10}}\right) \left(\frac{2 \text{ mol H}_2\text{O}}{1 \text{ mol P}_4\text{O}_{10}}\right)$$

$$\times \left(\frac{18.02 \text{ g H}_2\text{O}}{1 \text{ mol H}_2\text{O}}\right)$$

$$= 0.127 \text{ g H}_2\text{O}$$

11. PH_3

See section 23.5 of your text.

12. PH_3

See section 23.5 of your text.

SELF-TEST

I. We suggest that you go through your lecture notes and compile a list of chemical reactions, formulas, and names of compounds. Check your material to make sure that you did not copy any material incorrectly and then prepare your own self-test. Write a series of incomplete reactions with either the reactants or products given and a list of formulas without names or vice versa. The next day complete the reactions and give either the name or the formula of the compound without referring to any material. You may wish to exchange tests with your friends.

If your instructor has emphasized the necessity of learning all chemical reactions in Chapter 23 of your text, complete the following test in 10 minutes:

1. Draw the resonance structures of nitric acid.

2. Complete the following reactions:
   (a) Sb_2S_3 (s) + O_2 (g) →
   (b) NO_2 (g) + H_2O (l) →
   (c) Ca_3N_2 (s) + H_2O (l) →
   (d) As_4O_6 (s) + C (s) →
   (e) PBr_3 (l) + H_2O (l) →
   (f) PBr_3 (l) + O_2 (g) →
   (g) PI_3 (s) + I_2 (s) →

3. Hypophosphorous acid is a
   (a) monoprotic acid   (c) triprotic acid
   (b) diprotic acid   (d) base

THE NONMETALS, PART IV: CARBON, SILICON, BORON, AND THE NOBLE GASES

CHAPTER 24

OBJECTIVES

I. You should be able to demonstrate your knowledge of the following terms by defining them, describing them, or giving specific examples of them:

acetylide [24.3]
boranes [24.8]
catenation [24.1]
carbonyls [24.4]
carbide [24.3]
freons [24.5]
hydrocarbon [24.2]
noble gas [24.9]
silane [24.3]
three-center bond [24.7]

II. You should be familiar with the physical properties of the group 0 elements, carbon, silicon, and boron. The elements discussed in this chapter are shown in the shaded area of Figure 24.1.

III. You should be familiar with the abundance of these elements in nature, the ways in which they can be obtained from natural sources, and their uses.

IV. You should know the fundamental chemistry of these elements, i.e., the types of compounds they form, how such compounds are prepared, and how such compounds react.

V. You should be able to apply the basic information of all
previous chapters of your text toward an understanding of
the chemistry of these elements.

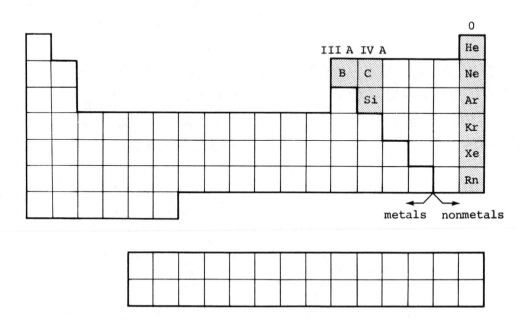

Figure 24.1   Periodic Table Showing Those Nonmetals Which Are
Discussed in This Chapter

EXERCISES   I. Write the formula or symbol for each compound or element
described below.

1. the acidic anhydride of boric acid

2. silicon dioxide

3. hydrogen cyanide

4. the group III A element which is classified as a
nonmetal

5. boron triiodide

6. potassium thiocyanate

7. the acidic anhydride of carbonic acid

8. xenon hexafluoride

9. diborane

10. the most abundant noble gas

11. the gas removed from air by photosynthesis

12. carbon disulfide

13. the most reactive noble gas

14. calcium carbide

15. the element which exhibits the most extensive catenation

16. the noble gas from which the first chemical compound was prepared

II. Answer each of the following:

1. How are pure neon, argon, krypton, and xenon obtained commercially?

2. What is the commercial source of pure helium?

3. Why have XeF_2, XeF_4, and XeF_6 been prepared, but not XeF_3 and XeF_5?

4. In addition to the xenon fluoride compounds mentioned in problem 3 above, some krypton fluoride compounds have been prepared. What are the chances of preparing compounds containing fluorine and either He, Ne, or Ar?

5. Draw Lewis structures for the following molecules and predict their shape:
   (a) SiO_2
   (b) Si_2H_6 or H_3SiSiH_3
   (c) HSCN
   (d) NCN^{2-}
   (e) SiO_4^{4-}

6. Use VSEPR theory to predict the shape of the following molecules:
   (a) CCl_4
   (b) BF_3
   (c) XeF_4
   (d) SiF_6^{2-}

7. Why is so little known about the chemistry of radon?

8. Why is the melting point of boron so much higher than the other III A elements?

9. Balance the following equations. If no reaction occurs write *NR*.
   (a) $CaC_2(s) + H_2O \rightarrow$
   (b) $CaCO_3(s) \xrightarrow{heat}$
   (c) $C_6H_{14}(l) + O_2(g) \rightarrow$

10. Which of the following elements forms a three-center bond in some of its compounds?
    (a) Xe   (c) Si
    (b) Rn   (d) B

ANSWERS TO EXERCISES

I. Formulas of compounds and characteristics of elements

1. B_2O_3

The reaction of B_2O_3 with water to form boric acid is

$$B_2O_3 + 3H_2O \rightarrow 2H_3BO_3$$

2. SiO_2

3. HCN

A very toxic gas.

4. B

Boron is classified as a nonmetal.

5. BI_3

Most covalent boron compounds have a total of six electrons in the valence shell of boron.

6. KSCN

7. CO_2

Carbon dioxide dissolves in water to form carbonic acid:

$$CO_2 + H_2O \rightarrow H_2CO_3$$

8. XeF_6

9. B_2H_6

10. Ar   Argon is approximately 0.9% of the earth's atmosphere.

11. CO_2

12. CS_2

13. Rn   Since all known isotopes of radon are radioactive and disintegrate quickly, little is known about its chemistry.

14. CaC_2

15. C   Boron and silicon also exhibit some catenation.

16. Xe   After radon, xenon is the most reactive noble gas.

II. Problems

1.   These gases are obtained by fractional distillation of liquefied air.

2.   Helium is obtained from natural gas deposits.

3.   The compounds XeF_3 and XeF_5 would each have an odd number of valence electrons. Most molecules that are easily prepared have an even number of valence electrons.

4.   The atoms of the elements of group 0 increase in size from He to Xe. The valence electrons are also less tightly held from He to Xe, i.e., the ionization potentials decrease down the group.

5.

(a) $\ddot{O}{=}Si{=}\ddot{O}$ Each silicon atom has a tetrahedral geometry.

(b)
$$\begin{array}{c} H \\ H\text{----}Si\text{---}Si\text{----}H \\ H \end{array}$$

The SCN portion of the molecule is linear; the HSC portion is angular.

(c) H
      $\ddot{S}{-}C{\equiv}N{:}$

The ion is linear and has two other linear resonance forms:

(d) $\ddot{:}N{=}C{=}\ddot{N}{:}^{2-}$

$\ddot{:}\ddot{N}{-}C{\equiv}N{:}^{2-}$ and $:N{\equiv}C{-}\ddot{\ddot{N}}{:}^{2-}$

(e)   The ion is tetrahedral

$$
\left[\begin{array}{c}
\ddot{\overset{\displaystyle ..}{O}}: \\
| \\
:\ddot{O}-Si-\ddot{O}: \\
| \\
:\ddot{O}:
\end{array} \right]^{4-}
$$

6. (a)   tetrahedral
   (b)   triangular planar
   (c)   square planar
   (d)   octahedral

7.   All isotopes are short-lived and radioactive ones.

8.   Boron has too few electrons to form a network but the
                  electrons appear to form three-center bonds of great
                  stability.

9. (a) $CaC_2 + H_2O \rightarrow CaO + H\text{-}C{\equiv}C\text{-}H$

   (b) $CaCO_3 \xrightarrow{\text{heat}} CaO + CO_2$

   (c) $2C_6H_{14} + 38O_2 \rightarrow 12CO_2 + 14H_2O$

10. boron   See problem 8 in this section.

SELF-TEST   Complete this test in 20 minutes.

1. Draw the structure of diamond.

2. Draw the structure of the carbonate ion.

3. Write the equation for the reaction of aluminum carbide with water.

4. Draw the structure of XeF_4.

5. Draw the structure of a three-carbon hydrocarbon.

METALS AND METALLURGY

CHAPTER
25

OBJECTIVES   I. You should be able to demonstrate your knowledge of the
             following terms by defining them, describing them, or
             giving specific examples of them:

    alkali metals [25.8]
    alkaline earth metals [25.9]
    amalgam [25.5]
    band theory [25.1]
    basic oxygen process [25.7]
    Bayer method [25.5]
    conduction band [25.1]
    conductor [25.1]
    degenerate [25.1]
    extrinsic semiconductor [25.2]
    flotation [25.5]
    flux [25.6]
    forbidden zone [25.1]
    gangue [25.4]
    Goldschmidt process [25.6]
    Hall process [25.6]
    insulator [25.1]
    intrinsic semiconductor [25.1]
    Kroll process [25.6]
    lanthanides [25.10, 25.11]
    lanthanide contraction [25.10]
    leaching [25.5]
    matte [25.6]
    metallurgy [25.5]
    n-type semiconductor [25.2]
    Parks process [25.7]
    p-type semiconductor [25.2]
    refining [25.5, 25.7]

roasting [25.5]
semiconductor [25.1, 25.2]
smelting [25.6]
transition metals [25.10]
valence band [25.1]
Van Arkel process [25.7]
zone refining [25.7]

II. You should be familiar with the chemical and physical properties of the metals shown in the shaded area of the periodic table in Figure 25.1 of the study guide.

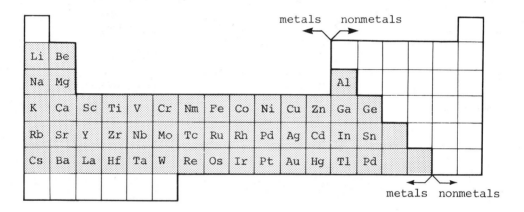

Figure 25.1 Periodic Chart Showing Those Metals Which Are Discussed in This Chapter

III. You should be familiar with the natural sources of the most important metals and the methods or treating, reducing, and refining them.

IV. You should be able to apply the basic information contained in previous chapters of your text toward an understanding of the chemistry of metals.

V. You should understand metallic bonding and semiconductors.

UNITS,   Carefully study your class notes and make sure you know
SYMBOLS,   the symbols for all the metals listed in Figure 25.1 of
MATHEMATICS   the study guide.

EXERCISES   I. Answer each of the following:

1. Which of the following metals does not have an
   important oxidation state of 1+?
   (a) Cu   (d) Fe
   (b) Na   (e) Tl
   (c) Au

2. Which of the following is not used as a desiccant?
   (a) NaCl   (d) $CaSO_4$
   (b) $CaCl_2$ (e) $Ba(ClO_4)_2$
   (c) $Mg(ClO_4)_2$

3. Which of the following is the softest metal?
   (a) Li   (d) Ba
   (b) Cs   (e) Na
   (c) Be

4. Aluminum is produced by
   (a) the Hall process   (d) the Bessemer process
   (b) zone refining   (e) the Van Arkel process
   (c) hydrogen reduction

5. The most stable oxidation state of the lanthanide
   elements is
   (a) 1+   (d) 5+
   (b) 2+   (e) 7+
   (c) 3+

6. Which of the following statements is not true of Be with
   respect to other elements of the group II A?
   (a) It has the highest melting point.
   (b) It is the most dense.
   (c) It has the highest first ionization potential.
   (d) It is the hardest.
   (e) It has the smallest atomic radius.

7. Which of the following transition metals has the
   greatest number of possible oxidation states?
   (a) Y, yttrium   (d) Zn, zinc
   (b) Sc, scandium   (e) Au, gold
   (c) Ru, ruthenium

8. Which of the following has the smallest atomic radius?
   (a) Mn, manganese   (d) Sc, scandium
   (b) Tc, technetium   (e) Zn, zinc
   (c) Re, rhenium

9. Common impurities in pig iron obtained from a blast
   furnace include all but one of the following. Identify
   the element that is not a contaminant.
   (a) carbon   (d) sulfur
   (b) silicon   (e) titanium
   (c) phosphorus

10. Which of the following is commonly found free in nature?
    (a) Fe   (d) Li
    (b) W   (e) Au
    (c) Ba

11. Which of the following metals has the lowest melting
    point?
    (a) Ag   (d) Pd
    (b) Cd   (e) Rh
    (c) In

12. Which of the following does not contain an unpaired
    s electron in the valence shell of the ground state
    atom?
    (a) Rb, rubidium   (d) Sn, tin
    (b) Cu, copper   (e) Au, gold
    (c) Mo, molybdenum

13. Which of the following compounds is the least soluble
    in water?
    (a) $Be(OH)_2$ (d) $Ba(OH)_2$
    (b) $Ca(OH)_2$ (e) $Mg(OH)_2$
    (c) $Sr(OH)_2$

14. Which of the following elements has common oxidation
    states of 2+, 4+, and 7+?
    (a) Mn   (d) La
    (b) Ni   (e) Cs
    (c) Pt

15. Which of the following elements has common oxidation
    states of 1+ and 3+?
    (a) Au   (d) K
    (b) Cd   (e) Hg
    (c) Sc

16. A metal in which the unfilled conduction band overlaps
    the filled valence band is
    (a) an intrinsic semiconductor
    (b) an extrinsic semiconductor
    (c) an *n*-type semiconductor
    (d) a conductor
    (e) an insulator

17. Metals with larger atomic radii tend to be
    (a) of lower density
    (b) less malleable
    (c) less ductile
    (d) less reactive
    (e) all of the above

18. The metal which can be separated from bauxite by the
    Bayer process is
    (a) aluminum   (d) iron
    (b) bismuth   (e) silver
    (c) copper

19. The metal which cannot be reduced by the Goldschmidt
    process is
    (a) aluminum   (d) manganese
    (b) barium   (e) all of these can be
    (c) chromium   reduced

20. Which of the following is not a refining process used
    in metallurgy?
    (a) Bessemer process   (d) Van Arkel process
    (b) Parkes process   (e) zone refining
    (c) the roasting process

II. Answer the following:

1. Complete the following equations. If no reaction occurs,
   write *NR* in the space provided for products.

   (a) $Al(s) + Cl_2(g) \rightarrow$

(b) $CaO(s) + H_2O(l) \rightarrow$

(c) $ZnS(s) + O_2(g) \xrightarrow{heat}$

(d) $Ce(s) + Br_2(g) \rightarrow$

(e) $Ba(s) + H_2O(l) \rightarrow$

(f) $Ni(s) + Cl_2(g) \rightarrow$

(g) $WO_3(s) + H_2(g) \xrightarrow{heat}$

(h) $Al_4C_3(s) + H_2O(l) \rightarrow$

(i) $Tl(s) + H^+(aq) \rightarrow$

(j) $Ca(s) + Cl_2(g) \rightarrow$

(k) $Pb(s) + H^+(aq) \rightarrow$

(l) $V(s) + H_2O(l) \rightarrow$

(m) $Na(s) + O_2(g) \rightarrow$

(n) $Al(s) + O_2(g) \xrightarrow{heat}$

(o) $Fe(s) + H^+(aq) \rightarrow$

2. Write the name of each of the following compounds in the space provided:

    (a) $Be(OH)_2$

_____   (b) $LiNO_3$

_____   (c) PbS

_____   (d) MoO_3

_____   (e) MnO_2

_____   (f) Li_2O

_____   (g) UF_4

_____   (h) Cu_2S

_____   (i) $TiCl_3$

_____   (j) Hg_2Cl_2

ANSWERS TO
EXERCISES

I. Properties of metals

1. (d), Fe   Each element of group I A has an oxidation state of 1+.
                  The 1+ oxidation state is an important oxidation state
                  of the transition metals Cu, Au, Ag, and Hg; compounds
                  of Cu^+, Ag^+, Au^+, and Hg^{2+} are common. Thallium is the
                  only group III A metal that has an oxidation state of 1+.

2. (a), NaCl   The ions of group II A metals and many compounds contain-
                  ing such ions hydrate readily.

3. (b), Cs   The softness of the metals of groups I A and II A
                  increases down the group with increasing atomic number.
                  The metals of group II A are harder than the metals of
                  group I A.

4. (a), Hall process

5. (c), 3+

6. (b),   Both barium and strontium are more dense than beryllium.
   It is the
   most dense.

7. (c), Ru   An element of the transition series that contains between
                  four and six electrons in the outer d shell can form a
                  variety of oxidation states.

8. (a), manganese Atomic size initially decreases and then increases across
                  a period of transition elements from left to right; the
                  minimum is reached near the center of the period. Atomic
                  size increases with increasing atomic number down a group.

9. (e), titanium

10. (e), Au

11. (c), In

12. (d), Sn

13. (a), $Be(OH)_2$

14. (a), Mn

15. (a), Au

16. (d), conductor

17. (a), lower density

18. (a), aluminum

19. (e), all can be reduced

20. (c), roasting

II. Reactions of metals and compounds containing metals

1. (a) $2Al(s) + 3Cl_2(g) \rightarrow 2AlCl_3(s)$

  (b) $CaO(s) + H_2O(l) \rightarrow Ca(OH)_2(aq)$

  (c) $2ZnS(s) + 3O_2(g) \xrightarrow{heat} 2ZnO(s) + 2SO_2(g)$

  (d) $2Ce(s) + 3Br_2(g) \rightarrow 2CeBr_3(s)$

  (e) $Ba(s) + 2H_2O(l) \rightarrow Ba(OH)_2(aq) + H_2(g)$

  (f) $Ni(s) + Cl_2(g) \rightarrow NiCl_2(s)$

  (g) $WO_3(s) + 3H_2(g) \xrightarrow{heat} W(s) + 3H_2O(g)$

  (h) $Al_4C_3(s) + 12H_2O(l) \rightarrow 4Al(OH)_3(s) + 3CH_4(g)$

  (i) $2Tl(s) + 2H^+(aq) \rightarrow 2Tl^+(aq) + H_2(g)$

  (j) $Ca(s) + Cl_2(g) \rightarrow CaCl_2(s)$

  (k) $Pb(s) + 2H^+(aq) \rightarrow Pb^{2+}(aq) + H_2(g)$

  (l) $V(s) + H_2O(l) \rightarrow NR$

  (m) $2Na(s) + O_2(g) \rightarrow Na_2O_2(s)$

  (n) $4Al(s) + 3O_2(g) \xrightarrow{heat} 2Al_2O_3(s)$

  (o) $Fe(s) + 2H^+(aq) \rightarrow Fe^{2+}(aq) + H_2(g)$

2. (a)   beryllium hydroxide
  (b)   lithium nitrate
  (c)   lead(II) sulfide
  (d)   molybdenum(VI) oxide
  (e)   manganese(IV) oxide, sometimes called manganese dioxide
  (f)   lithium oxide
  (g)   uranium(IV) fluoride
  (h)   copper(I) sulfide, also called cuprous sulfide
  (i)   titanium(III) chloride
  (j)   mercury(I) chloride, also called mercurous chloride

SELF-TEST   Complete the test in 10 minutes:

1. In the spaces provided, write the formulas of the products obtained from the following reactions:

   (a) $Mg(s) + H_2O(g) \rightarrow$

   (b) $Na(s) + O_2(g) \rightarrow$

   (c) $Li(s) + N_2(g) \rightarrow$

   (d) $HgS(s) + O_2(g) \xrightarrow{\text{heat}}$

   (e) $K(s) + O_2(g) \rightarrow$

   (f) $Ba(s) + P_4(l) \xrightarrow{\text{heat}}$

   (g) $Ce(s) + S(l) \xrightarrow{\text{heat}}$

2. Which of the following metals could be added to pure silicon to produce an *n*-type semiconductor?
   (a) gallium   (d) indium
   (b) germanium   (e) tin
   (c) arsenic

3. Which of the following metals would have the highest melting point?
   (a) potassium   (d) tantalum
   (b) vanadium   (e) mercury
   (c) zinc

4. Which of the following is not a process involving electrolysis?
   (a) producing sodium in a Downs cell
   (b) producing aluminum from molten cryolite
   (c) using the Hall process
   (d) electrolytically refining copper
   (e) zone refining silicon rods

5. Which of the following does not react directly with nitrogen to form a nitride?
   (a) potassium   (d) aluminum
   (b) calcium   (e) silicon
   (c) titanium

COMPLEX COMPOUNDS

CHAPTER 26

OBJECTIVES

I. You should be able to demonstrate your knowledge of the following terms by defining them, describing them, or giving specific examples of them:

bidentate [26.1]
chelate [26.1]
coordination isomer [26.4]
coordination number [26.1]
crystal field theory [26.5]
enantiomorphs [26.4]
Ewens-Bassett number [26.3]
geometric isomer [26.4]
hydrate isomer [26.4]
inert [26.2]
ionization isomer [26.4]
isomers [26.4]
ligand [26.1]
linkage isomer [26.4]
molecular orbital theory [26.5]
optical isomers [26.4]
porphyrin [26.1]
Stock number [26.3]
sterioisomerism [26.4]
structural isomer [26.4]
unidentate [26.1]

II. Given the names or chemical formulas of complex compounds, you should be able to write the corresponding chemical formulas or names.

III. You should be able to draw and identify the various types of isomers.

IV. You should be able to discuss the bonding in complex compounds.

EXERCISES   I. Answer each of the following with *true* or *false*. If the statement is false, correct it.

1. In general, Δ_o is larger for low-spin octahedral complexes of a metal than for high-spin octahedral complexes of that metal.

2. The ethylenediaminetetraacetate ion, EDTA, is a sexadentate chelate.

3. The formation of a complex can prevent a metal ion from disproportionating.

4. Tetrahedral complexes are six-coordinate complexes.

5. Both low-spin and high-spin octahedral complexes can exist for d^8 and d^9 ions.

6. Tetrahedral complexes never exist as *cis-trans* isomers.

7. Tetrahedral complexes never exist as optical isomers.

8. With the addition of Ag^+, 2 moles of chloride can be precipitated as AgCl from a solution containing 1 mole of $[Cr(H_2O)_5Cl]Cl_2 \cdot H_2O$.

9. Labile complexes rapidly undergo reactions in which the ligands are replaced.

10. Both chlorophyll and hemoglobin are metal ion complexes of porphyrins.

II. Write the name of each of the following complex compounds in the space provided:

1. $[Co(NH_3)_6]Cl_3$ _____

2. $K_4[Fe(CN)_6]$ _____

3. $[Co(NH_3)_5(SO_4)]Br$ _____

4. $[Co(NH_3)_5Br]SO_4$ _____

5. $[Pt(NH_3)_4Cl_2]PtCl_4$ _____

6. $[Co(H_2O)_6]Cl_2$ _____

7. $Ir(NH_3)_3Cl_3$ _____

8. $K_3[Fe(CN)_6]$ _____

9. $NH_4[Cr(SCN)_4(NO)_2]$ _____

10. $Pt(NH_3)_2Cl_2$ _____

11. $[Ni(H_2O)_5Cl]Cl$ _____

12. $[Cr(NH_3)_6][Cr(CN)_6]$ _____

III. Do the following:

1. Using Figure 26.1, identify the type of isomers represented by each of the following pairs:

(a) $[Co(en)_2ClBr]Cl$ and $[Co(en)_2Cl_2]Br$

(b) $[Pd(dipy)(SCN)_2]$ and $[Pd(dipy)(NCS)_2]$

(c)

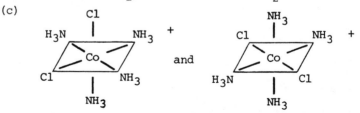

(d) $[Pt(NH_3)_4(OH)_2]SO_4$ and $[Pt(NH_3)_4(SO_4)OH]OH$

(e) $[Co(NH_3)_3(H_2O)_2Cl]Br_2$ and $[Co(NH_3)_3(H_2O)ClBr]Br \cdot H_2O$

(f)

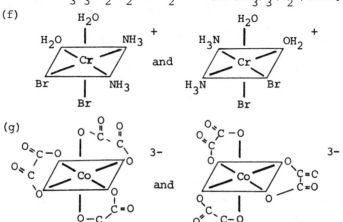

(g)

(h)

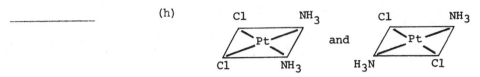

Figure 26.1   Isomerism in Complex Compounds

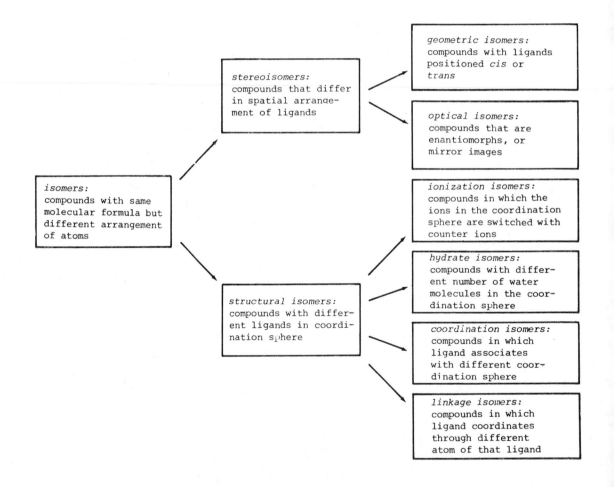

2. Which of the following complexes cannot have an optical isomer?

   (a) *trans*-Pt(NH$_3$)$_2$Cl$_2$, square planar

   (b)

   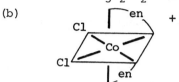

   (c)

   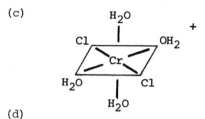

   (d)

   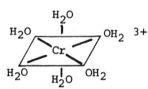

3. Which of the complexes in problem 2 above can exist as *cis-trans* isomers?

4. An elemental analysis of an unknown compound indicates an empirical formula of CrN$_4$Cl$_3$H$_{12}$. When dissolved in water, the compound forms two ions, and 1 mole of chloride ion can be precipitated for each mole of compound. The molecular weight of the compound is found to be 226. The compound exists in two isomeric forms. Draw the isomers.

5. The compound *cis*-Pt(NH$_3$)$_2$Cl$_2$ is an active anti-tumor agent. Does it have an optical isomer?

6. A compound when analyzed shows 65.0 percent Pt, 23.6 percent Cl, 9.3 percent N, and 2.0 percent H. The compound has a molecular weight of 600 and forms two ions in solution. No chloride can be removed from solution by the addition of Ag$^+$, and no ammonia can be evolved. Draw two possible isomers of the compound and identify the type of such isomers.

7. Determine whether each of the following transition metal ions is a d^0, d^1, d^2, d^3, ... d^9, or d^{10} ion:

(a) V^{5+}, vanadium ion

(b) $[Fe(CN)_6]^{4-}$, hexacyanoferrate(II) or hexacyanoferrate(4-)

(c) hexaaquamanganese(III) ion or hexaaquamanganese(3+)

(d) $[Ni(H_2O)_5Cl]Cl$, pentaaquachloronickel(II) chloride or pentaaquachloronickel(1+) chloride

(e) Ni^{3+}

(f) $NH_4[Cr(SCN)_4(NO)_2]$, ammonium dinitrosyltetrathiocyanatochromate(III) or ammonium dinitrosyltetrathiocyanatochromate(1-)

8. Using crystal field theory, diagram the electronic arrangement of $Fe(H_2O)_6^{2+}$ and $Fe(CN)_6^{4-}$. One of the complexes is low-spin the other is high-spin; make sure they are properly identified.

ANSWERS TO EXERCISES

I. Properties of complex compounds

1. True

For a high-spin octahedral complex the electrons are distributed in the t_{2g} and e_g orbitals such that the number of unpaired electrons is a maximum. Such a distribution occurs when the pairing energy is greater than Δ_o. For a low-spin octahedral complex the pairing energy is less than Δ_o and thus the lower-energy orbitals are filled completely before the higher-energy orbitals. For example, in an octahedral complex the high-spin state of a d^6 metal ion complex is

$$\underline{\uparrow} \qquad \underline{\uparrow} \quad e_g \qquad \uparrow$$
$$\underline{\uparrow\downarrow} \quad \underline{\uparrow} \quad \underline{\uparrow} \quad t_{2g} \qquad \Delta_o \downarrow$$

and the low-spin state of that d^6 metal ion complex is

$$\underline{\quad} \qquad \underline{\quad} \quad e_g \qquad \uparrow$$
$$\underline{\uparrow\downarrow} \quad \underline{\uparrow\downarrow} \quad \underline{\uparrow\downarrow} \quad t_{2g} \qquad \Delta_o \downarrow$$

2. True

See section 26.1 of your text.

3. True

See section 26.1 of your text.

4. False

Tetrahedral complexes are four-coordinate complexes.

5. False   In an octahedral complex the configuration of a d^8 ion is

$$\underline{\uparrow}\ \underline{\uparrow}\qquad e_g$$
$$\underline{\uparrow\downarrow}\ \underline{\uparrow\downarrow}\ \underline{\uparrow\downarrow}\qquad t_{2g}$$
$$d^8$$

and that of a d^9 ion is

$$\underline{\uparrow\downarrow}\ \underline{\uparrow}\qquad e_g$$
$$\underline{\uparrow\downarrow}\ \underline{\uparrow\downarrow}\ \underline{\uparrow\downarrow}\qquad t_{2g}$$
$$d^9$$

6. True   Each ligand is 109° from each of the other three ligands;
            therefore, *cis* and *trans* isomers cannot exist.

7. False   Optical isomers are possible and have been separated for
            some compounds. For instance, salts of the borosalicyl-
            aldehydro complex cannot be superimposed:

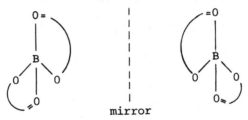

                                mirror

8. True   The two chlorides outside the brackets are not in the first
            coordination sphere of the metal ion. They exist as ions
            in solution and can be precipitated as AgCl with the
            addition of Ag^+ to the solution.

9. True

10. True

        II. Names of complex compounds

            Refer to Table 26.1 of the study guide or Section 26.3 of
            your text if you have any difficulty with this section.
            After you have checked your answers, write the formula of
            each compound in the space provided.

TABLE 24.1   Rules for naming complex compounds

1. If the complex compound is a salt, the cation is named first whether or not it is the complex ion.

2. The ligand constituents of the complex compound or complex ion are named in alphabetical order.

3. Anionic ligands are given -o endings; examples are: OH^-, hydroxo; O^{2-}, oxo; S^{2-}, thio; Cl^-, chloro; F^-, fluoro, CO_3^{2-}, carbonato; CN^-, cyano; CNO^-, cyanato; $C_2O_4^{2-}$, oxalato; NO_3^-, nitrato; NO_2^-, nitro; SO_4^{2-}, sulfato; and $S_2O_3^{2-}$, thiosulfato.

4. The names of neutral ligands are not changed. Exceptions to this rule are: H_2O, aquo; NH_3, ammine; CO, carbonyl; and NO, nitrosyl.

5. The number of ligands of a particular type is indicated by a prefix: di-, tri-, tetra-, penta-, and hexa- (for two to six). For complicated ligands (such as ethylenediamine), the prefixes bis, tris-, and tetrakis- (two to four) are employed.

6. The oxidation number of the central ion is indicated by a Roman numeral, which is set off by parentheses and placed after the name of the complex.

7. If the complex is an anion, the ending -ate is employed. If the complex is a cation or a neutral molecule, the name is not changed.

1. Hexaamminecobalt(III) chloride

        or

Hexaamminecobalt(3+) chloride

2. Potassium hexacyanoferrate(II)

        or

Potassium hexacyanoferrate(4-)

3. Pentaamminesulfatocobalt(III) bromide

        or

Pentaamminesulfatocobalt(1+) bromide

4. Pentaamminebromocobalt(III) sulfate

      or

  Pentaamminebromocobalt(2+) sulfate

5. Tetraamminedichloroplatinum(IV) tetrachloroplatinate(II)

      or

  Tetraamminedichloroplatinum(2+) tetrachloroplatinate(2-)

6. Hexaaquacobalt(II) chloride

      or

  Hexaaquacobalt(2+) chloride

7. Triamminetrichloroiridium(III)

      or

  Triamminetrichloroiridium

8. Potassium hexacyanoferrate(III)

      or

  Potassium hexacyanoferrate(3-)

9. Ammonium dinitrosyltetrathiocyanatochromate(III)

      or

  Ammonium dinitrosyltetrathiocyanatochromate(1-)

10. Diamminedichloroplatinum(II)

      or

  Diamminedichloroplatinum

11. Pentaaquachloronickel(II) chloride

      or

  Pentaaquachloronickel(1+) chloride

12. Hexaamminechromium(III) hexacyanochromate(III)

      or

  Hexaamminechromium(3+) hexacyanochromate(3-)

III. Isomers

See Section 26.4 of your text.

1. (a) ioniza-
       tion
       isomers

The Cl and Br are switched. In the first compound Br is in the first coordination sphere and in the second compound it is the counter ion.

   (b) linkage
       isomers

The SCN is bonded to the metal through the S atom in the first compound and through the N atom in the second compound.

   (c) geometric
       isomers

The chlorines are *cis* in the first compound and *trans* in the second.

   (d) ioniza-
       tion
       isomers

   (e) hydrate
       isomers

   (f) optical
       isomers

   (g) optical
       isomers

   (h) geometric
       isomers

2. (a), (c),
   and (d)

Mirror images of these complexes can be superimposed on the originals.

3. (a)

    cis   *trans*

(b)

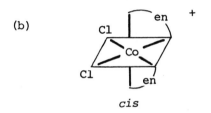

cis

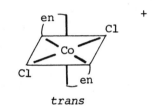

trans

(c)

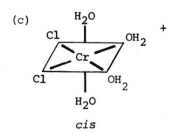

cis

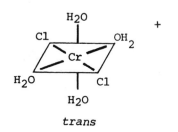

trans

4.

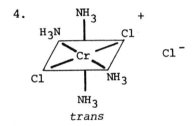

trans

The nitrogen-hydrogen ratio indicates the possibility of NH_3 in the compound and a formula of $Cr(NH_3)_4Cl_3$. Chromium usually forms octahedral complexes. Since 1 mole of chloride ion can be precipitated for each mole of compound, a chloride ion is free in solution; i.e., it is not in the first coordination sphere of Cr. The complex may exist in *cis* and *trans* forms:

cis-$[Cr(NH_3)_4Cl_2]Cl$

and

trans-$[Cr(NH_3)_4Cl_2]Cl$

cis

5. No

The compound

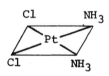

is square planer. Its mirror image can be superimposed on the original.

6. $[Pt(NH_3)_4][PtCl_4]$

and

$[Pt(NH_3)_3Cl][Pt(NH_3)Cl_3]$

From the data on percent composition an empirical formula of $PtCl_2N_2H_6$, or $Pt(NH_3)_2Cl_2$, can be determined. The compound has a molecular weight of 600; thus, the molecular formula must be $Pt_2(NH_3)_4Cl_4$. The compound is neutral; thus, the charge of the anion must equal the charge of the cation.

7. (a) d^0 (d) d^8
   (b) d^6 (e) d^7
   (c) d^4 (f) d^3

8.

Iron(II) is a d^6 ion. In $Fe(H_2O)_6^{2+}$ the iron(III) is in a high-spin state and in $Fe(CN)_6^{4-}$ it is in a low-spin state:

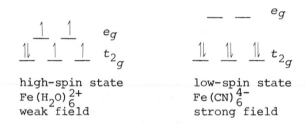

high-spin state
$Fe(H_2O)_6^{2+}$
weak field

low-spin state
$Fe(CN)_6^{4-}$
strong field

The crystal field approach indicates the unpaired, nonbonding electrons *and* explains the difference in energy of the d orbitals.

SELF-TEST

Complete the test in 20 minutes:

1. Write the name of each of the following compounds in the space provided:

   (a) $K_2[CuCl_4]$ _____

   (b) $K_2[PtCl_4]$ _____

   (c) $K_4[Fe(CN)_6]$ _____

   (d) $Pt(NH_3)_2Cl_4$ _____

   (e) $[Co(en)_2(SCN)Cl]Cl$ _____

   (f) $[Co(NH_3)_4Br_2]Br$ _____

   (g) $[Cu(NH_3)_4]_3[CrCl_6]_2$ _____

2. Draw the *cis* and *trans* isomers of the diamminedioxalatochromate(III) ion.

3. Is Cr(III) a d^0, d^1, d^2, d^3, d^4, d^5, d^6, d^7, d^8, d^9, or d^{10} ion?

4. Draw the electronic arrangement of molybdenum in ammonium aquapentachloromolybdate(III) as predicted by the crystal field theory.

5. Draw the structure of the anion in potassium hexacyanotungstenate(III).

NUCLEAR CHEMISTRY

CHAPTER
27

OBJECTIVES

I. You should be able to demonstrate your knowledge of the
following terms by defining them, describing them, or
giving specific examples of them:

activity [27.5]
activation analysis [27.10]
alpha decay [27.3]
antiparticle [27.3]
electron capture [27.3]
fast neutron [27.7]
fission [27.3]
gamma radiation [27.3]
Geiger-Müller counter [27.5]
geological dating [27.6]
half-life [27.5]
isotope dilution [27.10]
linear accelerator [27.7]
magic number [27.1]
mass number, A [27.1]
moderator [27.7, 27.8]
neutrino [27.3]
neutron-capture reaction [27.7]
nuclear fission [27.8]
nuclear fusion [27.9]
nuclear reactor [27.8]
nucleon [27.1]
nuclide [27.1]
particle-particle reaction [27.7]
plasma [27.9]
positron emission, β^+ decay [27.3]
radioactive decay series [27.6]
radiocarbon dating [27.5]

scintillation counter [27.5]
thermonuclear reaction [27.8]
transuranium element [27.7]
Wilson cloud chamber [27.5]
zone of stability [27.1]

II. You should be able to write balanced equations for nuclear reactions including those undergoing radioactive decay by alpha decay, beta decay, electron capture, and spontaneous fission; and also those nuclear fission and fusion reactions brought about in particle accelerators and nuclear power plants.

III. You should be able to determine the amount of energy released in a nuclear reaction by using the change in mass which occurs during the reaction.

IV. You should be able to use the rate law and values of $t_{1/2}$ to calculate the amount of a nuclide remaining after a specific time or to calculate the age of the sample.

V. You should be able to calculate binding energies of nuclei.

VI. You should understand and be able to use the curie (the unit of activity) in calculations.

VII. You should understand the practical implications of nuclear chemistry including the biological effects of radiation, uses in medical diagnostics, power generation, dating, activation analysis, and fundamental chemical equilibrium and kinetic studies.

UNITS,
SYMBOLS,
MATHEMATICS

The following symbols were used in this chapter; you should be familiar with them.

- A is the symbol for the mass number of an element. The mass number equals the total number of nucleons in the nucleus.

- α is the symbol used to represent an alpha particle, the helium nucleus.

- β^- is the symbol used to represent a beta particle, an electron.

- β^+ is the symbol used to represent a positron, the anti-particle of the β particle.

- C is the symbol for the curie, the unit used to measure the activity of a radioactive source. One curie equals 3.70×10^{10} nuclear disintegrations in one second.

- mC equals 10^{-3} curies or 3.70×10^{7} nuclear disintegrations in one second.

- μC equals 10^{-6} curies or 3.70×10^{4} nuclear disintegrations in one second.

- γ is the symbol used to represent gamma radiation.

- k is the symbol used to represent the rate constant of a nuclear decay reaction.

- MeV is the abbreviation for megaelectron volt, the commonly used unit of energy in nuclear chemistry.

- n or $_{0}^{1}n$ is the symbol used to represent a neutron.

- ν is the symbol used to represent a neutrino.

- $\bar{\nu}$ is the symbol used for the antineutrino.

- $t_{\frac{1}{2}}$ is the symbol used to represent the half-life of an unstable nuclide.
- Z is the symbol for the atomic number of an element. The atomic number equals the number of protons in the nucleus.

EXERCISES   I. Choose the correct answer for each of the following.

1. Which of the following symbols represents high energy electromagnetic radiation?
   (a) α   (c) γ
   (b) β   (d) δ

2. Which of the following nuclei has the highest binding energy?

   (a) $_{2}^{4}HE$ (c) $_{8}^{16}O$

   (b) $_{92}^{238}U$ (d) $_{20}^{40}Ca$

3. Which of the following symbols represents the particle in the list with the highest mass?
   (a) α   (c) n
   (b) γ   (d) π

4. Which of the following is the symbol used to represent one of the antiparticles?
   (a) α
   (b) β^+
   (c) $_{-1}^{0}e$
   (d) ν

5. Emissions from radioactive decay can never be detected with a
   (a) cyclotron
   (b) scintillation counter
   (c) Geiger-Müller counter
   (d) Wilson cloud chamber

6. For the nuclear reaction indicated by the notation $_4^9Be(p,\alpha)_3^6Li$,
   (a) $_4^9Be$ is the reacting nucleus bombarded by $_1^1H$
   (b) a $_2^4He$ nucleus is produced
   (c) $_3^6Li$ is a product
   (d) all of the above

7. Which of the following nuclei would be the least stable?
   (a) $_2^4He$
   (b) $_8^{16}O$
   (c) $_8^{20}O$
   (d) $_{20}^{40}Ca$

8. Which of the following nuclides has the highest atomic number?
   (a) $_7^{12}N$
   (b) $_8^{13}O$
   (c) $_6^{14}C$
   (d) $_7^{16}N$

10. A curie, C, is defined as 3.70×10^{10} disintegrations per second. One millicurie is

    (a) 3.70×10^7 disintegrations per second
    (b) 3.70×10^{10} disintegrations per millisecond
    (c) the activity of a curie sample after a millihalf-life
    (d) all of the above

11. Nuclides with proton-to-neutron ratios higher than those within the zone of stability probably decay by

    (a) β^- emission
    (b) α emission
    (c) electron capture
    (d) β^+ emission

12. The decay of $^{15}_{8}O$ to $^{15}_{7}N$ involves the release of 1.74 MeV of energy. The energy of the positron is probably

    (a) greater than 1.74 MeV
    (b) less than 1.74 MeV
    (c) equal to 1.74 MeV
    (d) equal to that of the neutrino which is also emitted

13. During the electron capture decay of a nucleus

    (a) an inner electron falls into the nucleus
    (b) the number of neutrons in the nucleus increases
    (c) X rays are emitted
    (d) all of the above

14. Radioactive decay processes are

    (a) first-order reactions
    (b) all very dangerous
    (c) always emitting X rays
    (d) none of these

15. Element number 110 will be called

    (a) ununnilium, Uun
    (b) unilium, Un
    (c) duounnilium, Dun
    (d) none of these

16. Moderators are used in nuclear reactors to

    (a) slow down neutrons
    (b) produce rational debates
    (c) prevent neutrons from being captured
    (d) all of the above

17. Practical nuclear fusion will require

    (a) the use of light elements
    (b) very high temperatures
    (c) generation of a plasma
    (d) all of these

18. Radioactive nuclides are used

    (a) in microwave ovens
    (b) to take X rays of bones
    (c) to preserve foods
    (d) all of the above

19. Activation analysis is used

    (a) to determine the quantity of an element in a sample
    (b) to initiate nuclear fission
    (c) to initiate nuclear fusion
    (d) none of these

20. Radioactive nuclides cause biological damage primarily by the

    (a) transmutation of the nuclei which are part of the biological system
    (b) forming of free radicals within the biological system
    (c) displacing normal nuclei from the biological system
    (d) none of these

II. Complete the following nuclear reactions by writing the missing product in the blank.

1. $^{218}_{84}Po \rightarrow ^{4}_{2}He + $ _____

2. $^{254}_{102}No \rightarrow ^{4}_{2}He + $ _____

3. $^{238}_{92}U + ^{1}_{0}n \rightarrow \gamma + $ _____

4. $^{238}_{92}U + ^{12}_{6}C \rightarrow 6\ ^{1}_{0}n + $ _____

5. $^{207}_{84}Po \rightarrow ^{0}_{1}e + $ _____

6. $^{228}_{88}Ra \rightarrow ^{0}_{-1}e + $ _____

7. $^{14}_{6}C \rightarrow ^{0}_{-1}e + $ _____

8. $^{238}_{92}U \rightarrow ^{4}_{2}He + $ _____

9. $^{87}_{36}Kr \rightarrow ^{0}_{-1}e + $ _____

10. $^{87}_{36}Kr \rightarrow ^{1}_{0}n + $ _____

11. $^{13}_{7}N \rightarrow ^{13}_{6}C + $ _____

12. $^{39}_{19}K + ^{1}_{0}n \rightarrow 2\ ^{1}_{0}n + $ _____

13. $^{238}_{92}U \rightarrow ^{234}_{90}Th +$ _____

14. $^{235}_{92}U + ^{1}_{0}n \rightarrow ^{139}_{56}Ba + 3\ ^{1}_{0}n +$ _____

15. $^{2}_{1}H + ^{3}_{1}H \rightarrow ^{4}_{2}He +$ _____

16. $^{0}_{-1}e + ^{55}_{26}Fe \rightarrow$ _____

III. Write equations for the following induced nuclear reactions. All are used to prepare unstable isotopes.

1. $^{114}_{48}Cd\,(n,p)$

2. $^{114}_{48}Cd\,(n,\gamma)$

3. $^{197}_{79}Au\,(p,3n)$

4. $^{60}_{28}Ni\,(\alpha,n)$

5. $^{63}_{29}Cu\,(p,n)$

6. $^{114}_{48}Cd\,(d,p)$

7. $^{90}_{40}Zr\,(\alpha,4n)$

8. $^{92}_{42}Mo\,(p,n)$

9. $^{207}_{82}Pb\,(\alpha,3n)$

10. $^{209}_{83}Bi\,(d,3n)$

IV. Work the following problems:

1. In the electron capture reaction

$$^{0}_{-1}e + ^{7}_{4}Be \overset{ec}{\rightarrow} ^{7}_{3}Li$$

how much energy in MeV is released? The mass of $^{7}_{4}Be$ is 7.0169 u and that of $^{7}_{3}Li$ is 7.0160 u.

2. Radon in a tube has been used in cervical cancer therapy. The half-life of radioactive radon is 3.8 days. If 11 micrograms of radon are sealed in a tube, how many micrograms remain after 21 days?

3. How long will it take before the amount of radon in the tube described in problem 2 above is reduced to one millionth of a microgram?

4. The $^{14}_{6}C$ activity of a fiber from an Egyptian mummy shroud is 7.50 disintegrations per minute per gram of carbon. How old is the fiber? The half-life of $^{14}_{6}C$ is 5770 years, and the $^{14}_{6}C$ activity of a piece of wood from the outer layer of a freshly cut tree is 15.2 disintegrations per minute per gram of carbon.

5. Use information from problem 4 above to determine the age of the oldest sample that can be dated by the $^{14}_{6}C$ technique. Assume that less than one disintegration per minute per gram of carbon cannot be detected with the equipment used with this technique.

6. A sample of $^{147}_{59}Pr$ is prepared and placed in a scintillation counter. The initial counting rate is 200/min. After 36 min the rate is 25 counts/min. What is the half-life of $^{147}_{59}Pr$?

7. How many grams of $^{60}_{27}Co$ will give 75×10^{-3} curies, or 75 millicuries, of radiation? The half-life of $^{60}_{27}Co$ is 5.2 years.

ANSWERS TO EXERCISES

I. Multiple choice questions

1. (c) γ

2. (d) $^{238}_{92}U$

3. (a) α

4. (b) β^{+}

5. (a) cyclotron

6. (d) all of the above

7. (c) $^{20}_{8}O$

8. (b) $^{13}_{8}O$

9. (d) $^{16}_{7}N$

10. (a) 3.70×10^{7} disintegrations per second

11. (d) β^+ emission

12. (b) less than 1.74 MeV--the neutrino carries away the rest of the energy

13. (d) all of the above

14. (a) first-order reactions

15. (a) ununnilium, Uun

16. (a) slow down neutrons

17. (d) all of these

18. (c) to preserve foods

19. (a) to determine the quantity of an element in a sample

20. (b) forming free radicals within the biological system

II. Balancing nuclear equations

1. $^{214}_{82}Pb$ An alpha particle is a helium nucleus, ^4_2He. Thus, when an alpha particle is emitted, two protons and two neutrons are lost from the nucleus of the atom.

2. $^{250}_{100}Fm$

3. $^{239}_{92}U$

4. $^{244}_{98}Cf$ The capture of a neutron, ^1_0n, by an atom increases the mass number of that atom by one unit. The loss of a neutron by an atom decreases the mass number of that atom by one unit.

5. $^{207}_{83}Bi$ A positron, ^0_1e, is the product of a transformation of a nuclear proton into a nuclear neutron.

6. $^{228}_{89}Ac$ The beta particle, $^0_{-1}e$, is a negatively charged, low-mass particle. It is a product of the transformation of a nuclear neutron into a nuclear proton.

7. $^{14}_7N$

8. $^{234}_{90}Th$

9. $^{87}_{37}Rb$

10. $^{86}_{36}Kr$

11. ^0_1e

12. $^{38}_{19}K$

13. $^{4}_{2}He$

14. $^{94}_{36}Kr$ This is the ^{235}U atomic bomb reaction.

15. $^{1}_{0}n$ This is a typical fusion reaction.

16. $^{55}_{25}Mn$

III. Nuclear bombardment reactions

1. $^{114}_{48}Cd + ^{1}_{0}n \rightarrow ^{114}_{47}Ag + ^{1}_{1}H$

2. $^{114}_{48}Cd + ^{1}_{0}n \rightarrow ^{115}_{48}Cd + \gamma$

3. $^{197}_{79}Au + ^{1}_{1}H \rightarrow ^{195}_{80}Hg + 3^{1}_{0}n$

4. $^{60}_{28}Ni + ^{4}_{2}He \rightarrow ^{63}_{30}Zn + ^{1}_{0}n$

5. $^{63}_{29}Cu + ^{1}_{1}H \rightarrow ^{63}_{30}Zn + ^{1}_{0}n$

6. $^{114}_{48}Cd + ^{2}_{1}H \rightarrow ^{115}_{48}Cd + ^{1}_{1}H$

7. $^{90}_{40}Zr + ^{4}_{2}He \rightarrow ^{90}_{42}Mo + 4^{1}_{0}n$

8. $^{92}_{42}Mo + ^{1}_{1}H \rightarrow ^{92}_{43}Tc + ^{1}_{0}n$

9. $^{207}_{82}Pb + ^{4}_{2}He \rightarrow ^{208}_{84}Po + 3^{1}_{0}n$

10. $^{209}_{83}Bi + ^{2}_{1}H \rightarrow ^{208}_{84}Po + 3^{1}_{0}n$

IV. Radioactive decay problems

1. 0.8 MeV   Determine the loss of mass:

         mass of reactant $^{7}_{4}Be$ = 7.0169 u

         mass of product $^{7}_{3}Li$ = 7.0160 u

         loss of mass = 0.0009 u

Calculate the energy equivalent of this mass difference by means of Einstein's equation:

$E = mc^2$

     = 0.0009 u (931 MeV/u)

     = 0.8 MeV

Thus, the energy released is 0.8 MeV.

2. 0.24 µg   We first calculate the rate constant for the radioactive
                decay of radon:

$$k = \frac{0.693}{t_{\frac{1}{2}}}$$

$$= \frac{0.693}{3.8 \text{ days}}$$

$$= 0.182/\text{day}$$

Then calculate the fraction of radon remaining after 21
days:

$$\log\left(\frac{N_0}{N}\right) = \frac{kt}{2.303}$$

$$= \frac{(0.182/\text{day})(21 \text{ days})}{2.30}$$

$$= 1.66$$

and

$$\frac{N_0}{N} = 10^{1.66}$$

$$= 10^{0.66} \times 10^1$$

$$= 4.6 \times 10^1, \text{ or } 46$$

Since $N_0 = 11$ µg, the amount of radon remaining after 21
days is

$$\frac{N_0}{N} = 46$$

$$N = \frac{N_0}{46}$$

$$= \frac{11 \text{ µg}}{46}$$

$$= 0.24 \text{ µg}$$

3. 90 days   Since $N_0 = 11$ µg and $N = 1 \times 10^{-6}$ µg, the length of time,
                t, can be computed directly:

$$\log\left(\frac{N_0}{N}\right) = \frac{kt}{2.303}$$

$$t = \left(\frac{2.303}{k}\right) \log\left(\frac{N_0}{N}\right)$$

$$= \left(\frac{2.3}{0.18/\text{day}}\right) \log\left(\frac{1.1}{1 \times 10^{-6}}\right)$$

$$= 9 \times 10^1 \text{ days, or } 90 \text{ days}$$

4. 6.45×10^3 years

Determine the value of k from the half-life:

$$k = \frac{0.693}{t_{\frac{1}{2}}}$$

$$= \frac{0.693}{5770 \text{ years}}$$

$$= 1.20 \times 10^{-4}/\text{year}$$

The number of disintegrations per minute is proportional to the number of atoms present; therefore, substitute the values for the number of disintegrations per minute into the fraction N_0/N:

$$\log \left(\frac{N_0}{N}\right) = \frac{kt}{2.303}$$

$$t = \left(\frac{2.303}{k}\right) \log \left(\frac{N_0}{N}\right)$$

$$= \left(\frac{2.303}{1.20 \times 10^{-4}/\text{year}}\right) \log \left(\frac{15.2 \text{ disint./min}}{7.00 \text{ disint./min}}\right)$$

$$= 6.45 \times 10^3 \text{ years}$$

5. 2×10^4 years

Determine the time, t:

$$\log \left(\frac{N_0}{N}\right) = \frac{kt}{2.303}$$

$$t = \left(\frac{2.303}{k}\right) \log \left(\frac{N_0}{N}\right)$$

$$= \left(\frac{2.3}{1.2 \times 10^{-4}/\text{year}}\right) \log \left(\frac{15 \text{ disint./min}}{1 \text{ disint./min}}\right)$$

$$= 2 \times 10^4 \text{ years}$$

The technique is actually limited to less than 20,000 years due to other restrictions.

6. 12 min

In one half-life, half the original material disintegrates, and the rate drops to 100 counts per minute. During the second half-life, the rate drops to 50 counts per minute, and during the third half-life it drops to 25 counts per minute. Therefore, 3 half-lives elapse between the initial and the final counts. Thus, the half-life is 1/3 of the elapsed time, or 36 min/3 = 12 min.

7. 6.6×10^{-5} g

$^{60}_{27}Co$

Convert the activity from millicuries to disintegrations/sec:

$$\text{activity} = (75 \text{ mC}) \left(\frac{3.70 \times 10^7 \text{ disint./sec}}{1\text{C}} \right) \left(\frac{1\text{C}}{1000 \text{ mC}} \right)$$

$$= 2.78 \times 10^6 \text{ disint./sec, or}$$

$$= 2.78 \times 10^6 \text{ atom/sec}$$

Since the rate constant is expressed in /years, we convert the activity from atoms/sec to atoms/year:

$$? \text{ activity} = \left(\frac{2.78 \times 10^6 \text{ atoms}}{1 \text{ sec}} \right) \left(\frac{60 \text{ sec}}{1 \text{ min}} \right) \left(\frac{60 \text{ min}}{1 \text{ hr}} \right) \left(\frac{24 \text{ hr}}{1 \text{ day}} \right) \left(\frac{365 \text{ days}}{1 \text{ year}} \right)$$

$$= 8.77 \times 10^{13} \text{ atoms/year}$$

Then calculate the rate constant, k:

$$k = \frac{0.693}{t_{\frac{1}{2}}}$$

$$= \frac{0.693}{5.2 \text{ years}}$$

$$= 0.133/\text{year}$$

Since activity = kN, we can calculate the number of atoms, N:

$$\text{activity} = kN$$

$$N = \frac{\text{activity}}{k}$$

$$= \frac{8.77 \times 10^{13} \text{ atoms/year}}{0.133 \text{ year}}$$

$$= 6.59 \times 10^{14} \text{ atoms}$$

Finally we calculate the number of grams:

$$? \text{ g } ^{60}_{27}Co = 6.59 \times 10^{14} \text{ atoms } ^{60}_{27}Co \left(\frac{1 \text{ mol } ^{60}_{27}Co}{6.022 \times 10^{23} \text{ atoms}} \right) \left(\frac{60 \text{ g } ^{60}_{27}Co}{1 \text{ mol } ^{60}_{27}Co} \right)$$

$$= 6.6 \times 10^{-5} \text{ g } ^{60}_{27}Co$$

SELF-TEST   Complete the test in 20 minutes:

1. A sample contains $^{35}_{16}S$ as the only radioactive species. How many moles of $^{35}_{16}S$ are in a sample that has an activity of 1.01×10^3 disintegrations per minute. The half-life of $^{35}_{16}S$ is 86.6 days.

2. The half-life of $^{31}_{14}Si$ is 2.6 hours. How many grams of $^{31}_{14}Si$ must be prepared if 1.0×10^{-12} g will be needed in an experiment 13 hours later?

3. Calculate the binding energy of an atom of $^{208}_{82}Pb$, which has a mass of 208.060 u. The masses of the proton, neutron, and electron are 1.007277 u, 1.008665 u, and 0.0005486 u, respectively.

4. Complete the following nuclear reactions by filling in the blank.

   (a) $^{124}_{54}Xe\,(n,\gamma)$ _____

   (b) $^{125}_{54}Xe \overset{ec}{\rightarrow}$ _____

   (c) $^{32}_{16}S + ^{2}_{1}H \rightarrow ^{33}_{17}Cl$ _____

   (d) $^{123}_{51}Sb\,(\alpha,2n)$ _____

   (e) _____ $(\alpha,5n)\ ^{123}_{53}I$

ORGANIC CHEMISTRY

CHAPTER
28

OBJECTIVES

I. You should be able to demonstrate your knowledge of the
following terms by defining them, describing them, or
giving specific examples of them:

addition reaction [28.5]
alcohol [28.6]
aldehyde [28.7]
alkane [28.1]
alkene [28.2]
alkyl [28.1]
alkyne [28.3]
amide [28.9]
amine [28.9]
aromatic [28.4]
aryl [28.4]
carbocation [28.5]
carbonyl [28.7]
carboxyl [28.8]
cis [28.2]
conjugated [28.10]
copolymer [28.10]
dehydrogenation [28.7]
ester [28.8]
ether [28.6]
Friedel-Crafts synthesis [28.5]
functional group [28.6]
geometric isomer [28.2]
Grignard reagent [28.7]
homologous [28.1]
ketone [28.7]
Markovnikov's rule [28.5]
meta [28.4]

monomer [28.10]
nucleophilic substitution [28.6]
olefin [28.2]
ortho [28.4]
para [28.4]
polymer [28.10]
primary [28.1, 28.6, 28.9]
saturated hydrocarbon [28.2]
secondary [28.1, 28.6, 28.9]
stereoisomer [28.2]
structural isomer [28.1]
substitution reaction [28.5]
tertiary [28.1, 28.6, 28.9]
trans [28.2]
unsaturated hydrocarbon [28.2]

II. Given the structures or names of simple organic compounds, you should be able to write the corresponding names or structures.

III. You should be familiar with the properties and reactions of the major classes of organic molecules.

IV. You should be able to identify isomers.

EXERCISES   I. Write the name of each of the following compounds in the space provided:

butane   1. $CH_3CH_2CH_2CH_3$

2 methyl butane   2. $CH_3CH_2CHCH_3$
$$\quad\quad\quad\quad\quad\quad |$$
$$\quad\quad\quad\quad\quad CH_3$$

3 ethyle methyl pentane   3.

$$CH_3$$
$$|$$
$$CH_2$$
$$|$$
$$CH_3-CH_2-C-CH_2-CH_3$$
$$|$$
$$CH_3$$

2 methyl pentane   4.

$$CH_3$$
$$|$$
$$CH_3-CH_2-CH_2-CH-CH_3$$

3 dimethyl hexane   5.

$$CH_3$$
$$|$$
$$CH_3-CH_2-CH_2-C-CH_3$$
$$|$$
$$CH_2$$
$$|$$
$$CH_3$$

2,3,5 trimethy 2 ethyl hexane   6.

$$\quad\quad\quad\quad\quad\quad CH_3$$
$$\quad\quad\quad\quad\quad\quad |$$
$$\quad\quad CH_3 \quad\quad CH_2$$
$$\quad\quad | \quad\quad\quad\quad |$$
$$CH_3-CH-CH_2-C-CH-CH_3$$
$$\quad\quad\quad\quad\quad\quad | \quad |$$
$$\quad\quad\quad\quad\quad CH_3 CH_3$$

cyclobutane   7. H_2C-CH_2
$$\quad\quad\quad\quad | \quad\quad |$$
$$\quad\quad\quad H_2C-CH_2$$

methyl cyclobutane   8.

$$CH_3$$
$$|$$
$$H_2C-CH$$
$$| \quad\quad |$$
$$H_2C-CH_2$$

2 butene   9. $CH_3-CH=CH-CH_3$

2 pentane   10. $CH_3-CH_2-CH=CH-CH_3$

4 methyl 2 pentene   11.

$$CH_3$$
$$|$$
$$CH_3-CH-CH=CH-CH_3$$

cyclobutene   12. H_2C-CH_2
$$\quad\quad\quad\quad\quad | \quad\quad |$$
$$\quad\quad\quad\quad HC=CH$$

3 methyl butyne   13.

$$CH_3$$
$$|$$
$$HC\equiv C-CH-CH_3$$

methylbenzene
~~Toluene~~   **14.**

orthdinitrobenzene **15.**

para di chloro benzene. **16.** Cl—⟨benzene⟩—Cl

meta dibromo benzene. **17.**

2 bromo butane **18.** $CH_3—CH—CH_2—CH_3$
                              $|$
                              Br

2 bromobutene **19.** $CH_2{=}C—CH_2—CH_3$
                             $|$
                             Br

3 bromobutene **20.** $CH_2{=}CH—CH—CH_3$
                             $|$
                             Br

1,4 dibromobutene **21.** $Br_2C{=}CH—CH_2—CH_2Br$

2 butanol **22.** $CH_3—CH—CH_2—CH_3$
                          $|$
                          OH

cyclohexanol **23.**

2,3 butanol **24.** $CH_3—CH—CH—CH_3$
                           $|$ $|$
                           OH OH

phenol
para clorhexanol **25.** Cl—⟨ring⟩—OH

butanal **26.**

$CH_3—CH_2—CH_2—\overset{\displaystyle O}{\overset{\|}{C}}—H$

3 chlorbutanal
2 chlor **27.**

$CH_3—\overset{Cl}{\underset{|}{CH}}—CH_2—\overset{\displaystyle O}{\overset{\|}{C}}—H$

2 pentone **28.**

$CH_3—\overset{\displaystyle O}{\overset{\|}{C}}—CH_2—CH_2—CH_3$

2 3 ethyl 2 pentone **29.**

$CH_3—CH_2—\underset{\displaystyle\underset{|}{\underset{CH_2}{|}}}{CH}—\overset{\displaystyle O}{\overset{\|}{C}}—CH_3$
                                     CH_3

propanoic acid,   30.

$$CH_3—CH_2—\overset{\overset{\text{O}}{\|}}{C}—OH$$

mbromobenzoic acid.   31.

$$\overset{\overset{\text{O}}{\|}}{C}—OH$$ with Br

_____   32.

$$\overset{\overset{\text{O}}{\|}}{C}—OCH_3$$

_____   33.

$$ClCH_2—\overset{\overset{\text{O}}{\|}}{C}—O—CH_2CH_3$$

_____   34.

$$CH_3CH_2—\overset{\overset{\text{H}}{|}}{N}—CH_2CH_3$$

_____   35.

—NH$_2$ with Cl

_____   36.

$$H_2C—CH_2$$
$$H_2C \quad CH_2$$
$$\overset{\overset{\|}{\text{O}}}{C}$$

_____   37.

$$CH_3—\overset{\overset{\text{O}}{\|}}{C}—O—CH_2—\overset{\overset{CH_3}{|}}{\underset{\underset{CH_3}{|}}{CH}}$$

_____   38.

$$CH_2=\overset{\overset{CH_3}{|}}{C}—\overset{\overset{CH_2\,|\,CH_3}{}}{\underset{\underset{CH_3}{|}}{C}}=\overset{}{C}—CH_3$$

_____   39.

_____   40.

$$CH_3—\overset{\overset{CH_3}{|}}{\underset{\underset{CH_3}{|}}{C}}—C≡CH$$

II. Complete the following reactions. If no reaction occurs, write *NR* in the space provided for products.

1. $CH_3CH_2—CH{=}CH—CH_3 \xrightarrow{MnO_4^-}$

2. $CH_3CH_2—CH{=}CH—CH_3 \xrightarrow{H_2/Pt}$

3. $CH_3—CH_2—CH{=}CH—CH_3 \xrightarrow{Br_2}$

4. 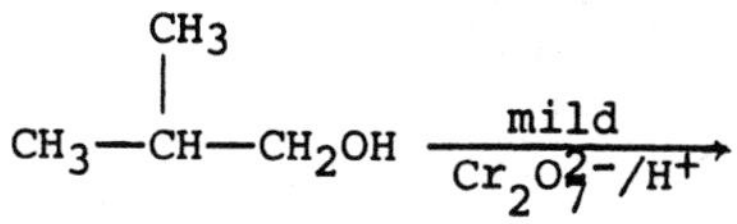
$$CH_3—\overset{\overset{\displaystyle CH_3}{|}}{CH}—CH_2OH \xrightarrow[Cr_2O_7^{2-}/H^+]{mild}$$

5.
$\xrightarrow{Br_2/FeBr_3}$

6.
$$CH_3—\overset{\overset{\displaystyle O}{\|}}{C}—H \xrightarrow{hot\ MnO_4^-}$$

7.
$$CH_3—CH_2—\overset{\overset{\displaystyle O}{\|}}{C}—H \xrightarrow{H_2/catalyst}$$

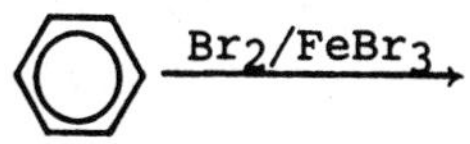

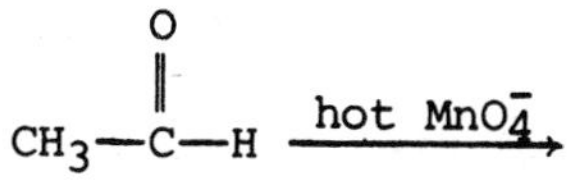

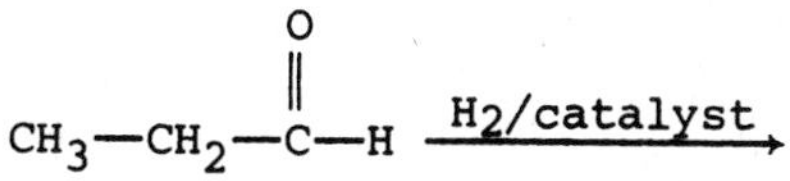

8. 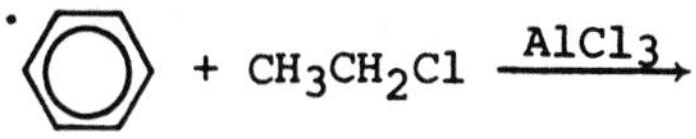
$+ CH_3CH_2Cl \xrightarrow{AlCl_3}$

9. 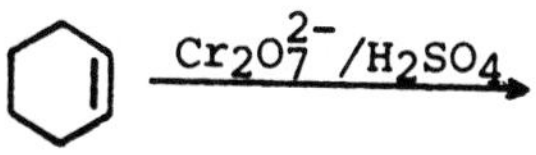
$\xrightarrow{Cr_2O_7^{2-}/H_2SO_4}$

III. Answer each of the following:

1. The molecule
$$CH_3—\overset{\overset{\displaystyle O}{\|}}{C}—OCH_2—CH_2—\overset{\overset{\displaystyle CH_3}{|}}{\underset{\underset{\displaystyle CH_3}{|}}{CH}}$$

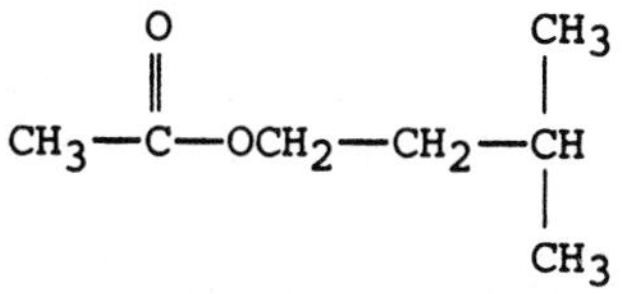

is an ester that has a banana-like odor. Write a reaction for the preparation of the ester from a carboxylic acid and an alcohol. Name the carboxylic acid, alcohol, and ester.

2. Teflon is prepared by polymerizing tetrafluoriethene
   (also called tetrafluoroethylene). Draw the structure
   of Teflon.

3. The compound commonly known as DDT has the following
   structure. Are there any stereoisomers or optical
   isomers of the molecule?

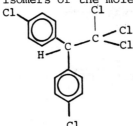

4. Draw and name all structural isomers with the formula
   C_7H_{16}.

5. Polyvinyl chloride, abbreviated PVC, has the formula

   From what single molecule can it be formed?

6. Identify the oxidation states of carbon in each compound
   of the following series:

   CH_4, H_3C—OH, H_2C═O, H—$C\underset{OH}{\overset{O}{<}}$, O═C═O

   Name each compound.

7. In a certain reaction benzaldehyde, which has the formula

   and an almond-like aroma, is converted to benzoic acid.
   Is the reaction an oxidation, a reduction, or a
   displacement reaction?

8. Draw all possible structural isomers of C_3H_6O. Name the
   isomers. You do not have sufficient information to name
   all of them.

9. Arrange the following compounds in order of increasing boiling point.

$$C_5H_{12}, \quad C_5H_{11}OH, \quad C_4H_9C{\overset{H}{\underset{O}{\Big\langle}}}, \quad C_4H_9C{\overset{O}{\underset{OH}{\Big\langle}}}, \quad C_7H_{16}$$

ANSWERS TO EXERCISES

I. Names of organic compounds

1. *n*-butane

The names of some straight-chain alkanes are given in Table 28.1 of the study guide.

2. methylbutane

Substituents on the longest chain are given special radical names (see Table 28.2 of the study guide).

3. 3-ethyl-3-methylpentane

When a substituent can be at one of several places on the longest chain, the positions on the chain are numbered and the substitution position is indicated with the appropriate number.

4. 2-methylpentane

The numbering of the longest chain is always begun at the end of the chain that will give the lowest number to the first substituted position.

TABLE 28.1   Names of Some Straight-Chain Compounds

Name of Compound	Formula
methane	CH_4
ethane	C_2H_6
propane	C_3H_8
butane	C_4H_{10}
pentane	C_5H_{12}
hexane	C_6H_{14}
heptane	C_7H_{16}
octane	C_8H_{18}
nonane	C_9H_{20}
decane	$C_{10}H_{22}$
hexadecane	$C_{16}H_{34}$
heptadecane	$C_{17}H_{36}$

TABLE 28.2   Names of Simple Radicals

Formula	Name
CH_3—	methyl
CH_3CH_2—	ethyl
$CH_3CH_2CH_2$—	*normal*-propyl or *n*-propyl
CH_3CH— 　　$\|$ 　　CH_3	isopropyl
$CH_3CH_2CH_2CH_2$—	*normal*-butyl or *n*-butyl
CH_3CHCH_2— 　　$\|$ 　　CH_3	isobutyl
CH_3CH_2CH— 　　　$\|$ 　　　CH_3	*secondary*-butyl or *sec*-butyl
CH_3 　　$\|$ CH_3C— 　　$\|$ 　　CH_3	*tertiary*-butyl or *tert*-butyl
⬡—	phenyl
Br—	bromo
Cl—	chloro
O_2N—	nitro

5. 3,3-dimethyl-   The longest chain is circled and the numbering is included:
   hexane

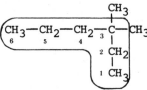

6. 2,3,5-trimethyl-
   3-ethylhexane

7. cyclobutane

The names of some cycloalkanes are given in Table 28.3 of the study guide.

TABLE 28.3   Names of Some Cycloalkanes

Formula	Compound	Formula	Compound
	cyclo-propane		cyclo-hexane
	cyclo-butane		cyclo-heptane

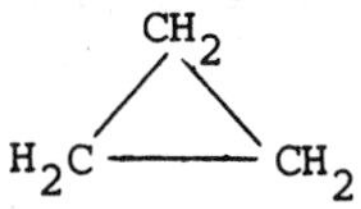

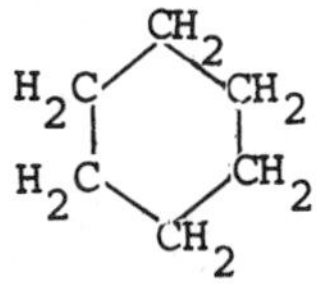

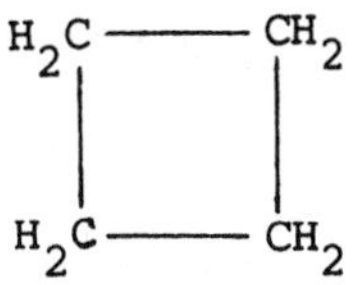

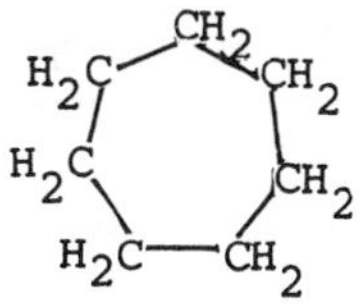

8. methylcyclo-butane

The names of the substituents in Table 28.2 of the study guide are used to name such compounds, but the naming becomes more complex when more than one substituent is added because of the possibility of isomers.

9. 2-butene

The name of an alkene is derived from the name of the corresponding alkane by changing the ending from *-ane* to *-ene*. When necessary, a number is used to show the double-bond position. Because of the restricted rotation about a double bond, *cis* and *trans* isomers exist:

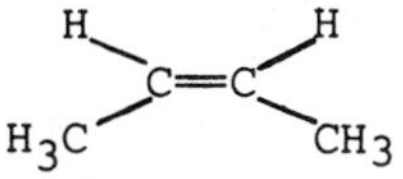

cis-2-butene 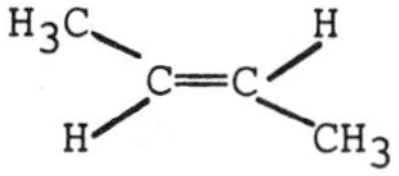   *trans*-2-butene

10. 2-pentene

The numbering always starts at the end of the chain that gives the lowest numbers to the substituents. Therefore, the compound is 2-pentene, not 3-pentene. For this compound *cis* and *trans* isomers exist:

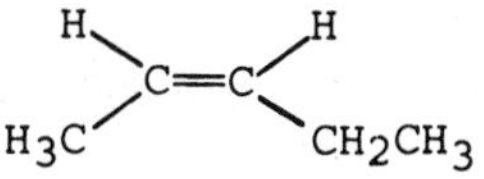

cis-2-pentene 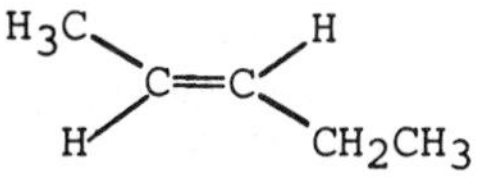   *trans*-2-pentene

11. 4-methyl-
    2-pentene

For this compound *cis* and *trans* isomers exist:

$$
\begin{array}{ccc}
\text{H} & & \text{H} \\
& \text{C}=\text{C} & \\
\text{CH}_3 & & \text{CH(CH}_3\text{)CH}_3
\end{array}
\qquad
\begin{array}{ccc}
\text{H}_3\text{C} & & \text{H} \\
& \text{C}=\text{C} & \\
\text{H} & & \text{CH(CH}_3\text{)CH}_3
\end{array}
$$

cis-4-methyl-2-pentene   *trans*-4-methyl-2-pentene

12. cyclobutene

The name of a cycloalkene is derived from the name of the corresponding cycloalkane by changing the ending from *-ane* to *-ene*.

13. methyl-1-
    butyne

No number is needed for the methyl group because there is only one possible position. The name of a compound containing a triple bond is derived from the name of the corresponding alkane by changing the ending from *-ane* to *-yne*.

14. methylbenzene   Another acceptable name is toluene.

15. *o*-dinitro-
    benzene

The prefixes *ortho-* (*o*-), *meta-* (*m*-), and *para-* (*p*-) are used to designate relative positions of two substituents on a ring.

16. *p*-dichloro-
    benzene

17. *m*-dibromo-
    benzene

18. 2-bromobutane

19. 2-bromo-1-
    butene

20. 3-bromo-1-
    butene

21. 1,1,4-tribromo-
    1-butene

22. 2-butanol

A hydrocarbon containing an OH functional group is an alcohol. The name of an alcohol is derived from the name of the corresponding alkane by changing the ending from *-ane* to *-anol*. When necessary, a number is used to indicate the position of the OH group.

23. cyclohexanol   The name of a cycloalcohol is derived from the name of the corresponding cycloalkane by changing the ending *-ane* to *-anol*.

24. 2,3-butanediol A polyhydroxy alcohol contains more than one OH group. The number of OH groups in a polyhydroxy alcohol is indicated in the name by addition of the appropriate ending to the name of the corresponding alkane: *diol*, *triol*, and so on. The position of an OH group is indicated in the name by a number.

25. *p*-chlorophenol

26. butanal   The name of an aldehyde is derived from the name of the corresponding alkane by changing the ending -e to *-al*.

27. 3-chlorobuta-   For aldehydes the numbering of carbon atoms always starts
    nal   with the carbon atom that is double bonded to oxygen.

28. 2-pentanone   The name of a ketone is derived from the name of the corresponding alkane by changing the ending *-ane* to *-one*. Numbering of carbon atoms begins at the end of the chain that gives the lowest number of the carbon of the carbonyl group, C═O.

29. 3-ethyl-2-pentanone

30. propanoic   The name of a carboxylic acid is derived from the parent
    acid   hydrocarbon by deletion of the final -e, addition of the ending *-oic*, and addition of the separate word *acid*. This compound is also called propanoic acid.

31. *m*-bromobenzoic acid

32. methyl   The name of an ester reflects the alcohol and acid from
    benzoate   which the ester is derived: the name of the hydrocarbon radical attached to the OH of the alcohol (see Table 28.2 of the study guide) is used to indicate the parent alcohol, and the ending *-ate* added to the base of the parent acid is used to indicate the parent acid.

33. ethyl chloroacetate

34. diethylamine   The name of an amine is formed by the addition of the names of the radicals (see Table 28.2 of the study guide) that are attached to the N atom to the word *amine*.

35. *m*-chloroani-   The molecule ⬡—NH_2 is called aniline.
    line

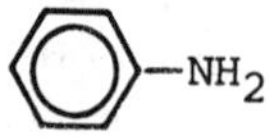

36. cyclopentanone

37. isopropyl acetate

38. 2,3,4-trimethyl-1,3-hexadiene

39. triphenylamine

40. 3,3-dimethyl-1-butyne

II. Reactions of organic compounds

1. an oxidation reaction: CH_3CH_2—CH=CH—CH_3 $\xrightarrow[\text{heat}]{MnO_4^-}$ CH_3CH_2—$\overset{\overset{O}{\|}}{C}$—$OH$ + CH_3—$\overset{\overset{O}{\|}}{C}$—$OH$

   propanoic acid   ethanoic acid

2. an addition reaction: CH_3CH_2—CH=CH—CH_3 $\xrightarrow{H_2/Pt}$ $CH_3(CH_2)_3CH_3$

   pentane

3. an addition reaction:

   CH_3—CH_2—CH=CH—CH_3 $\xrightarrow{Br_2}$ CH_3—CH_2—$\overset{\overset{Br}{|}}{CH}$—$\overset{\overset{Br}{|}}{CH}$—$CH_3$

   2,3-dibromopentane

4. an oxidation reaction:

   CH_3—$\overset{\overset{CH_3}{|}}{CH}$—$CH_2OH$ $\xrightarrow[\text{Cr}_2\text{O}_7^{2-}/\text{H}^+]{\text{mild}}$ CH_3—$\overset{\overset{CH_3}{|}}{CH}$—$\overset{\overset{O}{\|}}{C}$—$H$

   methylpropanal

5. a substitution reaction:

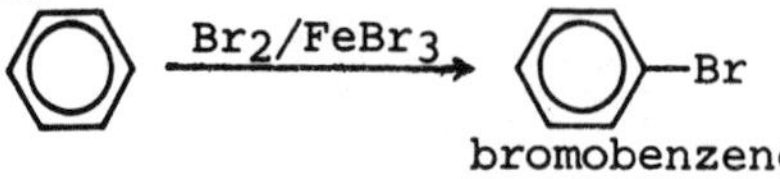

   ⬡ $\xrightarrow{Br_2/FeBr_3}$ ⬡—Br

   bromobenzene

6. an oxidation reaction:

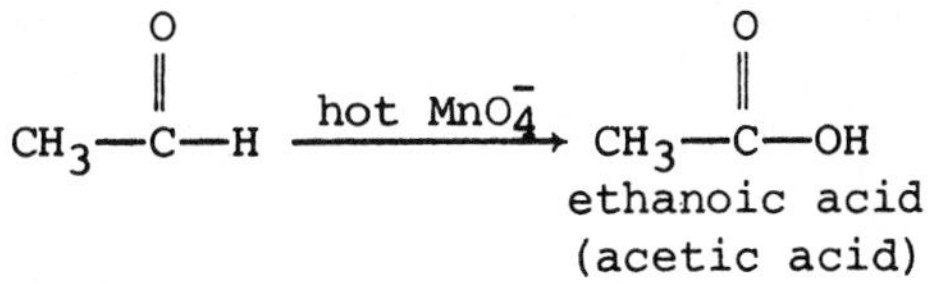

$$CH_3-\overset{\overset{\displaystyle O}{\|}}{C}-H \xrightarrow{\text{hot } MnO_4^-} CH_3-\overset{\overset{\displaystyle O}{\|}}{C}-OH$$

ethanoic acid
(acetic acid)

7. a reduction reaction:

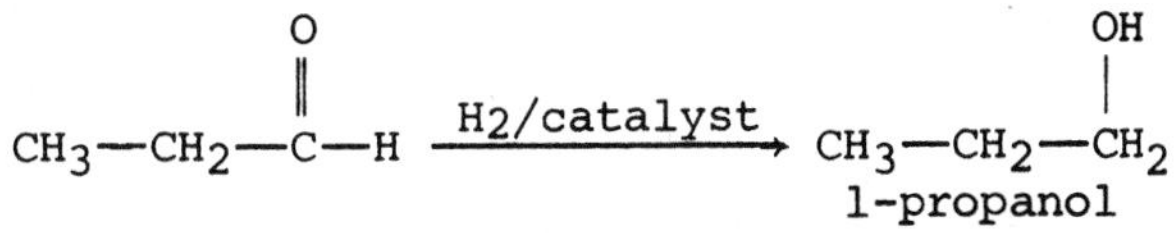

$$CH_3-CH_2-\overset{\overset{\displaystyle O}{\|}}{C}-H \xrightarrow{\text{H}_2/\text{catalyst}} CH_3-CH_2-\overset{\overset{\displaystyle OH}{|}}{CH_2}$$

1-propanol

8. a Friedel-Crafts reaction:

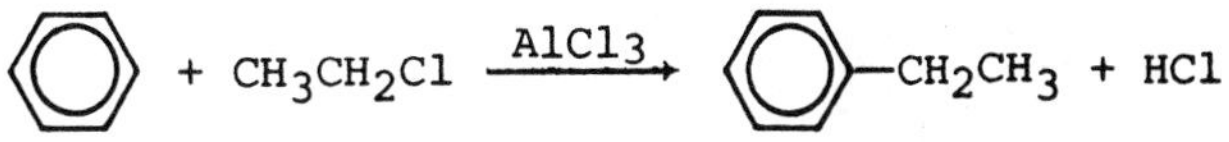

⬡ + CH_3CH_2Cl $\xrightarrow{AlCl_3}$ ⬡$-CH_2CH_3$ + HCl

9. an oxidation reaction:

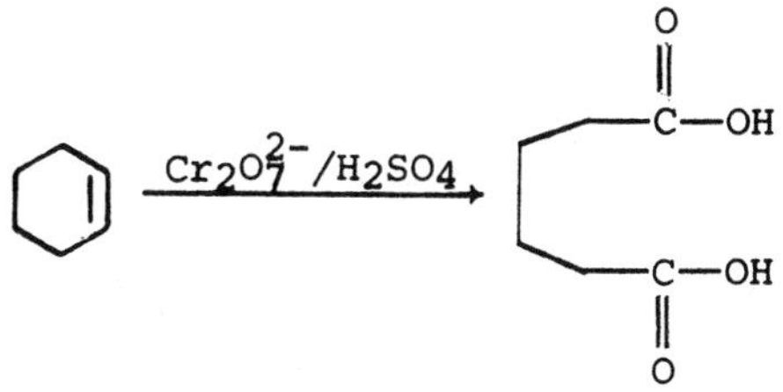

⬡ $\xrightarrow{Cr_2O_7^{2-}/\text{H}_2SO_4}$

$$\begin{array}{c}\overset{\overset{\displaystyle O}{\|}}{C}-OH\\|\\\overset{\overset{\displaystyle }{}}{C}-OH\\\|\\O\end{array}$$

or drawn differently

$$HO-\overset{\overset{\displaystyle O}{\|}}{C}-\overset{\overset{\displaystyle H}{|}}{\underset{\underset{\displaystyle H}{|}}{C}}-\overset{\overset{\displaystyle H}{|}}{\underset{\underset{\displaystyle H}{|}}{C}}-\overset{\overset{\displaystyle H}{|}}{\underset{\underset{\displaystyle H}{|}}{C}}-\overset{\overset{\displaystyle H}{|}}{\underset{\underset{\displaystyle H}{|}}{C}}-\overset{\overset{\displaystyle O}{\|}}{C}-OH$$

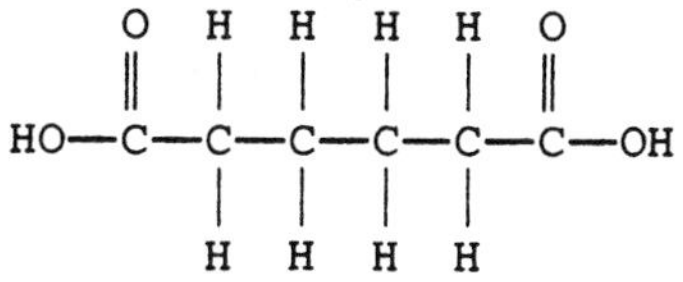

III. Properties of organic compounds

1.

$$CH_3-\overset{\overset{\displaystyle O}{\|}}{C}-OH \;+\; HO-CH_2-CH_2-\overset{\overset{\displaystyle CH_3}{|}}{\underset{\underset{\displaystyle CH_3}{|}}{CH}} \;\rightarrow\; CH_3-\overset{\overset{\displaystyle O}{\|}}{C}-OCH_2-CH_2-\overset{\overset{\displaystyle CH_3}{|}}{\underset{\underset{\displaystyle CH_3}{|}}{CH}}$$

ethanoic acid   3-methyl-1-butanol   isopentylacetate
(acetic acid)

2.

$$-\overset{\overset{\displaystyle F}{|}}{\underset{\underset{\displaystyle F}{|}}{C}}-\overset{\overset{\displaystyle F}{|}}{\underset{\underset{\displaystyle F}{|}}{C}}-\overset{\overset{\displaystyle F}{|}}{\underset{\underset{\displaystyle F}{|}}{C}}-\overset{\overset{\displaystyle F}{|}}{\underset{\underset{\displaystyle F}{|}}{C}}-$$

The compound has the structure of polyethene, but the H atoms are replaced by F atoms.

3. no

4.

$CH_3-CH_2-CH_2-CH_2-CH_2-CH_2-CH_3$ heptane

$CH_3-\overset{\overset{\displaystyle }{}}{\underset{\underset{\displaystyle CH_3}{|}}{CH}}-CH_2-CH_2-CH_2-CH_3$ 2-methylhexane

$CH_3-CH_2-\overset{\overset{\displaystyle }{}}{\underset{\underset{\displaystyle CH_3}{|}}{CH}}-CH_2-CH_2-CH_3$ 3-methylhexane

$CH_3-\overset{\overset{\displaystyle CH_3}{|}}{\underset{\underset{\displaystyle CH_3}{|}}{C}}-CH_2-CH_2-CH_2$ 2,2-dimethylpentane

$CH_3-\overset{\overset{\displaystyle }{}}{\underset{\underset{\displaystyle CH_3}{|}}{CH}}-\overset{\overset{\displaystyle }{}}{\underset{\underset{\displaystyle CH_3}{|}}{CH}}-CH_2-CH_3$ 2,3-dimethylpentane

$CH_3-\overset{\overset{\displaystyle }{}}{\underset{\underset{\displaystyle CH_3}{|}}{CH}}-CH_2-\overset{\overset{\displaystyle }{}}{\underset{\underset{\displaystyle CH_3}{|}}{CH}}-CH_3$ 2,4-dimethylpentane

$CH_3-CH_2-\overset{\overset{\displaystyle CH_3}{|}}{\underset{\underset{\displaystyle CH_3}{|}}{C}}-CH_2-CH_3$ 3,3-dimethylpentane

$CH_3-CH_2-\overset{\overset{\displaystyle }{}}{\underset{\underset{\underset{\underset{\displaystyle CH_3}{|}}{CH_3}}{|}}{\underset{\displaystyle CH_2}{}}}CH-CH_2-CH_3$

ethylpentane

$CH_3-\overset{\overset{\displaystyle CH_3}{|}}{\underset{\underset{\displaystyle CH_3}{|}}{C}}-\overset{\overset{\displaystyle CH_3}{|}}{CH}-CH_3$ 2,2,3-trimethylbutane

5. $HC\!\!=\!\!CH_2$

6.

methane	methanol	formaldehyde	formic acid	carbon dioxide
4-	2-	0	2+	4+
CH_4	$H_3C\!-\!OH$	$H_2C\!\!=\!\!O$	$H\!-\!C\overset{O}{\underset{OH}{\diagup\!\!\!\diagdown}}$	$O\!\!=\!\!C\!\!=\!\!O$

7. oxidation   Note the change in the oxidation state of the carbon atom that is double bonded to the oxygen:

$$\underset{0}{\bigcirc\!\!-\!\!\overset{\overset{O}{\|}}{C}\!\!-\!\!H} \quad \rightarrow \quad \underset{2+}{\bigcirc\!\!-\!\!\overset{\overset{O}{\|}}{C}\!\!-\!\!OH}$$

8.   You should have been able to name the compounds in the left column, but you may not have been able to name those in the right column:

$$\underset{HO}{\overset{H}{\diagdown}}C\!\!\overset{CH_2}{\underset{CH_2}{\diagup\!\!\diagdown}}$$

cyclopropanol

$$CH_3\!-\!CH_2\!-\!\overset{\overset{O}{\|}}{C}\!-\!H$$

propanal

$$CH_3\!-\!\overset{\overset{O}{\|}}{C}\!-\!CH_3$$

propanone or
acetone

$$\overset{OH}{\underset{|}{}}\\ CH_3\!-\!CH\!\!=\!\!CH$$

1-propen-1-ol

$$\overset{OH}{\underset{|}{}}\\ CH_2\!-\!CH\!\!=\!\!CH_2$$

2-propen-1-ol

$$\overset{OH}{\underset{|}{}}\\ CH_3\!-\!C\!\!=\!\!CH_2$$

1-propen-2-ol

$$CH_3\!-\!O\!-\!CH\!\!=\!\!CH_2$$

3-oxa-1-butene or
methylvinyl ether

$$\overset{O\!-\!CH_2}{\underset{H_2C\!-\!CH_2}{|\quad|}}$$

trimethylene oxide

9.

	Molecular Weight	Boiling Point (°C)
C_5H_{12}	72	36
C_7H_{16}	100	99
$C_4H_9C\underset{O}{\overset{H}{\lessgtr}}$	86	103
$C_5H_{11}OH$	88	138
$C_4H_9C\underset{OH}{\overset{O}{\lessgtr}}$	102	187

Boiling point usually increases with increasing molecular weight. The polar aldehyde has a higher boiling point than the nonpolar alkane with a similar molecular weight. Due to hydrogen bonding, alcohols have higher boiling points. Carboxylic acids have even higher boiling points than alcohols because pairs of carboxylic acid molecules are held together by two hydrogen bonds.

SELF-TEST   I. Complete the test in 20 minutes:

1. Write the name of each of the following compounds in the space provided:

_____ (a)

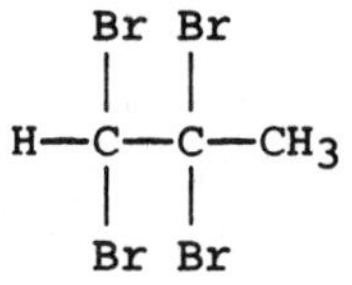

$$H-\underset{\underset{Br}{|}}{\overset{\overset{Br}{|}}{C}}-\underset{\underset{Br}{|}}{\overset{\overset{Br}{|}}{C}}-CH_3$$

_____ (b)

O_2N- ⬡ $\overset{NO_2}{\underset{NO_2}{}}$ $-CH_3$

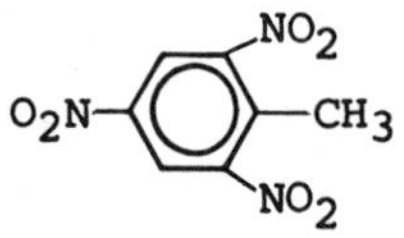

_____ (c) $CH_3-CH_2-C{\equiv}CH$

_____ (d)

$$CH_3-\underset{\underset{CH_3}{|}}{CH}-CH_2-CH_2-\underset{\underset{CH_3}{|}}{\overset{\overset{CH_3}{|}}{C}}-CH_2-CH_3$$

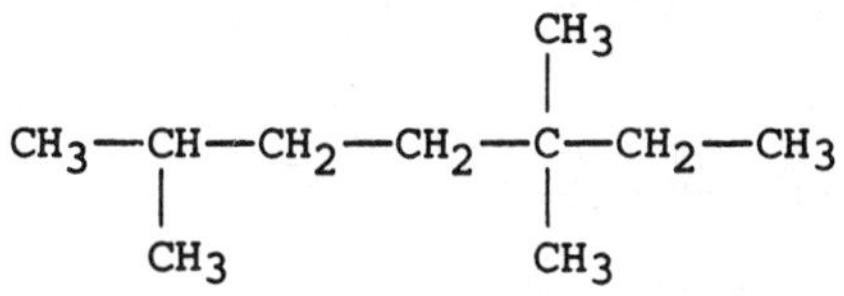

(e)

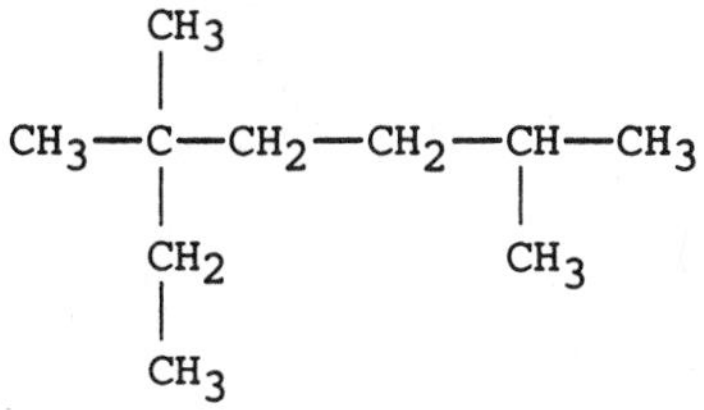

$$CH_3-\underset{\underset{\underset{CH_3}{|}}{\underset{CH_2}{|}}}{\overset{\overset{CH_3}{|}}{C}}-CH_2-CH_2-\underset{\underset{CH_3}{|}}{CH}-CH_3$$

(f)

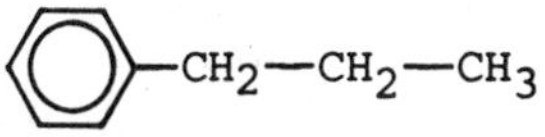

⬡$-CH_2-CH_2-CH_3$

(g) $CH_2\!\!=\!\!CH-CH_2-CH\!\!=\!\!CH_2$

(h)

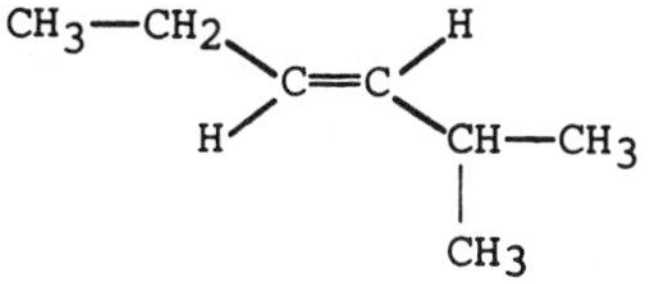

$$\underset{H}{\overset{CH_3-CH_2}{\diagdown}}C\!\!=\!\!C\underset{\underset{\underset{CH_3}{|}}{CH-CH_3}}{\overset{H}{\diagup}}$$

(i)

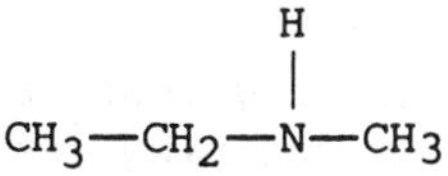

$$CH_3-CH_2-\underset{\overset{|}{}}{\overset{\overset{H}{|}}{N}}-CH_3$$

(j)

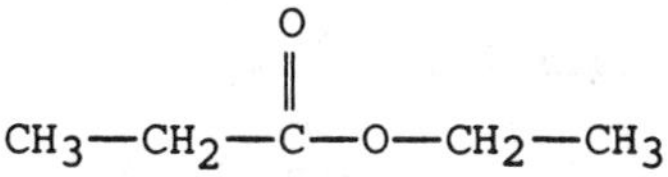

$$CH_3-CH_2-\overset{\overset{O}{\|}}{C}-O-CH_2-CH_3$$

(k) $CH_3-CH_2-CH_2-O-\underset{\underset{O}{\|}}{C}-CH_3$

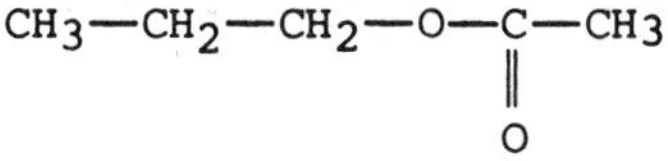

(l)

$$CH_3-CH_2-CH_2-CH_2-CH_2-\overset{\overset{O}{\|}}{C}-H$$

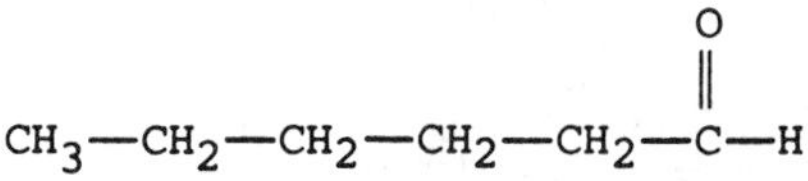

2. Write the formulas of the products obtained from each of the following reactions. If no reaction occurs, write *NR*.

(a) $CH_3CH_2-OH \xrightarrow[\text{heat}]{\text{Cu, O}_2}$

(b) $CH_2\!\!=\!\!CH_2 \xrightarrow{\text{HCN}}$

(c) $CH_3-CH_2-C\!\!\equiv\!\!N \xrightarrow[\text{catalyst}]{\text{H}_2}$

3. Draw all structural isomers of C_3H_6 and name each.

4. Draw all isomers of C_3H_8O and name each.

BIOCHEMISTRY

CHAPTER
29

OBJECTIVES   I. You should be able to demonstrate your knowledge of the
                     following terms by defining them or describing them:

                     α-amino acid [29.1]
                     adenine [29.4]
                     adenosine diphosphate [29.6]
                     adenosine triphosphate [29.6]
                     α-helix [29.4]
                     ADP [29.6]
                     anticodon [29.6]
                     anabolic [29.6]
                     ATP [29.6]
                     carbohydrates [29.2]
                     catabolic process [29.6]
                     chiral [29.1]
                     condon [29.4]
                     coenzyme [29.5]
                     cytosine [29.4]
                     deoxyribonucleic acid [29.4]
                     disaccharides [29.2]
                     DNA [29.4]
                     enantiomorph [29.1]
                     enzymes [29.5]
                     fatty acids [29.3]
                     guanine [29.4]
                     inhibition [29.5]
                     lipids [29.3]
                     metabolism [29.6]
                     monosaccharides [29.2]
                     nucleic acid [29.4]
                     peptide [29.1]
                     polypeptide [29.1]

primary structure [29.1]
protein [29.1]
quaternary structure [29.1]
ribonucleic acid [29.4]
RNA [29.4]
saponification [29.3]
secondary structure [29.1]
substrate [29.5]
tertiary structure [29.1]
thymine [29.4]
triglycerides [29.3]
turnover number [29.5]
uracil [29.4]
zwitterion [29.1]

II. You should be familiar with the basic structural characteristics of amino acids, carbohydrates, fats and oils,
and nucleic acids.

III. You should understand the function of enzymes.

IV. You should understand the process of protein synthesis.

UNITS,
SYMBOLS,
MATHEMATICS

The following symbols were used in this chapter. You
should be familiar with them.

- A is the abbreviation for adenine.
- C is the abbreviation for cytosine
- D- is a prefix used with the name of optically active
    compounds to designate the configuration around
    the chiral carbon.
- DNA is the abbreviation used for deoxyribonucleic acid.
- G is the abbreviation for guanine.
- L- is a prefix used with the name of optically active
    compounds to designate the configuration around the
    chiral carbon.
- RNA is the abbreviation used for ribonucleic acid.
- T is the abbreviation for thymine.
- U is the abbreviation for uracil.

- (+) is the symbol used with the names of optically
  active compounds to indicate that plane polarized
  light passing through the compound is rotated
  clockwise.
- (-) is the symbol used with the names of optically
  active compounds to indicate that plane polarized
  light passing through the compound is rotated
  counterclockwise.

EXERCISES   Choose the best answer to each of the following questions:

1. Which of the following is a dipeptide containing one
   glycine, $^{+}H_3N-CH_2-COO^{-}$?

a. $^{+}H_3N-CH_2-\overset{\displaystyle O}{\overset{\|}{C}}-\underset{\displaystyle H}{N}-CH_2-\overset{\displaystyle O}{\overset{\|}{C}}-\underset{\displaystyle H}{N}-\overset{\displaystyle CH_3}{\underset{\displaystyle H}{C}}-COO^{-}$

b. $^{+}H_3N-\overset{\displaystyle CH_3}{\underset{\displaystyle H}{C}}-\overset{\displaystyle O}{\underset{\displaystyle O}{C}}-\underset{\displaystyle H}{N}-\overset{\displaystyle CH_3}{\underset{\displaystyle H}{C}}-COO^{-}$

c. $^{+}H_3N-\overset{\displaystyle H}{\underset{\displaystyle CH_2}{C}}-\overset{\displaystyle O}{\overset{\|}{C}}-\underset{\displaystyle H}{N}-CH_2-COO^{-}$

(with benzene ring—OH below the CH_2)

d. $^{+}H_3N-CH_2-\overset{\displaystyle O}{\overset{\|}{C}}-\underset{\displaystyle H}{N}-CH_2-COO^{-}$

2. Which of the above is a tripeptide?

3. Which of the peptide residues in problem 1 could
   form the O—H---O type hydrogen bond to the glycine
   residue, $-\underset{\displaystyle H}{N}-CH_2-\overset{\displaystyle O}{\overset{\|}{C}}-$, of another protein?

4. How many optical isomers of glycine, $^+H_3N-CH_2-COO^-$, should exist?

a. 0   c. 4
b. 2   d. 16

5. The backbone or the primary structure of a protein is formed from

a. amino acids   c. enzymes
b. sugars   d. codon

6. How many chiral carbon atoms are there in isoleucine,

$$\begin{array}{c} COO^- \\ | \\ {}^+H_3N-C-H \\ | \\ H_3C-C-H \\ | \\ CH_2 \\ | \\ CH_3 \end{array}$$

a. 1   c. 3
b. 2   d. 4

7. How many chiral carbon atoms are there in leucine,

$$\begin{array}{c} COO^- \\ | \\ {}^+H_3N-C-H \\ | \\ H_3C-C-H \\ | \\ CH_2 \\ | \\ H_3C-C-CH_3 \\ | \\ H \end{array}$$

a. 1   c. 3
b. 2   d. 4

8. Which of the following is the zwitterion form of L-alanine?

a.
$$\begin{array}{c} COOH \\ | \\ H_2N-C-H \\ | \\ CH_3 \end{array}$$

c.
$$\begin{array}{c} COO^- \\ | \\ {}^+H_3N-C-H \\ | \\ CH_3 \end{array}$$

b.
$$\begin{array}{c} COO^- \\ | \\ H_2N-C-H \\ | \\ CH_3 \end{array}$$

d.
$$\begin{array}{c} COOH \\ | \\ {}^+H_3N-C-H \\ | \\ CH_3 \end{array}$$

9. Which of the following is the zwitterion form of L-aspartic acid?

a.

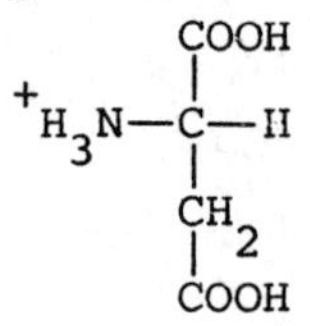

$$
\begin{array}{c}
COOH \\
| \\
{}^+H_3N-C-H \\
| \\
CH_2 \\
| \\
COOH
\end{array}
$$

c.

$$
\begin{array}{c}
COO^- \\
| \\
{}^+H_3N-C-H \\
| \\
CH_2 \\
| \\
COO^-
\end{array}
$$

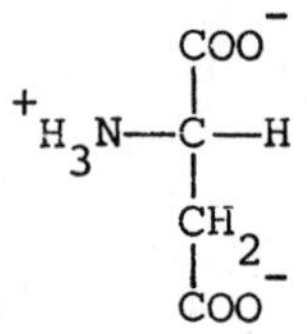

b.

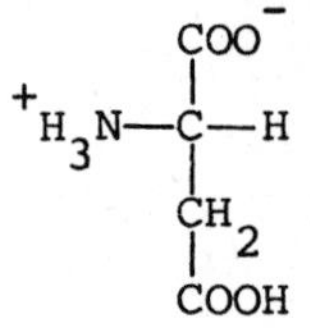

$$
\begin{array}{c}
COO^- \\
| \\
{}^+H_3N-C-H \\
| \\
CH_2 \\
| \\
COOH
\end{array}
$$

d.

$$
\begin{array}{c}
COO^- \\
| \\
H_2N-C-H \\
| \\
CH_2 \\
| \\
COO^-
\end{array}
$$

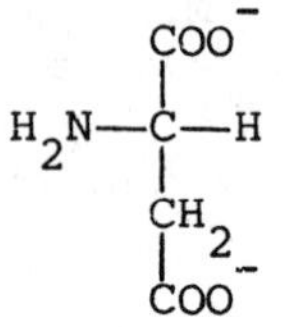

10. Which of the following is dextrorotatory, i.e., rotates plane polarized light clockwise?

a. L-(+)-alanine
b. D-(-)-alanine
c. L-(-)-glucose
d. none of the above

11. Which of the following amino acids is shown as an L-isomer?

a.

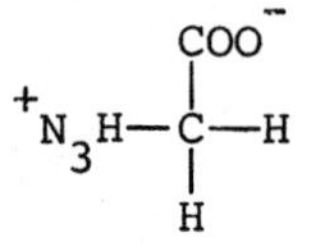

$$
\begin{array}{c}
COO^- \\
| \\
{}^+N_3H-C-H \\
| \\
H
\end{array}
$$

glycine

c.

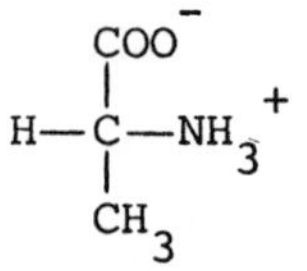

$$
\begin{array}{c}
COO^- \\
| \\
H-C-NH_3{}^+ \\
| \\
CH_3
\end{array}
$$

alanine

b.

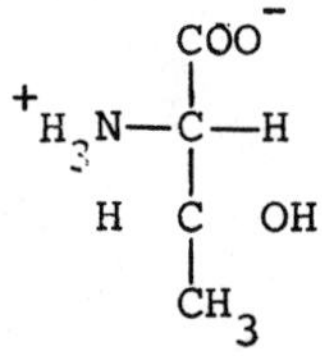

$$
\begin{array}{c}
COO^- \\
| \\
{}^+H_2N-C-H \\
| \\
H\ C\ OH \\
| \\
CH_3
\end{array}
$$

threonine

d.

$$
\begin{array}{c}
COO^- \\
| \\
H-C-NH_3{}^+ \\
| \\
CH_2OH
\end{array}
$$

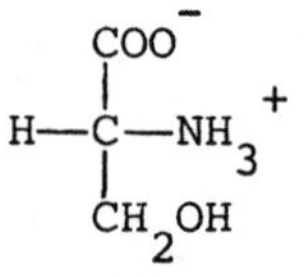

serine

12. How many stereoisomers of the five-membered fructose ring could exist? The structure of the ring is

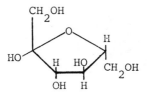

    a. 4
    b. 8
    c. 16
    d. 32

13. How many stereoisomers of palmitic acid, $CH_3(CH_2)_{14}COOH$, could exist?

    a. 0
    b. 14
    c. 256
    d. 16,384

14. How many stereoisomers could exist for the triglyceride formed from three moles of palmitic acid and one mole of glycerol?

    a. 2
    b. 42
    c. 512
    d. 49,152

15. The double helix of DNA is its

    (a) primary structure
    (b) secondary structure
    (c) tertiery structure
    (d) quarternary structure

16. The messenger RNA formed from the portion of DNA chain A-C-G is

    (a) U-G-C
    (b) G-C-U
    (c) U-T-C
    (d) T-G-U

17. The three-base sequence G-G-U in a messenger RNA chain

    (a) is a codon
    (b) specifies the amino acid glycine
    (c) represents guanine-guanine-uracil
    (d) all of the above

18. Messenger RNA carries information

    (a) for producing amino acid sequences in peptides
    (b) from anticodons in DNA
    (c) in triplet sets called codons
    (d) all of the above

19. The messenger RNA codon for the DNA sequence A-G-T is

    (a) A-G-U
    (b) T-C-A
    (c) U-G-A
    (d) U-C-A

20. The transfer RNA anticodon which is specific for tyrosine is G-U-A. What is the corresponding messenger RNA codon?

    (a) C-A-U
    (b) G-U-A
    (c) C-A-T
    (d) G-T-A

21. One of the sequences in one of the human hemoglobin chains is valine-leucine-serine. According to Table 29.1 of the study guide, the corresponding messenger RNA codon could have the base sequence

    (a) C-A-U-G-A-C-U-C-A
    (b) C-A-T-G-A-C-T-C-A
    (c) U-G-A-G-U-C-A-U-G
    (d) G-U-A-C-U-G-A-G-U

22. An enzyme is

    (a) a biological catalyst
    (b) a protein
    (c) a surface catalyst
    (d) all of the above

TABLE 29.1   Examples of some coding for amino acid
             sequences in protein

Amino acid	DNA anticodon	messenger RNA codon	transfer RNA anticodon
Alanine	C-G-T	G-C-A	C-G-U
Cystine	A-C-G	U-G-C	A-C-G
Glycine	C-C-A	G-G-U	C-C-A
Leucine	G-A-C	C-U-G	G-A-C
Lycine	T-T-T	A-A-A	U-U-U
Proline	G-G-A	C-C-U	G-G-A
Serine	T-C-A	A-G-U	U-C-A
Tyrosine	A-T-G	U-A-C	A-U-G
Valine	C-A-T	G-U-A	C-A-U

23. The number of moles of reactant transformed in one
    minute by an enzyme is

    (a) the turnover number
    (b) usually between 1 mol/min to 5 mol/min
    (c) at least 6×10^{23} molecules/min
    (d) all of the above

24. The process in which the activity of an enzyme is
    decreased by forming reversible enzyme-inhibitor
    combinations is called

    (a) competitive inhibition
    (b) product inhibition
    (c) feedback inhibition
    (d) non-competitive inhibition

25. Curve A in the following diagram represents the
    reaction coordinates for

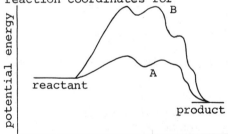

    (a) an uncatalyzed reaction
    (b) a catalyzed reaction
    (c) an enzymatic reaction which is more inhibited than
        that represented by Curve B
    (d) none of the above

26. The compounds whose reactions are catalyzed by enzymes are called

    (a) proteins
    (b) DNA
    (c) RNA
    (d) substrates

27. Photosynthesis is

    (a) a catabolic process
    (b) an anabolic process
    (c) used to produce ATP directly from CO_2, H_2O and light
    (d) all of the above

28. Energy obtained from oxidizing one mole of glucose to CO_2 and water is stored in the body

    (a) as heat
    (b) in ADP
    (c) in ATP
    (d) in DNA

29. During the process in which ATP goes to ADP

    (a) energy is released
    (b) energy is stored
    (c) water is released
    (d) the cell cools

30. The process in which fructose yields CO_2 and water in the body is called

    (a) a catabolic process
    (b) an anabolic process
    (c) a reduction
    (d) none of the above

31. The process in which DNA produces daughter DNA molecules is called

    (a) a catabolic process
    (b) an anabolic process
    (c) bisexual reproduction
    (d) coiling-decoiling

ANSWERS TO EXERCISES

Introduction to biochemistry

1. c

The glycine residue is shaded in the following structure

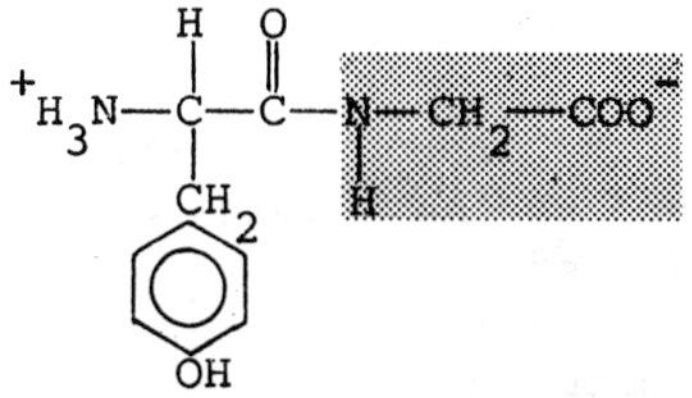

Answer "a" has two glycine residues shown shaded in the following structure

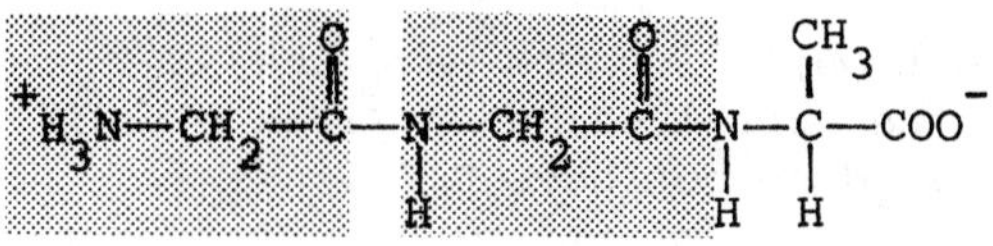

2. a

In addition to the two glycines shown in the shaded areas of the answer to problem 1, there is an L-alanine.

3. c

The bond could be formed between the OH group on the phenyl ring and the C=O group of glycine.

4. a

There are no chiral carbons. A carbon must be bonded to 4 distinctly different groups to be chiral.

5. a

Proteins are large chains of amino acids.

6. b

The two chiral carbons are shaded in the following structure. Each is bonded to 4 distinctly different groups.

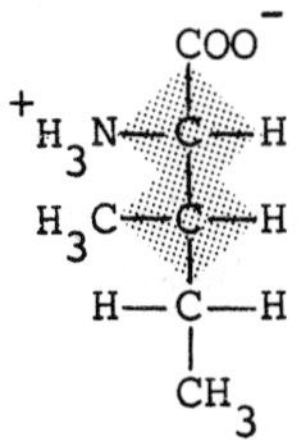

7. a   The chiral carbon is shaded in the structure below

$$
\begin{array}{c}
\text{COO}^{-} \\
| \\
^{+}\text{H}_3\text{N}-\overset{\displaystyle}{\text{C}}-\text{H} \\
| \\
\text{CH}_2 \\
| \\
\text{H}_3\text{C}-\overset{\displaystyle}{\underset{\displaystyle \text{H}}{\text{C}}}-\text{CH}_3
\end{array}
$$

8. c   In the zwitterion, the proton from the carboxylic acid group is transferred to the α-amine group. The net molecular charge is zero.

9. b   See the answer to problem 8.

10. a   The plus sign in parentheses indicates that plane polarized light is rotated clockwise.

11. b   By convention, L isomers have the NH_3^{+} group drawn to the left of the chiral carbon. Since glycine has no chiral carbon, it does not have D,L isomers.

12. c   There are four chiral carbon atoms. The number of sterioisomers is $2^4=16$.

13. a   There are no chiral carbons--no sterioisomers.

14. a   There is one chiral carbon and two sterioisomers. The chiral carbon is shaded in the following structure.

$$
\begin{array}{c}
\quad\quad\quad\quad\quad \text{O} \quad\quad \text{H} \\
\quad\quad\quad\quad\quad || \quad\quad | \\
\text{CH}_3(\text{CH}_2)_{14}\text{C}-\text{O}-\text{C}-\text{H} \\
\quad\quad\quad\quad\quad \text{O} \quad\quad | \\
\quad\quad\quad\quad\quad || \\
\text{CH}_3(\text{CH}_2)_{14}\text{C}-\text{O}-\text{C}-\text{H} \\
\quad\quad\quad\quad\quad \text{O} \quad\quad | \\
\quad\quad\quad\quad\quad || \\
\text{CH}_3(\text{CH}_2)_{14}\text{C}-\text{O}-\text{C}-\text{H} \\
\quad\quad\quad\quad\quad\quad\quad\quad\quad | \\
\quad\quad\quad\quad\quad\quad\quad\quad\quad \text{H}
\end{array}
$$

15. b   The spatial arrangement of a chain is its secondary structure. Here each chain is a helix.

16. a   The complementary pairs are:

DNA	messenger RNA
A (adenine)	U (uracil)
G (guanine)	C (cytosine)
C (cytosine)	G (guanine)

Therefore for an A-C-G anticodon, the corresponding codon becomes U-G-C. See the second line of Table 29.1 of the study guide.

17. d   See Table 29.1 of the study guide.

18. d   See Table 29.1 of the study guide.

19. d   See Table 29.1 of the study guide.

20. a   See Table 29.1 of the study guide.

21. d   See Table 29.1 of the study guide.

22. d

23. a   Turnover numbers are usually between 10,000 and 5,000,000 mol/min.

24. a

25. b   Catalyzed reactions have lower activation energies.

26. d

27. b   During photosynthesis carbohydrates are produced from CO_2 and H_2O.

28. c   See section 29.6 of your text.

29. a

30. a

31. b   New large molecules are synthesized.

SELF-TEST   Answer the following questions in 10 minutes.

1. Draw the structure of the amino acid, L-glutamine in
   the zwitterion form.   The R group is $H_2N-\underset{\underset{O}{\|}}{C}-CH_2-CH_2$

2. How many chiral carbons are in L-glutamine?

3 Draw the structure of D-glutamine in the zwitterion
   form.

4. How many codons are necessary to identify the amino acid
   sequence alanine-cytosine-glycine-cytosine-alanine?

5. The following molecule may be classified as

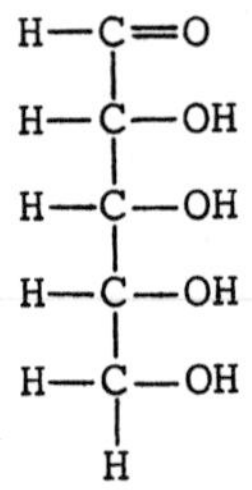

   ```
   H—C=O
       |
   H—C—OH
       |
   H—C—OH
       |
   H—C—OH
       |
   H—C—OH
       |
       H
   ```

   a. an amino acid   c. a ketone
   b. a peptide   d. a carbohydrate

6. The hydrogen bonding between bases on each strand of
   DNA accounts in large part for the shape of the helix.
   The base cytosine and the base guanine form three
   hydrogen bonds.   Draw a structure for the cytosine-
   guanine interaction.   The structure of cytosine is

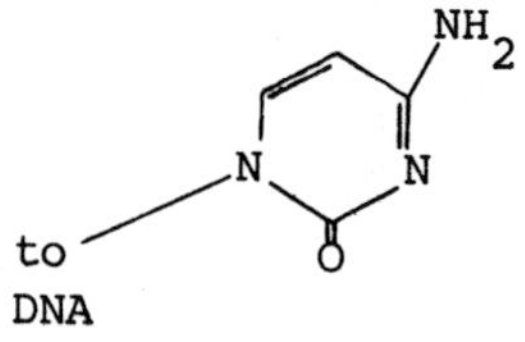

   to
   DNA

   The structure of guanine is

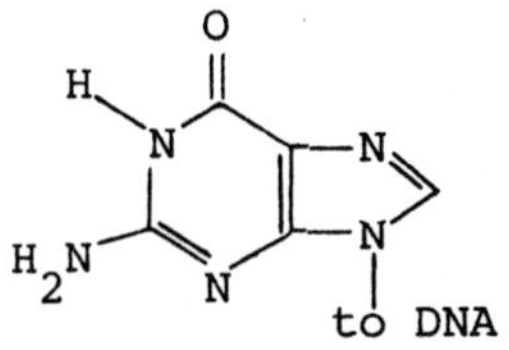

   to DNA

ANSWERS TO SELF-TESTS

CHAPTER 1   I. 1. (c)   5. (d)   9. (c)   13. (b)

2. (b)   6. (a)   10. (d)   14. (b)

3. (c)   7. (c)   11. (b)   15. (b)

4. (c)   8. (b)   12. (a)

II. 1. 1.0×10^{-2} cm

2. 1.24 g/cm^3

3. 8.5×10^{11} kg

CHAPTER 2

1. proton
   neutron
   electron
2. neutron
3. gamma
4. isotopes
5. 6.944 u

6. $^{12}_{6}C$

7. lead, Pb
8. fluorine, F
9. 27

10. 10
11. periods
12. ion
13. gamma
14. nucleus
15. 29

CHAPTER 3

1. hydrogen
   nitrogen
   oxygen
   the halogens:
     fluorine
     chlorine
     bromine
     iodine
     astatine
2. (b) C_2H_3

3. 3.03×10^{22} atoms
4. (b) CaO
5. $C_6H_5OCl_2$

CHAPTER 4

1. 0.744 g
2. 0.406 g H_2C_2

3. 56%

CHAPTER 5

1. $+3.80$ J
2. 0.0607 g Al
3. -206.5 kJ
4. $-270.$ kJ
5. -92.30 kJ
6. $+1381.8$ kJ

CHAPTER 6

1. (b)	6. (d)	11. (d)	16. (b)
2. (b)	7. (a)	12. (a)	17. (a)
3. (b)	8. (b)	13. (d)	18. (b)
4. (c)	9. (b)	14. (c)	19. (a)
5. (d)	10. (a)	15. (d)	20. (d)

CHAPTER 7

1. (d)
2. (a)
3. (b)
4. $+155$ kJ

5. Cu^+: $1s^2 2s^2 2p^6 3s^2 3p^6 3d^{10}$
   Cu^{2+}: $1s^2 2s^2 2p^6 3s^2 3p^6 3d^9$
   N^{3-}: $1s^2 2s^2 2p^6$
   Cl^-: $1s^2 2s^2 2p^6 3s^2 3p^6$

CHAPTER 8

1. (a)
   not (b): no octet on N
   not (c): no octet on N
   not (d): no octet on N
   not (e): too many electrons
   not (f): too many electrons

2. $\ddot{\text{S}}=\text{C}=\ddot{\text{N}}^{\ominus} \leftrightarrow \ :\text{S}\equiv\text{C}-\ddot{\text{N}}^{\oplus}\text{:}^{2\ominus} \leftrightarrow \ :\ddot{\text{S}}^{\ominus}-\text{C}\equiv\text{N}:$

3. (d) F
4. (c) Sb

CHAPTER 9

1.

Formula	Number of Electron Pairs		Shape of Molecule or Ion
	Bonding	Nonbonding	
SCl_2	2	2	angular
XeF_4	4	2	square planar
AlH_4^-	4	0	tetrahedral
$TeCl_4$	4	1	trigonal pyramidal or distorted tetrahedral
SeF_5^-	5	1	square pyramidal

2.

	Hybrid Orbitals	Geometric Shape
PCl_5	dsp^3	trigonal bipyramidal
PCl_4^+	sp^3	tetrahedral
PCl_6^-	d^2sp^3	octahedral

3.

Molecule	Total Electrons in Orbitals				Bond Order	Number of Unpaired Electrons
	σ	$\sigma*$	π	$\pi*$		
Be_2	2	2	0	0	0	0
B_2	2	2	2	0	1	2
N_2	4	2	4	0	3	0
O_2	4	2	4	2	2	2
NO^+	4	2	4	0	3	0

CHAPTER 10

I. 1. 16.0 g/mol
2. 105 g/mol
3. 12 mm
4. 800 mm
5. 0.900 g/L

6. 2 L NH_3,
0 L O_2,
8 L NO,
12 L H_2O
7. 43.9 g/mol

II. 1. H_2, 6
2. same, same
3. 1/17
4. intermolecular attractive forces, molecular volume

III.

a. pressure vs. volume

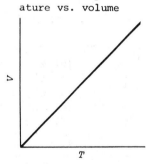

c. absolute temper-
ature vs. volume

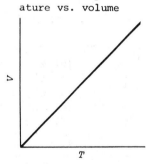

b. pressure vs. the
product of pres-
sure and volume

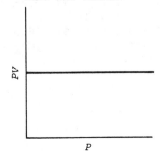

d. energy distribu-
tion of molecules

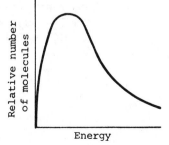

CHAPTER 11   **1.**   218 atm

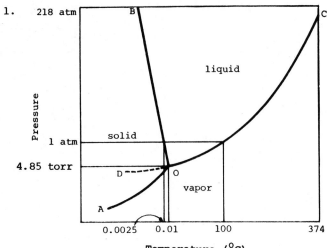

Pressure

1 atm   solid

4.85 torr   D------ O

A

liquid

vapor

0.0025   0.01   100   374

Temperature ($^\circ$C)

2. (a) 1   (b) 2
3. (a) I_2 (d) CO
    (b) K   (e) Si
    (c) C
4. 58.7 g/mol

CHAPTER 12   **1.** 11.7 m   **4.** 16.6 mL
                     2. 9.71 M   **5.** 125 g/mol
                     3. 6.00 M

CHAPTER 13   **1.** (a) $SO_2(g) + 2OH^-(aq) \rightarrow SO_3^{2-}(aq) + H_2O$
                         (b) $ZnSO_4(s) + Ba(OH)_2(aq) \rightarrow Zn(OH)_2(s) + BaSO_4(s)$
                         (c) $5Fe^{2+}(aq) + MnO_4^-(aq) + 8H^+ \rightarrow 5Fe^{3+}(aq) + Mn^{2+}(aq) + 4H_2O$
                         (d) $CO_2(g) + H_2O(1) \rightarrow H^+(aq) + HCO_3^-(aq)$
                         (e) $Na^+(aq) + SO_4^{2-}(aq) + Cu^{2+}(aq) + Cl^-(aq) \rightarrow NR$
                         (f) $K_2O(s) + H_2O(1) \rightarrow 2K^+(aq) + 2OH^-(aq)$

                     2. 4.30%

                     3. 123 g/eq

                     4. $2ClO_2 + 2OH^- \rightarrow ClO_2^- + ClO_3^- + H_2O$
                         $IO_4^- + H_2AsO_3^- \rightarrow IO_3^- + H_2AsO_4^-$
                         $4Sb + 4NO_3^- + 4H^+ \rightarrow Sb_4O_6 + 4NO + 2H_2O$
                         $O_3 + 6I^- + 6H^+ \rightarrow 3H_2O + 3I_2$
                         $2NO_3^- + 3H_2S + 2H^+ \rightarrow 2NO + 3S + 4H_2O$
                         $Cr_2O_7^{2-} + 6Cl^- + 14H^+ \rightarrow 2Cr^{3+} + 3Cl_2 + 7H_2O$

5.

Element	Oxidation number
(a) Cl	$4+ \rightarrow 3+$
Cl	$4+ \rightarrow 5+$
(b) I	$7+ \rightarrow 5+$
As	$3+ \rightarrow 5+$
(c) Sb	$0 \rightarrow 3+$
N	$5+ \rightarrow 2+$
(d) O	$0 \rightarrow 2-$
I	$1- \rightarrow 0$
(e) N	$5+ \rightarrow 2+$
S	$2- \rightarrow 0$
(f) Cr	$6+ \rightarrow 3+$
Cl	$1- \rightarrow 0$

6. 3.12 g/eq

CHAPTER 14

1. order of $HgCl_2 = 1$, order of $C_2O_4^{2-} = 2$,
   $k = 4.7 \times 10^{-3} M^{-2}$ min^{-1}
2. (a) remains the same   (d) decreases
   (b) increases   (e) increases
   (c) remains the same
3. 17°

CHAPTER 15

1. (a) increase   (d) increase
   (b) no change   (e) increase
   (c) no change
2. increase temperature, decrease pressure, remove SO_2
3. 133 atm^{-1}
4. 6.25×10^{-3} M

CHAPTER 16

1. NH_3, H^-
2. H_2S
3. $SOCl_2$ (SO^{2+} is the acid ion)
4. F^-
5. $\overset{..}{\underset{..}{S}}:$
6. $H\overset{..}{\underset{..}{S}}^-$, $HC_2H_3O_2$
7. HPO_4^{2-}

CHAPTER 17

1. 12.87
2. 2.0×10^{-6}
3. 0.18 M
4. 8.85
5. 5.2

CHAPTER 18

1. 1.2×10^{-4} g Ag^{2+}/250 mL
2. $PbSO_4$; 4.3×10^{-5} M
3. 6.7×10^{-16} M

CHAPTER 19

1. -246 kJ/mol
2. reaction is spontaneous: $K_P = 1.0 \times 10^{-5}$
3. -107.9 J/K
   - 58.5 kJ
4. 8.84

CHAPTER 20   I.   1. +1.22 V
2. 0.755 g
3. (a) Cu^{2+} (d) yes
   (b) $Cu^{2+} + 2e^- \rightarrow Cu°$ (e) negative
   (c) positive

II.   1. (b) 0.61 V   4. (c) Pt
2. (b) lowering the pH   5. (d) VO_2^{2+} and Ni(s)
3. (a) Ni   6. (c) VO^{2+}

CHAPTER 21   1. (a) $Mg(s) + H_2O(g) \rightarrow MgO(s) + H_2(g)$
(b) $H_2(g) + Cl_2(g) \rightarrow 2HCl(g)$
(c) $2Na(s) + 2H_2O(l) \rightarrow 2Na^+(aq) + 2OH^-(aq) + H_2(g)$
(d) $2OH^-(aq) + 2Cl_2(aq) \rightarrow ClO^-(aq) + Cl^-(aq) + H_2O(l)$
(e) $FeS(s) + 2H^+(aq) + 2Cl^-(aq) \rightarrow Fe^{2+}(aq) + 2Cl^-(aq) + H_2S(g$
(f) $2Br_2(g) + Ag_2O(s) + H_2O(l) \rightarrow 2AgBr(s) + 2HOBr(aq)$
(g) $PCl_3(l) + 3H_2O(l) \rightarrow 3HCl(aq) + H_3PO_3(aq)$
(h) $Cl^-(aq) + 3H_2O(l) \xrightarrow[\text{heat}]{\text{electrolysis}} ClO_3^-(aq) + 3H_2(g)$
(i) $Cl^-(aq) + H^+(aq) + H_2O(l) \rightarrow NR$
(j) $Cl_2(g) + 2I^-(aq) \rightarrow 2Cl^-(aq) + I_2(s)$

CHAPTER 22   1. (a) $FeS(s) + 2H^+(aq) + 2Cl^-(aq) \rightarrow Fe^{2+}(aq) + 2Cl^-(aq) + H_2S(g$
(b) $Ba^{2+}(aq) + SO_4^{2-}(aq) \rightarrow BaSO_4(s)$
(c) $C_{12}H_{22}O_{11}(s) + 11H_2SO_4(conc) \rightarrow 11H_2SO_4 \cdot H_2O + 12C(s)$
   sucrose
   or
   $C_{12}H_{22}O_{11}(s) + 6H_2SO_4 \rightarrow 5H_2SO_4 \cdot 2H_2O + H_2SO_4 \cdot H_2O + 12C$
   or any other balanced equation containing hydrated
   sulfuric acid species and 12 carbon atoms as products.
(d) $S_2O_3^{2-}(aq) + 2H^+(aq) \rightarrow S(s) + SO_2(g) + H_2O(l)$
(e) $2H_2O \xrightarrow{\text{electrolysis}} 2H_2(g) + O_2(g)$
(f) $2Na_2O_2(s) + 2H_2O(l) \rightarrow 4Na^+(aq) + 4OH^-(aq) + O_2(g)$
(g) $S(s) + O_2(g) \rightarrow SO_2(g)$
(h) $2H_2S(g) + 3O_2(g) \rightarrow 2H_2O(g) + 2SO_2(g)$
(i) $Se(s) + O_2(g) \rightarrow SeO_2(s)$

CHAPTER 23   I.   1.

2. (a) $2Sb_2S_3(s) + 9O_2(g) \xrightarrow{\text{heat}} Sb_4O_6(g) + 6SO_2(g)$
(b) $3NO_2(g) + H_2O(l) \rightarrow 2H^+(aq) + 2NO_3^-(aq) + NO(g)$
(c) $Ca_3N_2(s) + 6H_2O(l) \rightarrow 3Ca^{2+}(aq) + 6OH^-(aq) + 2NH_3(g)$
(d) $As_4O_6(s) + 6C(s) \xrightarrow{\text{heat}} As_4(g) + 6CO(g)$
(e) $PBr_3(l) + 3H_2O(l) \rightarrow H_3PO_3(aq) + 3HBr(aq)$
(f) $2PBr_3(l) + O_2(g) \rightarrow 2POBr_3(s)$
(g) $PI_3(s) + I_2(s) \rightarrow NR$
3. monoprotic acid

CHAPTER 24   1.

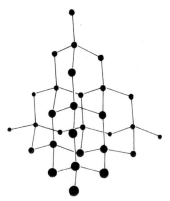

2.

:Ö: 2-
    |
    C=Ö:
    |
:Ö:

3. $Al_4C_3(s) + 12H_2O(l) \longrightarrow 4Al(OH)_3(s) + 3CH_4(g)$

4.

```
F           F
  \       /
    Xe
   ··
  /       \
F           F
```

5. $H_3C-CH_2-CH_3$ $H_2C=CH-CH_3$

$HC\equiv C-CH_3$ $H_2C=C=CH_2$

CHAPTER 25   1. (a) $MgO(s)$ $H_2(g)$ (e) $KO_2(s)$
                 (b) $Na_2O_2(s)$ (f) $Ba_3P_2(s)$
                 (c) $Li_3N(s)$ (g) $Ce_2S_3(s)$
                 (d) $Hg(g)$, $SO_2(g)$
              2. (c) arsenic
              3. (d) tantalum
              4. (e) zone refining silicon rods
              5. (a) potassium

CHAPTER 26   1. (a) potassium tetrachlorocuprate(II)
                 (b) potassium tetrachloroplatinate(II)
                 (c) potassium hexacyanoferrate(II)
                 (d) diamminetetrachloroplatinum(IV)
                 (e) chlorobis(ethylenediamine)thiocyanocobalt(III)
                     chloride
                 (f) tetraamminedibromocobalt(III) bromide
                 (g) tetraamminecopper(II) hexachlorochromate(III)

2.

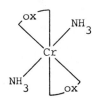   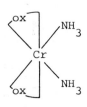

trans   *cis*

3. d^3

4.

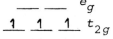

e_g

$\underline{1}$ $\underline{1}$ $\underline{1}$ t_{2g}

5.

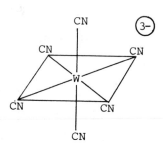

CHAPTER 27   1. 3.02×10^{16} mol
2. 3.2×10^{-11} g
3. 7.84 MeV
4. (a) $^{125}_{54}\text{Xe}$ (d) $^{125}_{53}\text{I}$

   (b) $^{125}_{53}\text{I}$ (e) $^{124}_{51}\text{Sb}$

   (c) $^{1}_{0}\text{n}$

CHAPTER 28   I. 1. (a) 1,1,2,2-tetrabromopropane
(b) 2,4,6-trinitrotoluene
(c) 1-butyne
(d) 2,5,5-trimethylheptane
(e) 2,5,5-trimethylheptane
(f) *n*-propylbenzene
(g) 1,4-pentadiene
(h) *trans*-2-methyl-3-hexene
(i) methylethylamine
(j) ethyl propionate
(k) *n*-propyl acetate
(l) *n*-hexanal

2. (a)

$$\text{CH}_3-\overset{\displaystyle \overset{\text{O}}{\|}}{\text{C}}-\text{H}$$

(b) $\text{CH}_3-\text{CH}_2-\text{C}{\equiv}\text{N}$

(c) $\text{CH}_3-\text{CH}_2-\text{CH}_2-\text{NH}_2$

3.

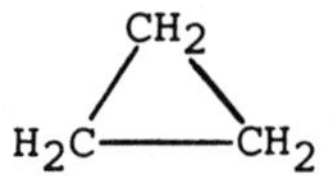

cyclopropane

$CH_2{-}CH{=}CH_2$

propene

4. 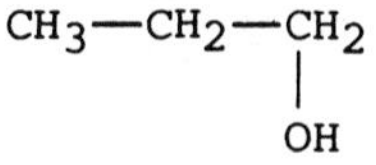 $CH_3{-}CH_2{-}CH_2$ $CH_3{-}CH{-}CH_3$ $CH_3{-}O{-}CH_2{-}CH_3$

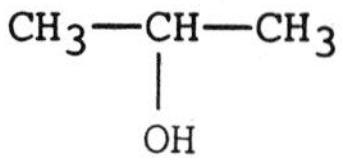

        OH OH

    1-propanol   2-propanol   ethylmethylether

CHAPTER 29   1. 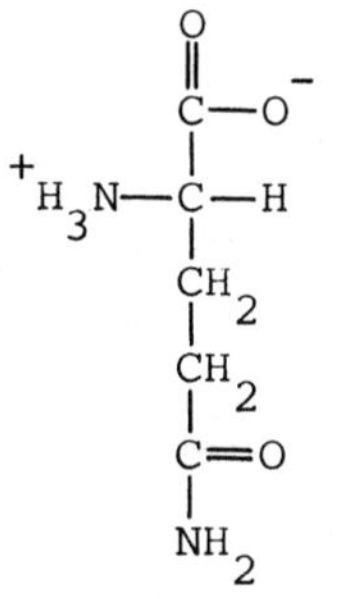   3.

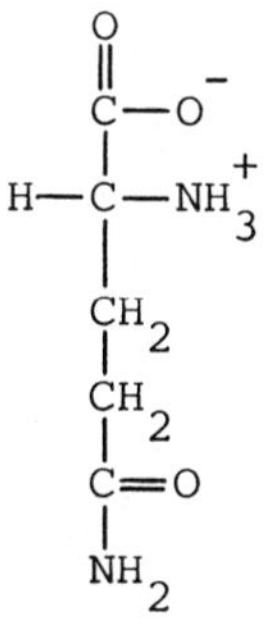

2. one

4. 15

5. (d) a carbohydrate

6.

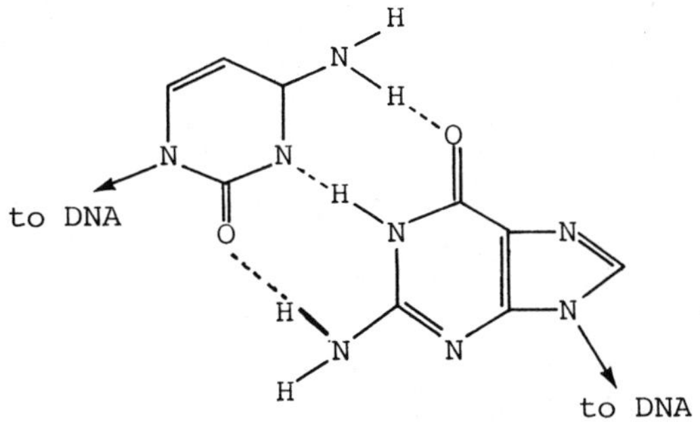